Solutions Manual

Calculus
Differentiation & Integration

Build Your Self-Confidence and Enjoyment of Math!
Problems Included

Aejeong Kang

MathRadar

Send all inquiries to:
MathRadar, LLC
5705 Spring Hill Dr.
Mckinney, Texas 75072

Visit www.mathradar.com for more information and a sneak preview of the MathRadar series of math books.

Send inquires via email: info@mathradar.com

Solutions Manual - Calculus (Differentiation & Integration)

ISBN-13: 978-0-9960450-0-1

ISBN-10: 0996045007

Printed in the United States of America.

Preface

I wrote these books because I am a mother and I have a strong academic background in mathematics. I have a BS degree in Mathematics and Master's degree in Mathematics as well. I have completed Ph.D. program in Biostatistics.

After receiving the big blessing of our first child, a daughter, I decided to forgo my personal career goals to become a full-time mother. When our daughter entered 7th grade, that meant lots of help with her study of math-my passion. However, I struggled to find good math books that would help her understand difficult concepts both clearly and quickly. About two years ago, I talked with my husband and my kids (now I have 2 children 8th grader, Nichole and 1st grader, Richard) about an idea that it would be better to write math books myself at least for my kids because I really want my kids study math with best books. After the conversation, I decided that the best way to help my children was by writing math books for them myself. They wholeheartedly agreed.

That's why I've been able to pour all my knowledge, energy, and soul into Mathradar Series. Because I'm a mom, I would do anything for my children. Thanks to my family's endless support, I wrote them ten books, designed for use in junior high, high-school, and advanced high-school mathematics.

And that would have been the end of my journey, but my husband and children insisted that I share my work outside of our family. They encouraged me to make my work available to other parents looking, as I was, for well-written, great mathematics books for their children.

So I finally decided to publish these books. I do so with the hope that they will help your children find success and confidence in learning and studying mathematics.

But I would never have begun or finished this project without the support of my family. Kyungwan, Nichole, and Richard, you are my world. Thank you.

Aejeong Kang

Introduction

After reading several pages of explanation/description about a certain mathematical concept, you still don't get it.

You have worked on many related problems to understand mathematical concepts, but you still feel completely lost in the mathematical jungle.

You bought a math book with good reviews, but it only offers short answers without detailed solutions. You feel confused and frustrated.

You've tried multiple learning math books, but you've still not getting good grades in math. It seems like math is just not for you.

If any one of these situation sound familiar, the MathRadar series will help you escape!

Everyone has different learning abilities and academic skill. MathRadar series is written and organized with emphasis on helping each individual study mathematics at his/her own pace. Each book consists of clean and concise summaries, callouts, additional supporting explanations, quick reminders and/or shortcuts to facilitate better understanding. Each concept is thoroughly explained with step-by-step instruction and detailed proofs.
With the numerous examples and exercises, students can check their comprehension levels with both basic and more advanced problems.

Carry the MathRadar series with you!
Work on them anytime and anywhere!
Finally, you can start to enjoy mathematics!

Whether you are struggling or advanced in your math skills, the MathRadar series books will build your self-confidence and enjoyment of math.

I hope Math Radar is what you need and will be a great tool for your hard work.
Your comments or suggestions are greatly appreciated.
Please visit my website at www. mathradar.com or email me at ae-jeong@mathradar.com
Thank you very much. And remember, math can be fun!

Chapter 1. The Concept of Limits

#1 Determine whether the following sequences have limits.

(1) $\dfrac{2}{1}, \dfrac{2}{2}, \dfrac{2}{3}, \dfrac{2}{4}, \dfrac{2}{5}, \cdots\cdots$

Since $a_n = \dfrac{2}{n}$, $\displaystyle\lim_{n\to\infty} a_n = \lim_{n\to\infty} \dfrac{2}{n} = 2\lim_{n\to\infty}\dfrac{1}{n} = 2\cdot 0 = 0$

(2) $0, \dfrac{1}{3}, \dfrac{2}{4}, \dfrac{3}{5}, \cdots\cdots$

Since $a_n = \dfrac{n-1}{n+1}$, $\displaystyle\lim_{n\to\infty} a_n = \lim_{n\to\infty}\dfrac{n-1}{n+1} = \lim_{n\to\infty}\left(\dfrac{1-\frac{1}{n}}{1+\frac{1}{n}}\right) = \dfrac{\lim\limits_{n\to\infty}\left(1-\frac{1}{n}\right)}{\lim\limits_{n\to\infty}\left(1+\frac{1}{n}\right)} = \dfrac{1-\lim\limits_{n\to\infty}\frac{1}{n}}{1+\lim\limits_{n\to\infty}\frac{1}{n}} = \dfrac{1-0}{1+0} = 1$

(3) $\dfrac{1^2}{2}, \dfrac{2^2}{4}, \dfrac{3^2}{6}, \dfrac{4^2}{8}, \cdots\cdots$

Since $a_n = \dfrac{n^2}{2n}$, $\displaystyle\lim_{n\to\infty} a_n = \lim_{n\to\infty}\dfrac{n^2}{2n} = \lim_{n\to\infty}\dfrac{n}{2} = \dfrac{1}{2}\lim_{n\to\infty} n = \dfrac{1}{2}\cdot\infty = \infty$ (No limit exists.)

(4) $\sin\dfrac{\pi}{2}, \sin\pi, \sin\dfrac{3\pi}{2}, \sin 2\pi, \cdots\cdots$

The sequence is $1, 0, -1, 0, 1, 0, \cdots\cdots$ (No limit exists.)

(5) 0.9, 0.99, 0.999, 0.9999, $\cdots\cdots$

Since $a_n = 1 - \dfrac{1}{10^n}$, $\displaystyle\lim_{n\to\infty} a_n = \lim_{n\to\infty}\left(1 - \dfrac{1}{10^n}\right) = \lim_{n\to\infty} 1 - \lim_{n\to\infty}\dfrac{1}{10^n} = 1 - 0 = 1$

(6) $-1, -4, -9, -16, \cdots\cdots$

Since $a_n = -n^2$, $\displaystyle\lim_{n\to\infty} a_n = \lim_{n\to\infty} -n^2 = -\lim_{n\to\infty} n^2 = -\infty$ (No limit exists.)

(7) $\{\log_2 n - \log_2(n+1)\}$

Note that $\log_2 n - \log_2(n+1) = \log_2\dfrac{n}{n+1}$

$n = 1;\ \dfrac{n}{n+1} = \dfrac{1}{2}$

$n = 2;\ \dfrac{n}{n+1} = \dfrac{2}{3}$

$n = 3;\ \dfrac{n}{n+1} = \dfrac{3}{4}$

$\vdots$

$\therefore \left\{\dfrac{n}{n+1}\right\}$ converges to 1. $\therefore \displaystyle\lim_{n\to\infty} a_n = \lim_{n\to\infty}\left\{\log_2\dfrac{n}{n+1}\right\} = \log_2 1 = 0$

(8) $\{(-2)^{n-1}\}$

$n = 1, 2, 3, \cdots\cdots \Rightarrow 1, -2, 4, -8, \cdots\cdots$ (No limit exists.)

(9) $\{\sin\dfrac{n\pi}{2}\cos\dfrac{n\pi}{2}\}$

$n = 1, 2, 3, \cdots\cdots \Rightarrow 1\cdot 0,\ 0\cdot(-1),\ (-1)\cdot 0,\ 0\cdot 1,\ \cdots\cdots\ ;\ 0, 0, 0, 0, \cdots\cdots$

The sequence converges to 0.

(10) $\{\tan(n\pi + (-1)^n \frac{\pi}{4}\}$

$n = 1, 2, 3, \cdots\cdots \Rightarrow -1,\ 1, -1,\ 1, \cdots\cdots$ (No limit exists.)

#2 Evaluate the following limits.

(1) $\displaystyle\lim_{n\to\infty} \frac{3n-2}{5n+7}$

$$\lim_{n\to\infty} \frac{3n-2}{5n+7} = \lim_{n\to\infty} \frac{3-\frac{2}{n}}{5+\frac{7}{n}} = \frac{3-0}{5+0} = \frac{3}{5}$$

(2) $\displaystyle\lim_{n\to\infty} \frac{6n^2+n+3}{5n^2-3n-4}$

$$\lim_{n\to\infty} \frac{6n^2+n+3}{5n^2-3n-4} = \lim_{n\to\infty} \frac{6+\frac{1}{n}+\frac{3}{n^2}}{5-\frac{3}{n}-\frac{4}{n^2}} = \frac{6+0+0}{5-0-0} = \frac{6}{5}$$

(3) $\displaystyle\lim_{n\to\infty} \frac{n^2+n-2}{4n^3-1}$

$$\lim_{n\to\infty} \frac{n^2+n-2}{4n^3-1} = \lim_{n\to\infty} \frac{\frac{1}{n}+\frac{1}{n^2}-\frac{2}{n^3}}{4-\frac{1}{n^3}} = \frac{0+0+0}{4-0} = 0$$

(4) $\displaystyle\lim_{n\to\infty} \frac{2n^3}{n^2+1}$

$$\lim_{n\to\infty} \frac{2n^3}{n^2+1} = \lim_{n\to\infty} \frac{2n}{1+\frac{1}{n^2}} = \infty \quad \text{(No limit exists.)}$$

(5) $\displaystyle\lim_{n\to\infty}\{\log(n+3) - \log n\}$

$$\lim_{n\to\infty}\{\log(n+3) - \log n\} = \lim_{n\to\infty} \log\left(\frac{n+3}{n}\right) = \lim_{n\to\infty} \log\left(1+\frac{3}{n}\right) = \log 1 = 0$$

(6) $\displaystyle\lim_{n\to\infty} \frac{(n+1)(3n-2)}{(n-1)(2n+1)}$

Since the degrees of the numerator and denominator are the same, the limit is the ratio of the coefficients of them. $\displaystyle\lim_{n\to\infty} \frac{(n+1)(3n-2)}{(n-1)(2n+1)} = \frac{3}{2}$

(7) $\displaystyle\lim_{n\to\infty} \frac{-2n^3-1}{n^2+1}$

Since the degree of the numerator is greater than the degree of the denominator,

$$\lim_{n\to\infty} \frac{-2n^3-1}{n^2+1} = -\infty$$

(8) $\displaystyle\lim_{n\to\infty}\{\log(10n^2 - 3n) - \log(n^2 + 2)\}$

$$\lim_{n\to\infty}\{\log(10n^2 - 3n) - \log(n^2 + 2)\} = \lim_{n\to\infty}\left\{\log\left(\frac{10n^2-3n}{n^2+2}\right)\right\} = \log 10 = 1$$

(9) $\lim\limits_{n\to\infty}\left\{\left(1-\frac{1}{2^2}\right)\left(1-\frac{1}{3^2}\right)\left(1-\frac{1}{4^2}\right)\cdots\cdots\left(1-\frac{1}{n^2}\right)\right\}$

$\lim\limits_{n\to\infty}\left\{\left(1-\frac{1}{2^2}\right)\left(1-\frac{1}{3^2}\right)\left(1-\frac{1}{4^2}\right)\cdots\cdots\left(1-\frac{1}{n^2}\right)\right\}$

$=\lim\limits_{n\to\infty}\left\{\left(1-\frac{1}{2}\right)\left(1+\frac{1}{2}\right)\left(1-\frac{1}{3}\right)\left(1+\frac{1}{3}\right)\left(1-\frac{1}{4}\right)\left(1+\frac{1}{4}\right)\cdots\cdots\left(1-\frac{1}{n}\right)\left(1+\frac{1}{n}\right)\right\}$

$=\lim\limits_{n\to\infty}\left(\frac{1}{2}\cdot\frac{\cancel{3}}{\cancel{2}}\cdot\frac{\cancel{2}}{\cancel{3}}\cdot\frac{\cancel{4}}{\cancel{3}}\cdot\frac{\cancel{3}}{\cancel{4}}\cdot\frac{\cancel{5}}{\cancel{4}}\cdots\cdots\frac{n\cancel{-1}}{\cancel{n}}\cdot\frac{n+1}{n}\right)$

$=\lim\limits_{n\to\infty}\left(\frac{1}{2}\cdot\frac{n+1}{n}\right)=\frac{1}{2}$

(10) $\lim\limits_{n\to\infty}\dfrac{1^2+2^2+3^2+\cdots\cdots+n^2}{n^3+1}$

Note that: $1+2+3+\cdots\cdots+n=\sum\limits_{k=1}^{n}k=\dfrac{n(n+1)}{2}$

$$1^2+2^2+3^2+\cdots\cdots+n^2=\sum\limits_{k=1}^{n}k^2=\dfrac{n(n+1)(2n+1)}{6}$$

$\lim\limits_{n\to\infty}\dfrac{1^2+2^2+3^2+\cdots\cdots+n^2}{n^3+1}=\lim\limits_{n\to\infty}\dfrac{\frac{n(n+1)(2n+1)}{6}}{n^3+1}=\lim\limits_{n\to\infty}\dfrac{n(n+1)(2n+1)}{6n^3+6}=\dfrac{2}{6}=\dfrac{1}{3}$

(11) $\lim\limits_{n\to\infty}\dfrac{1\cdot2+2\cdot3+3\cdot4+\cdots\cdots+n(n+1)}{n(1+2+3+\cdots\cdots+n)}$

Note that: $1\cdot2+2\cdot3+3\cdot4+\cdots\cdots+n(n+1)=\sum\limits_{k=1}^{n}k(k+1)=\sum\limits_{k=1}^{n}k^2+\sum\limits_{k=1}^{n}k$

$$=\dfrac{n(n+1)(2n+1)}{6}+\dfrac{n(n+1)}{2}=\dfrac{n(n+1)(n+2)}{3}$$

$\lim\limits_{n\to\infty}\dfrac{1\cdot2+2\cdot3+3\cdot4+\cdots\cdots+n(n+1)}{n(1+2+3+\cdots\cdots+n)}=\lim\limits_{n\to\infty}\dfrac{\frac{n(n+1)(n+2)}{3}}{n\cdot\frac{n(n+1)}{2}}=\dfrac{\frac{1}{3}}{\frac{1}{2}}=\dfrac{2}{3}$

(12) $\lim\limits_{n\to\infty}(\sqrt{n+2}-\sqrt{n-2})$

$\lim\limits_{n\to\infty}(\sqrt{n+2}-\sqrt{n-2})=\lim\limits_{n\to\infty}\dfrac{(\sqrt{n+2}-\sqrt{n-2})(\sqrt{n+2}+\sqrt{n-2})}{\sqrt{n+2}+\sqrt{n-2}}=\lim\limits_{n\to\infty}\dfrac{(n+2)-(n-2)}{\sqrt{n+2}+\sqrt{n-2}}$

$$=\lim\limits_{n\to\infty}\dfrac{4}{\sqrt{n+2}+\sqrt{n-2}}=0$$

(13) $\lim\limits_{n\to\infty}\dfrac{1}{\sqrt{n+1}-\sqrt{n}}$

$\lim\limits_{n\to\infty}\dfrac{1}{\sqrt{n+1}-\sqrt{n}}=\lim\limits_{n\to\infty}\dfrac{\sqrt{n+1}+\sqrt{n}}{(\sqrt{n+1}-\sqrt{n})(\sqrt{n+1}+\sqrt{n})}=\lim\limits_{n\to\infty}\dfrac{\sqrt{n+1}+\sqrt{n}}{n+1-n}=\lim\limits_{n\to\infty}(\sqrt{n+1}+\sqrt{n})=\infty$

(14) $\lim\limits_{n\to\infty}(\sqrt{n^2+n}-n)$

$\lim\limits_{n\to\infty}(\sqrt{n^2+n}-n)=\lim\limits_{n\to\infty}\dfrac{(\sqrt{n^2+n}-n)(\sqrt{n^2+n}+n)}{\sqrt{n^2+n}+n}=\lim\limits_{n\to\infty}\dfrac{n^2+n-n^2}{\sqrt{n^2+n}+n}=\lim\limits_{n\to\infty}\dfrac{n}{\sqrt{n^2+n}+n}=\lim\limits_{n\to\infty}\dfrac{1}{\sqrt{\frac{n^2+n}{n^2}}+1}$

$$=\lim\limits_{n\to\infty}\dfrac{1}{\sqrt{1+\frac{1}{n}}+1}=\dfrac{1}{2}$$

(15) $\lim\limits_{n\to\infty}(1+2n^2-3n^3)=\lim\limits_{n\to\infty}n^3\left(\dfrac{1}{n^3}+\dfrac{2}{n}-3\right)=-\infty$

(16) $\lim\limits_{n\to\infty}(3n^3 - 2n^2 - n + 1)$

$$\lim_{n\to\infty}(3n^3 - 2n^2 - n + 1) = \lim_{n\to\infty} n^3\left(3 - \frac{2}{n} - \frac{1}{n^2} + \frac{1}{n^3}\right) = \infty$$

(17) $\lim\limits_{n\to\infty}\sqrt{n}(\sqrt{n+1} - \sqrt{n-1})$

$$\lim_{n\to\infty}\sqrt{n}(\sqrt{n+1} - \sqrt{n-1}) = \lim_{n\to\infty}\frac{\sqrt{n}(\sqrt{n+1}-\sqrt{n-1})(\sqrt{n+1}+\sqrt{n-1})}{\sqrt{n+1}+\sqrt{n-1}} = \lim_{n\to\infty}\frac{\sqrt{n}(n+1-(n-1))}{\sqrt{n+1}+\sqrt{n-1}}$$

$$= \lim_{n\to\infty}\frac{2\sqrt{n}}{\sqrt{n+1}+\sqrt{n-1}} = \lim_{n\to\infty}\frac{2}{\sqrt{1+\frac{1}{n}}+\sqrt{1-\frac{1}{n}}} = \frac{2}{2} = 1$$

(18) $\lim\limits_{n\to\infty}(\sqrt{1+2+3+\cdots\cdots+(n+1)} - \sqrt{1+2+3+\cdots\cdots+n})$

$$\lim_{n\to\infty}(\sqrt{1+2+3+\cdots\cdots+(n+1)} - \sqrt{1+2+3+\cdots\cdots+n})$$

$$= \lim_{n\to\infty}\left(\sqrt{\frac{(n+1)(n+2)}{2}} - \sqrt{\frac{n(n+1)}{2}}\right) = \lim_{n\to\infty}\frac{\sqrt{n+1}}{\sqrt{2}}\left(\sqrt{n+2}-\sqrt{n}\right) = \lim_{n\to\infty}\frac{\sqrt{n+1}}{\sqrt{2}}\cdot\frac{(\sqrt{n+2}-\sqrt{n})(\sqrt{n+2}+\sqrt{n})}{\sqrt{n+2}+\sqrt{n}}$$

$$= \lim_{n\to\infty}\frac{\sqrt{n+1}}{\sqrt{2}}\cdot\frac{(n+2-n)}{(\sqrt{n+2}+\sqrt{n})} = \lim_{n\to\infty}\frac{\sqrt{2}\sqrt{n+1}}{\sqrt{n+2}+\sqrt{n}} = \lim_{n\to\infty}\frac{\sqrt{2}\sqrt{1+\frac{1}{n}}}{\sqrt{1+\frac{2}{n}}+1} = \frac{\sqrt{2}}{2}$$

(19) $\lim\limits_{n\to\infty}\dfrac{a^{n+1}+b^{n+1}}{a^n+b^n}\ (a>b>0)$

$$\lim_{n\to\infty}\frac{a^{n+1}+b^{n+1}}{a^n+b^n} = \lim_{n\to\infty}\frac{a+\frac{b^{n+1}}{a^n}}{1+\frac{b^n}{a^n}} = \lim_{n\to\infty}\frac{a+b\left(\frac{b}{a}\right)^n}{1+\left(\frac{b}{a}\right)^n} = \frac{a}{1} = a \quad \left(\because\ \text{Since } a>b>0,\ \lim_{n\to\infty}\left(\frac{b}{a}\right)^n = 0\right)$$

(20) $\lim\limits_{n\to\infty}\dfrac{1+2+2^2+\cdots\cdots+2^n}{2^n}$

$$\lim_{n\to\infty}\frac{1+2+2^2+\cdots\cdots+2^n}{2^n} = \lim_{n\to\infty}\frac{1}{2^n}(1+2+2^2+\cdots\cdots+2^n) = \lim_{n\to\infty}\frac{1}{2^n}\left(\frac{1-2^{n+1}}{1-2}\right)$$

$$= \lim_{n\to\infty}\frac{1}{2^n}(2^{n+1}-1) = \lim_{n\to\infty}\left(2-\frac{1}{2^n}\right) = 2 \quad \left(\because\ \text{Since } \frac{1}{2}<1,\ \lim_{n\to\infty}\left(\frac{1}{2}\right)^n = 0\right)$$

(21) $\lim\limits_{n\to\infty}(\sqrt{n^2+2n} + \sqrt{n^2+4n} + \cdots\cdots + \sqrt{n^2+20n} - 10n)$

$$\lim_{n\to\infty}(\sqrt{n^2+2n} + \sqrt{n^2+4n} + \cdots\cdots + \sqrt{n^2+20n} - 10n)$$

$$\lim_{n\to\infty}\{(\sqrt{n^2+2n} - n) + (\sqrt{n^2+4n} - n) + \cdots\cdots + (\sqrt{n^2+20n} - n)\}$$

$$= \lim_{n\to\infty}\left\{\left(\frac{n^2+2n-n^2}{\sqrt{n^2+2n}+n}\right) + \left(\frac{n^2+4n-n^2}{\sqrt{n^2+4n}+n}\right) + \cdots\cdots + \left(\frac{n^2+20n-n^2}{\sqrt{n^2+20n}+n}\right)\right\}$$

$$= \lim_{n\to\infty}\left\{\left(\frac{2n}{\sqrt{n^2+2n}+n}\right) + \left(\frac{4n}{\sqrt{n^2+4n}+n}\right) + \cdots\cdots + \left(\frac{20n}{\sqrt{n^2+20n}+n}\right)\right\}$$

$$= \lim_{n\to\infty}\left\{\left(\frac{2}{\sqrt{1+\frac{2}{n}}+1}\right) + \left(\frac{4}{\sqrt{1+\frac{4}{n}}+1}\right) + \cdots\cdots + \left(\frac{20}{\sqrt{1+\frac{20}{n}}+1}\right)\right\}$$

$$= \frac{2}{2} + \frac{4}{2} + \frac{6}{2} + \cdots\cdots + \frac{20}{2} = 1 + 2 + 3 + \cdots\cdots + 10 = \frac{10(10+1)}{2} = 55$$

(22) $\lim\limits_{n\to\infty} \dfrac{\sqrt{n^2-2014}-n}{n-\sqrt{n^2-2013}}$

$$\lim\limits_{n\to\infty} \frac{\sqrt{n^2-2014}-n}{n-\sqrt{n^2-2013}} = \lim\limits_{n\to\infty} \frac{\left(\sqrt{n^2-2014}-n\right)\left(\sqrt{n^2-2014}+n\right)\left(n+\sqrt{n^2-2013}\right)}{\left(n-\sqrt{n^2-2013}\right)\left(\sqrt{n^2-2014}+n\right)\left(n+\sqrt{n^2-2013}\right)}$$

$$= \lim\limits_{n\to\infty} \frac{\left(n^2-2014-n^2\right)\left(n+\sqrt{n^2-2013}\right)}{\left(n^2-n^2+2013\right)\left(\sqrt{n^2-2014}+n\right)} = \lim\limits_{n\to\infty} \frac{-2014\left(n+\sqrt{n^2-2013}\right)}{2013\left(\sqrt{n^2-2014}+n\right)}$$

$$= \lim\limits_{n\to\infty} \frac{-2014\left(1+\sqrt{1-\frac{2013}{n^2}}\right)}{2013\left(\sqrt{1-\frac{2014}{n^2}}+1\right)} = \frac{-2014}{2013} \cdot \frac{2}{2} = -\frac{2014}{2013}$$

(23) $\lim\limits_{n\to\infty} \dfrac{1^2-2^2+3^2-4^2+\cdots\cdots+(2n-1)^2-(2n)^2}{1-n^2}$

$$\lim\limits_{n\to\infty} \frac{1^2-2^2+3^2-4^2+\cdots\cdots+(2n-1)^2-(2n)^2}{1-n^2}$$

$$= \lim\limits_{n\to\infty} \frac{(1-2)(1+2)+(3-4)(3+4)+\cdots\cdots+(2n-1-2n)(2n-1+2n)}{1-n^2}$$

$$= \lim\limits_{n\to\infty} \frac{-(1+2)-(3+4)-\cdots\cdots-(2n-1+2n)}{1-n^2}$$

$$= -\lim\limits_{n\to\infty} \frac{1+2+3+4+\cdots\cdots+2n-1+2n}{1-n^2} = -\lim\limits_{n\to\infty} \frac{\frac{2n(2n+1)}{2}}{1-n^2} = -\lim\limits_{n\to\infty} \frac{n(2n+1)}{1-n^2} = \lim\limits_{n\to\infty} \frac{2n^2+n}{n^2-1} = 2$$

#3 Calculate the limits of the following sequences:

(1) $a_n = \dfrac{1}{1+r^n}$ $(r \neq -1)$

i) When $|r| < 1$, $\lim\limits_{n\to\infty} r^n = 0$ $\quad\therefore\ \lim\limits_{n\to\infty} \dfrac{1}{1+r^n} = 1$

ii) When $r = 1$, $\lim\limits_{n\to\infty} r^n = 1$ $\quad\therefore\ \lim\limits_{n\to\infty} \dfrac{1}{1+r^n} = \dfrac{1}{2}$

iii) When $|r| > 1$, $\lim\limits_{n\to\infty} |r^n| = \infty$ $\quad\therefore\ \lim\limits_{n\to\infty} \dfrac{1}{1+r^n} = 0$

(2) $a_n = \dfrac{r^{n+1}-1}{r^n+1}$ $(r > 0)$

i) When $0 < r < 1$, $\lim\limits_{n\to\infty} r^n = 0$; $\lim\limits_{n\to\infty} r^{n+1} = 0$ $\quad\therefore\ \lim\limits_{n\to\infty} \dfrac{r^{n+1}-1}{r^n+1} = \dfrac{-1}{1} = -1$

ii) When $r = 1$, $\lim\limits_{n\to\infty} r^n = \lim\limits_{n\to\infty} 1^n = 1$; $\lim\limits_{n\to\infty} r^{n+1} = \lim\limits_{n\to\infty} 1^{n+1} = 1$ $\quad\therefore\ \lim\limits_{n\to\infty} \dfrac{r^{n+1}-1}{r^n+1} = 0$

iii) When $r > 1$, $\lim\limits_{n\to\infty} r^n = \infty$ $\quad\therefore\ \lim\limits_{n\to\infty} \dfrac{r^{n+1}-1}{r^n+1} = \lim\limits_{n\to\infty} \dfrac{r-\frac{1}{r^n}}{1+\frac{1}{r^n}} = r$

#4 Find the value.

(1) For two convergent sequences $\{a_n\}$ and $\{b_n\}$, $\lim\limits_{n\to\infty}(a_n + b_n) = 4$ and $\lim\limits_{n\to\infty}(a_n b_n) = 3$.

Find the value of $\lim\limits_{n\to\infty} a_n^2 + \lim\limits_{n\to\infty} b_n^2$.

Let $\lim\limits_{n\to\infty} a_n = a$ and $\lim\limits_{n\to\infty} b_n = b$

Note that $\lim\limits_{n\to\infty}(a_n + b_n) = \lim\limits_{n\to\infty} a_n + \lim\limits_{n\to\infty} b_n = a + b = 4$

$\lim\limits_{n\to\infty}(a_n b_n) = \lim\limits_{n\to\infty} a_n \lim\limits_{n\to\infty} b_n = ab = 3$

$\therefore \lim\limits_{n\to\infty} a_n^2 + \lim\limits_{n\to\infty} b_n^2 = (\lim\limits_{n\to\infty} a_n)^2 + (\lim\limits_{n\to\infty} b_n)^2 = a^2 + b^2 = (a + b)^2 - 2ab = 4^2 - 2 \cdot 3 = 10$

(2) For a sequence $\{a_n\}$ such that $\lim\limits_{n\to\infty}(2n - 1)a_n = 5$, find the value of $\lim\limits_{n\to\infty} na_n$.

$\lim\limits_{n\to\infty} na_n = \lim\limits_{n\to\infty}\left\{(2n - 1)a_n \cdot \frac{n}{2n-1}\right\} = \lim\limits_{n\to\infty}\{(2n - 1)a_n\} \cdot \lim\limits_{n\to\infty}\left(\frac{n}{2n-1}\right) = 5 \cdot \frac{1}{2} = \frac{5}{2}$

(3) For a sequence $\{a_n\}$ such that $\lim\limits_{n\to\infty} \frac{2a_n+3}{a_n-1} = 4$, find the value of $\lim\limits_{n\to\infty} a_n$.

Let $\frac{2a_n+3}{a_n-1} = b_n$

Then, $2a_n + 3 = (a_n - 1)b_n$; $a_n(b_n - 2) = 3 + b_n$ $\cdots\cdots$ ①

Since $\lim\limits_{n\to\infty} b_n = 4$, $b_n - 2 \neq 0$ when $n \to \infty$

By ①, $a_n = \frac{b_n+3}{b_n-2}$

$\therefore \lim\limits_{n\to\infty} a_n = \lim\limits_{n\to\infty} \frac{b_n+3}{b_n-2} = \frac{4+3}{4-2} = \frac{7}{2}$

(4) Find the value of n such that $n^{n^{n^{\cdot^{\cdot^{n}}}}} = \frac{2}{3}$.

$n^{n^{n^{\cdot^{\cdot^{n}}}}} = \frac{2}{3} \Rightarrow \log n^{n^{n^{\cdot^{\cdot^{n}}}}} = \log\frac{2}{3}$

$\therefore n^{n^{\cdot^{\cdot^{n}}}} \log n = \log\frac{2}{3}$ $\qquad \therefore \frac{2}{3}\log n = \log\frac{2}{3}$ $\qquad \therefore \log n = \frac{3}{2}\log\frac{2}{3} = \log\left(\frac{2}{3}\right)^{\frac{3}{2}}$

$\therefore n = \left(\frac{2}{3}\right)^{\frac{3}{2}} = \sqrt{\left(\frac{2}{3}\right)^3} = \frac{2\sqrt{2}}{3\sqrt{3}} = \frac{2\sqrt{6}}{9}$

(5) For a positive integer n, $A_n = \left\{(x, y) \mid 0 \leq x \leq \frac{1}{2} + \frac{1}{2^2} + \frac{1}{2^3} + \cdots\cdots + \frac{1}{2^n}, \ 0 \leq y \leq 1\right\}$.

When a_n is the maximum value of $2x + y$, find the value of $\lim\limits_{n\to\infty} a_n$.

Since $\frac{1}{2} + \frac{1}{2^2} + \frac{1}{2^3} + \cdots\cdots + \frac{1}{2^n} = \frac{\left(\frac{1}{2}\right)\left\{1-\left(\frac{1}{2}\right)^n\right\}}{1-\left(\frac{1}{2}\right)} = 1 - \frac{1}{2^n}, \ 0 \leq x \leq 1 - \frac{1}{2^n}$

$\therefore 0 \leq 2x \leq 2 - \frac{1}{2^{n-1}}$

Since $0 \leq y \leq 1$, $0 \leq 2x + y \leq 3 - \frac{1}{2^{n-1}}$

Since a_n is the maximum value of $2x + y$, $a_n = 3 - \dfrac{1}{2^{n-1}}$

$\therefore \displaystyle\lim_{n\to\infty} a_n = \lim_{n\to\infty}\left(3 - \dfrac{1}{2^{n-1}}\right) = 3$

(6) For real numbers a, b and c such that $\displaystyle\lim_{n\to\infty}\dfrac{bn^3+cn-4}{an^2+3n-2} = 5$, **find the value of $a + b + c$.**

If $b \neq 0$, then $\displaystyle\lim_{n\to\infty}\dfrac{bn^3+cn-4}{an^2+3n-2} = \infty$

Thus, b must be zero to have the limit; i.e., $b = 0$

If $a \neq 0$, then $\displaystyle\lim_{n\to\infty}\dfrac{bn^3+cn-4}{an^2+3n-2} = \lim_{n\to\infty}\dfrac{cn-4}{an^2+3n-2} = 0$ which is not 5. Thus, $a = 0$

Therefore, $\displaystyle\lim_{n\to\infty}\dfrac{bn^3+cn-4}{an^2+3n-2} = \lim_{n\to\infty}\dfrac{cn-4}{3n-2} = \dfrac{c}{3} = 5$ and hence $c = 15$

$\therefore\ a + b + c = 0 + 0 + 15 = 15$

(7) For real numbers a and b such that $\displaystyle\lim_{n\to\infty}\left(\sqrt{n^2 + an + 3} - \sqrt{n^2 + bn + 2}\right) = 4$,

find the value of $a - b$.

$\displaystyle\lim_{n\to\infty}\left(\sqrt{n^2 + an + 3} - \sqrt{n^2 + bn + 2}\right) = \lim_{n\to\infty}\dfrac{(n^2+an+3)-(n^2+bn+2)}{\sqrt{n^2+an+3}+\sqrt{n^2+bn+2}}$

$= \displaystyle\lim_{n\to\infty}\dfrac{(a-b)n+1}{\sqrt{n^2+an+3}+\sqrt{n^2+bn+2}} = \lim_{n\to\infty}\dfrac{(a-b)+\frac{1}{n}}{\sqrt{1+\frac{a}{n}+\frac{3}{n^2}}+\sqrt{1+\frac{b}{n}+\frac{2}{n^2}}}$

$= \dfrac{a-b}{\sqrt{1}+\sqrt{1}} = \dfrac{a-b}{2}$

$\therefore\ \dfrac{a-b}{2} = 4 \qquad \therefore\ a - b = 8$

(8) For convergent sequence $\{a_n\}$ such that $\displaystyle\lim_{n\to\infty}\dfrac{3^n-2^{n+1}a_n}{3^n a_n+2^n} = 5$, **find the value of** $\displaystyle\lim_{n\to\infty} a_n$.

$\displaystyle\lim_{n\to\infty}\dfrac{3^n-2^{n+1}a_n}{3^n a_n+2^n} = \lim_{n\to\infty}\dfrac{1-2\left(\frac{2}{3}\right)^n a_n}{a_n+\left(\frac{2}{3}\right)^n} = \lim_{n\to\infty}\dfrac{1}{a_n} = 5$

$\therefore\ \displaystyle\lim_{n\to\infty} a_n = \dfrac{1}{5}$

(9) When an infinite geometric sequence $\left\{\left(\dfrac{a}{2}\right)^{n-1}(a + 2)\right\}$ **is convergent,**

find the range of the real number a.

The first term of the sequence $\left\{\left(\dfrac{a}{2}\right)^{n-1}(a + 2)\right\}$ is $a + 2$ and the common ratio of it is $\dfrac{a}{2}$.

To have a limit, $a + 2 = 0$ or $-1 < \dfrac{a}{2} \leq 1$ \qquad That is, $a = -2$ or $-2 < a \leq 2$

Therefore, $-2 \leq a \leq 2$

(10) For any positive integer n, $A = \begin{bmatrix} n^2 + n & 2 \\ n^2 & 3 \end{bmatrix}$ is a 2×2 matrix.

When a_n is the sum of all entries of the inverse matrix of A, find the value of $\lim\limits_{n \to \infty} a_n$.

$A = \begin{bmatrix} n^2 + n & 2 \\ n^2 & 3 \end{bmatrix} \Rightarrow A^{-1} = \frac{1}{3(n^2+n)-2n^2}\begin{bmatrix} 3 & -2 \\ -n^2 & n^2+n \end{bmatrix} = \frac{1}{n^2+3n}\begin{bmatrix} 3 & -2 \\ -n^2 & n^2+n \end{bmatrix}$

$\therefore \ a_n = \frac{1}{n^2+3n}(3 - 2 - n^2 + n^2 + n) = \frac{1+n}{n^2+3n}$

$\therefore \ \lim\limits_{n \to \infty} a_n = \lim\limits_{n \to \infty}\left(\frac{1+n}{n^2+3n}\right) = \lim\limits_{n \to \infty}\left(\frac{\frac{1}{n^2}+\frac{1}{n}}{1+\frac{3}{n}}\right) = \frac{0}{1} = 0$

(11) For any positive integer n, $f(n) = 1 + 2 + 3 + \cdots\cdots + n$.

Find the value of $\lim\limits_{n \to \infty} \frac{f(2n^2)}{\{f(n)\}^2}$.

$f(n) = 1 + 2 + 3 + \cdots\cdots + n = \frac{n(n+1)}{2}$

$f(2n^2) = \frac{2n^2(2n^2+1)}{2} = n^2(2n^2 + 1)$

$\{f(n)\}^2 = \left\{\frac{n(n+1)}{2}\right\}^2 = \frac{n^2(n+1)^2}{4}$

$\therefore \ \lim\limits_{n \to \infty} \frac{f(2n^2)}{\{f(n)\}^2} = \lim\limits_{n \to \infty} \frac{n^2(2n^2+1)}{\frac{n^2(n+1)^2}{4}} = \lim\limits_{n \to \infty} \frac{4n^2(2n^2+1)}{n^2(n+1)^2} = 8$

(12) For two points $P(n, f(n))$ and $Q(n + 1, f(n + 1))$ on a quadratic function $f(x) = 2x^2$, let a_n be the distance between P and Q.

Find the value of $\lim\limits_{n \to \infty} \frac{a_n}{n}$ (where n is a positive integer).

Since $P\left(n, f(n)\right) = P(n, 2n^2)$ and $Q\left(n + 1, f(n + 1)\right) = Q(n + 1, 2(n + 1)^2)$,

$a_n = \sqrt{(n + 1 - n)^2 + \{2(n + 1)^2 - 2n^2\}^2} = \sqrt{1 + (4n + 2)^2} = \sqrt{1 + 16n^2 + 16n + 4}$

$\quad = \sqrt{16n^2 + 16n + 5}$

$\therefore \ \lim\limits_{n \to \infty} \frac{a_n}{n} = \lim\limits_{n \to \infty} \frac{\sqrt{16n^2+16n+5}}{n} = \lim\limits_{n \to \infty} \frac{\sqrt{16+\frac{16}{n}+\frac{5}{n^2}}}{1} = \sqrt{16} = 4$

(13) For a sequence $\{a_n\}$, $a_n = n\left(\sqrt{\frac{n-1}{2n+1}} - a\right)$ $(n \geq 1, \ n; \text{positive integer})$,

find the value of the real number a so that $\lim\limits_{n \to \infty} a_n$ exists and calculate the limit.

Since $\lim\limits_{n \to \infty} n = \infty$, $\lim\limits_{n \to \infty}\left(\sqrt{\frac{n-1}{2n+1}} - a\right) = 0$ to have a limit.

Since $\lim\limits_{n \to \infty}\sqrt{\frac{n-1}{2n+1}} = \lim\limits_{n \to \infty}\sqrt{\frac{1-\frac{1}{n}}{2+\frac{1}{n}}} = \sqrt{\frac{1}{2}} = \frac{\sqrt{2}}{2}$, $a = \frac{\sqrt{2}}{2}$

$$\lim_{n\to\infty} a_n = \lim_{n\to\infty} n\left(\sqrt{\frac{n-1}{2n+1}} - \frac{\sqrt{2}}{2}\right) = \lim_{n\to\infty} n\left(\frac{\frac{n-1}{2n+1} - \frac{1}{2}}{\sqrt{\frac{n-1}{2n+1}} + \frac{\sqrt{2}}{2}}\right) = \lim_{n\to\infty} n\left(\frac{\frac{-3}{4n+2}}{\sqrt{\frac{n-1}{2n+1}} + \frac{\sqrt{2}}{2}}\right)$$

$$= \frac{\frac{-3}{4}}{\sqrt{\frac{1}{2}} + \frac{\sqrt{2}}{2}} = \frac{-3}{4\sqrt{2}} = \frac{-3\sqrt{2}}{8}$$

(14) For sequences $\{a_n\}$ and $\{b_n\}$ such that $\lim_{n\to\infty}(n-1)a_n = 5$ and $\lim_{n\to\infty}(2n^2-1)b_n = 10$, find the value of $\lim_{n\to\infty}\{(n+1)^3 a_n b_n\}$.

Let $(n-1)a_n = c_n$

Then, $a_n = \frac{c_n}{n-1}$, $n \neq 1$

Let $(2n^2 - 1)b_n = d_n$

Then, $b_n = \frac{d_n}{2n^2-1}$, $2n^2 \neq 1$

Since $\lim_{n\to\infty} c_n = 5$ and $\lim_{n\to\infty} d_n = 10$,

$$\lim_{n\to\infty}\{(n+1)^3 a_n b_n\} = \lim_{n\to\infty}\left\{(n+1)^3 \cdot \frac{c_n}{n-1} \cdot \frac{d_n}{2n^2-1}\right\} = \lim_{n\to\infty}\left\{\frac{(n+1)^3}{(n-1)(2n^2-1)}c_n d_n\right\}$$

$$= \lim_{n\to\infty}\left\{\frac{(n+1)^3}{(n-1)(2n^2-1)}\right\} \cdot \lim_{n\to\infty} c_n \cdot \lim_{n\to\infty} d_n = \frac{1}{2}\cdot 5 \cdot 10 = 25$$

(15) For a convergent sequence $\{a_n\}$, $\frac{a_n-4}{a_n+2} = \frac{1^3+2^3+3^3+\cdots\cdots+n^3}{n^4+1}$. Find the value of $\lim_{n\to\infty} a_n$.

$$\lim_{n\to\infty}\frac{a_n-4}{a_n+2} = \lim_{n\to\infty}\frac{1^3+2^3+3^3+\cdots\cdots+n^3}{n^4+1} = \lim_{n\to\infty}\left[\left(\frac{1}{n^4+1}\right)\left\{\frac{n(n+1)}{2}\right\}^2\right] = \frac{1}{4}$$

Let $\lim_{n\to\infty} a_n = a$. Then, $\lim_{n\to\infty}\frac{a_n-4}{a_n+2} = \frac{a-4}{a+2} = \frac{1}{4}$

$\therefore$ $4(a-4) = a+2$; $3a = 18$; $a = 6$

Therefore, $\lim_{n\to\infty} a_n = 6$

(16) When $\lim_{n\to\infty}(a_n - b_n) = 2$ and $\lim_{n\to\infty} a_n = \infty$, find the value of $\lim_{n\to\infty}\left(\frac{a_n^2}{b_n} - \frac{b_n^2}{a_n}\right)$. $(a_n b_n = 0)$

Let $a_n - b_n = c_n$

Then, $1 - \frac{b_n}{a_n} = \frac{c_n}{a_n}$

Since $\lim_{n\to\infty} c_n = 2$ and $\lim_{n\to\infty} a_n = \infty$, $\lim_{n\to\infty}\frac{c_n}{a_n} = 0$

$\therefore$ $\lim_{n\to\infty}\left(1 - \frac{b_n}{a_n}\right) = 0$; $\lim_{n\to\infty}\frac{b_n}{a_n} = 1$

$$\lim_{n\to\infty}\left(\frac{a_n^2}{b_n} - \frac{b_n^2}{a_n}\right) = \lim_{n\to\infty}\left(\frac{a_n^3 - b_n^3}{a_n b_n}\right) = \lim_{n\to\infty}\left\{\frac{(a_n-b_n)(a_n^2 + a_n b_n + b_n^2)}{a_n b_n}\right\}$$

$$= \lim_{n\to\infty} \left\{ (a_n - b_n)\left(\frac{a_n}{b_n} + 1 + \frac{b_n}{a_n}\right) \right\} = \lim_{n\to\infty} (a_n - b_n) \cdot \lim_{n\to\infty} \left(\frac{a_n}{b_n} + 1 + \frac{b_n}{a_n}\right)$$

$$= \lim_{n\to\infty} c_n \cdot (1 + 1 + 1) = 2 \cdot 3 = 6$$

(17) For a sequence $\{a_n\}$ such that $[\log_2 a_n] = 3n$ (n; positive integer),

find the value of $\lim\limits_{n\to\infty} \dfrac{\log_8 a_n}{3n}$. (Where $[x]$ is the greatest integer less than or equal to x.)

$[\log_2 a_n] = 3n \quad\Rightarrow\quad 3n \le \log_2 a_n < 3n + 1 \quad\Rightarrow\quad n \le \frac{1}{3}\log_2 a_n < n + \frac{1}{3}$

$\therefore\ n \le \log_{2^3} a_n < n + \frac{1}{3}$; $n \le \log_8 a_n < n + \frac{1}{3}$

$\therefore\ \dfrac{n}{3n} \le \dfrac{\log_8 a_n}{3n} < \dfrac{n+\frac{1}{3}}{3n}$; $\dfrac{1}{3} \le \dfrac{\log_8 a_n}{3n} < \dfrac{1}{3} + \dfrac{1}{9n}$

Since $\lim\limits_{n\to\infty} \dfrac{1}{3} = \dfrac{1}{3}$ and $\lim\limits_{n\to\infty}\left(\dfrac{1}{3} + \dfrac{1}{9n}\right) = \dfrac{1}{3}$, $\lim\limits_{n\to\infty} \dfrac{\log_8 a_n}{3n} = \dfrac{1}{3}$

(18) For real numbers a ($a < 0$) and b, $a + b = 1$.

For a matrix $A = \begin{bmatrix} 1 & a \\ 0 & b \end{bmatrix}$, $A^n = \begin{bmatrix} 1 & a_n \\ 0 & b_n \end{bmatrix}$ ($n = 1, 2, 3, \cdots$). Find the value of $\lim\limits_{n\to\infty} \dfrac{b_n}{a_n}$.

$A = \begin{bmatrix} 1 & a \\ 0 & b \end{bmatrix}$

$A^2 = \begin{bmatrix} 1 & a \\ 0 & b \end{bmatrix}\begin{bmatrix} 1 & a \\ 0 & b \end{bmatrix} = \begin{bmatrix} 1 & a + ab \\ 0 & b^2 \end{bmatrix} = \begin{bmatrix} 1 & a(1 + b) \\ 0 & b^2 \end{bmatrix}$

$A^3 = \begin{bmatrix} 1 & a \\ 0 & b \end{bmatrix}\begin{bmatrix} 1 & a(1+b) \\ 0 & b^2 \end{bmatrix} = \begin{bmatrix} 1 & a + ab + ab^2 \\ 0 & b^3 \end{bmatrix} = \begin{bmatrix} 1 & a(1 + b + b^2) \\ 0 & b^3 \end{bmatrix}$

$\vdots$

$A^n = \begin{bmatrix} 1 & a(1 + b + b^2 + \cdots + b^{n-1}) \\ 0 & b^n \end{bmatrix} = \begin{bmatrix} 1 & \frac{a(1-b^n)}{1-b} \\ 0 & b^n \end{bmatrix}$

$\therefore\ a_n = \dfrac{a(1-b^n)}{1-b}$ and $b_n = b^n$

Since $a < 0$ and $a + b = 1$, $a < 0$ and $b > 1$

Since $\lim\limits_{n\to\infty} b_n = \lim\limits_{n\to\infty} b^n = \infty$,

$$\lim_{n\to\infty} \frac{b_n}{a_n} = \lim_{n\to\infty} \frac{b^n}{\frac{a(1-b^n)}{1-b}} = \lim_{n\to\infty} \frac{1}{\frac{a}{1-b}\left(\frac{1}{b^n} - 1\right)} = -\frac{1}{\frac{a}{1-b}} = -\frac{1-b}{a} = \frac{b-1}{a} = \frac{-a}{a} = -1$$

#5 Find the limit.

(1) For a sequence $\{a_n\}$ such that $a_1 = 1$ and $a_{n+1} = \frac{1}{2}a_n + 3$ $(n \geq 1)$, find:

    1) a_n   **2) $\lim\limits_{n \to \infty} a_n$**

$a_1 = 1$

$a_{n+1} = \frac{1}{2}a_n + 3 \quad \Rightarrow \quad a_{n+1} = \frac{1}{2}(a_n + 6)$

$a_{n+1} - 6 = \frac{1}{2}(a_n + 6) - 6 = \frac{1}{2}(a_n - 6)$

$\therefore$ $\{a_n - 6\}$ is a geometric sequence with the common ratio $\frac{1}{2}$ and

    the first term $a_1 - 6 = 1 - 6 = -5$

1) Since $a_n - 6 = (-5) \cdot \left(\frac{1}{2}\right)^{n-1}$, $\quad a_n = -5\left(\frac{1}{2}\right)^{n-1} + 6$

2) $\lim\limits_{n \to \infty} a_n = \lim\limits_{n \to \infty}\left\{-5\left(\frac{1}{2}\right)^{n-1} + 6\right\} = 6$

(2) For a sequence $\{a_n\}$ such that $a_1 = 1$ and $a_{n+1} = 2a_n$, find the value of $\lim\limits_{n \to \infty}\frac{a_n}{3^{n-1}}$.

Since $\{a_n\}$ is a geometric sequence with common ratio r, $a_n = 1 \cdot 2^{n-1} = 2^{n-1}$

$\therefore \quad \lim\limits_{n \to \infty}\frac{a_n}{3^{n-1}} = \lim\limits_{n \to \infty}\frac{2^{n-1}}{3^{n-1}} = \lim\limits_{n \to \infty}\left(\frac{2}{3}\right)^{n-1} = 0$

(3) For a sequence $\{a_n\}$ such that $a_1 = 1$, $a_2 = 3$, and $2a_{n+2} - 3a_{n+1} + a_n = 0$ $(n \geq 1)$,

    find the value of $\lim\limits_{n \to \infty} a_n$.

$2a_{n+2} - 3a_{n+1} + a_n = 0 \quad \Rightarrow \quad 2(a_{n+2} - a_{n+1}) = a_{n+1} - a_n \quad \Rightarrow \quad a_{n+2} - a_{n+1} = \frac{1}{2}(a_{n+1} - a_n)$

$\therefore \quad a_n = a_1 + \sum_{k=1}^{n-1}(a_2 - a_1)\left(\frac{1}{2}\right)^{k-1} = 1 + \sum_{k=1}^{n-1}(3-1)\left(\frac{1}{2}\right)^{k-1} = 1 + \sum_{k=1}^{n-1}\left(\frac{1}{2}\right)^{k-2} = 1 + \frac{2\left\{1 - \left(\frac{1}{2}\right)^{n-1}\right\}}{1 - \left(\frac{1}{2}\right)}$

$= 1 + 4\left\{1 - \left(\frac{1}{2}\right)^{n-1}\right\} = 5 - 2^2\left(\frac{1}{2}\right)^{n-1} = 5 - \left(\frac{1}{2}\right)^{-2}\left(\frac{1}{2}\right)^{n-1} = 5 - \left(\frac{1}{2}\right)^{n-3}$

$\therefore \quad \lim\limits_{n \to \infty} a_n = \lim\limits_{n \to \infty}\left\{5 - \left(\frac{1}{2}\right)^{n-3}\right\} = 5$

(4) For a sequence $\{a_n\}$, let S_n be the partial finite sum of the first n terms.

    Find the value of $\lim\limits_{n \to \infty} a_n$ **1) when $S_1 = 10$ and $S_{n+1} = \frac{1}{2}S_n + 1$ $(n = 1, 2, 3, \cdots)$**

                             2) when $S_n = \frac{n(n+3)}{4(n+1)(n+2)}$

1) $S_{n+1} = \frac{1}{2}S_n + 1 \quad \Rightarrow \quad S_{n+1} - 2 = \frac{1}{2}(S_n - 2)$

    $\therefore$ $\{S_n - 2\}$ is a geometric sequence with first term $S_1 - 2 = 10 - 2 = 8$ and

    common ratio $\frac{1}{2}$. $\therefore$ $S_n - 2 = 8\left(\frac{1}{2}\right)^{n-1}$; $\quad S_n = 2 + 8\left(\frac{1}{2}\right)^{n-1}$

Note that $a_n = S_n - S_{n-1}$

$$= \left\{ 2 + 8\left(\tfrac{1}{2}\right)^{n-1} \right\} - \left\{ 2 + 8\left(\tfrac{1}{2}\right)^{n-2} \right\} = 8\left(\tfrac{1}{2}\right)^{n-2}\left(\tfrac{1}{2} - 1\right) = -8\left(\tfrac{1}{2}\right)^{n-1}, \ n \geq 2$$

$\therefore \ \lim\limits_{n\to\infty} a_n = \lim\limits_{n\to\infty} \left\{ -8\left(\tfrac{1}{2}\right)^{n-1} \right\} = 0$

2) When $n \geq 2$, $a_n = S_n - S_{n-1} = \dfrac{n(n+3)}{4(n+1)(n+2)} - \dfrac{(n-1)(n+2)}{4n(n+1)}$

$\therefore \ \lim\limits_{n\to\infty} a_n = \lim\limits_{n\to\infty} \left\{ \dfrac{n(n+3)}{4(n+1)(n+2)} - \dfrac{(n-1)(n+2)}{4n(n+1)} \right\}$

$\qquad = \lim\limits_{n\to\infty} \left\{ \dfrac{n(n+3)}{4(n+1)(n+2)} \right\} - \lim\limits_{n\to\infty} \left\{ \dfrac{(n-1)(n+2)}{4n(n+1)} \right\} = \dfrac{1}{4} - \dfrac{1}{4} = 0$

(5) For a positive integer n, let a_n be the decimal part of $\sqrt{n^2 + 3n}$.

Find the value of $\lim\limits_{n\to\infty} \dfrac{10}{a_n}$.

Note that $(n + 1)^2 < n^2 + 3n < (n + 2)^2$, $n \geq 2$

$\therefore \ n + 1 < \sqrt{n^2 + 3n} < n + 2, \ n \geq 2$

$\therefore \ \sqrt{n^2 + 3n} = (n + 1).\times\times\times$

That is, the integer part of $\sqrt{n^2 + 3n}$ is $n + 1$ and the decimal part of it is $\sqrt{n^2 + 3n} - (n + 1)$.

Thus, $a_n = \sqrt{n^2 + 3n} - (n + 1)$

$\therefore \ \lim\limits_{n\to\infty} \dfrac{10}{a_n} = \lim\limits_{n\to\infty} \left\{ \dfrac{10}{\sqrt{n^2+3n}-(n+1)} \right\} = \lim\limits_{n\to\infty} \left\{ \dfrac{10\left(\sqrt{n^2+3n}+(n+1)\right)}{\left(\sqrt{n^2+3n}-(n+1)\right)\left(\sqrt{n^2+3n}+(n+1)\right)} \right\}$

$\qquad = \lim\limits_{n\to\infty} \left\{ \dfrac{10\left(\sqrt{n^2+3n}+(n+1)\right)}{(n^2+3n)-(n+1)^2} \right\} = \lim\limits_{n\to\infty} \left\{ \dfrac{10\left(\sqrt{n^2+3n}+(n+1)\right)}{n-1} \right\}$

$\qquad = \lim\limits_{n\to\infty} \left\{ \dfrac{10\left(\sqrt{\frac{n^2+3n}{n^2}}+\frac{n+1}{n}\right)}{\frac{n-1}{n}} \right\} = \lim\limits_{n\to\infty} \left\{ \dfrac{10\left(\sqrt{1+\frac{3}{n}}+1+\frac{1}{n}\right)}{1-\frac{1}{n}} \right\} = \dfrac{10(\sqrt{1}+1)}{1} = 20$

(6) For a sequence $\{a_n\}$ such that $2n^2 - 1 < na_n < 2n^2 + 1 \ (n = 1, 2, 3, \cdots\cdots)$,

find the value of $\lim\limits_{n\to\infty} \dfrac{a_n+3n-1}{n+3}$.

$2n^2 - 1 < na_n < 2n^2 + 1 \ \Rightarrow \ \dfrac{2n^2-1}{n^2} < \dfrac{na_n}{n^2} < \dfrac{2n^2+1}{n^2} \ \ ; \ \ \dfrac{2n^2-1}{n^2} < \dfrac{a_n}{n} < \dfrac{2n^2+1}{n^2}$

Since $\lim\limits_{n\to\infty} \dfrac{2n^2-1}{n^2} = 2$ and $\lim\limits_{n\to\infty} \dfrac{2n^2+1}{n^2} = 2$, $\lim\limits_{n\to\infty} \dfrac{a_n}{n} = 2$

$\therefore \ \lim\limits_{n\to\infty} \dfrac{a_n+3n-1}{n+3} = \lim\limits_{n\to\infty} \dfrac{\frac{a_n}{n}+3-\frac{1}{n}}{1+\frac{3}{n}} = \dfrac{2+3-0}{1+0} = 5$

(6) If $\left\{a_n + \frac{n}{2n+1}\right\}$ is convergent, then $\{a_n{}^2\}$ is convergent.

True

If $\{a_n\}$ diverges, then $\left\{a_n + \frac{n}{2n+1}\right\}$ diverges. Thus, $\{a_n\}$ is convergent.

Let $\lim\limits_{n\to\infty} a_n = a$.

Then, $\lim\limits_{n\to\infty} a_n{}^2 = \left(\lim\limits_{n\to\infty} a_n\right)^2 = a^2$;i.e., $\{a_n{}^2\}$ converges to a^2.

Therefore, $\{a_n{}^2\}$ is convergent sequence.

(7) For two convergent sequences $\{a_n\}$ and $\{b_n\}$,

  if $\{a_n - b_n\}$ converges to a negative number, then $a_n < b_n$.

False $\because$ Let $\{a_n\}$; $2, 1, 1, 1, \cdots\cdots$

$\{b_n\}$; $1, 2, 2, 2, \cdots\cdots$

Then $\{a_n\}$ and $\{b_n\}$ are convergent sequences.

$\{a_n - b_n\}$ converges to -1.

But, $a_n \not< b_n$ when $n = 1$

(8) If $\lim\limits_{n\to\infty} a_{2n-1} = a$, $\lim\limits_{n\to\infty} a_{2n} = b$, and $a < b$, then $\{a_n\}$ is convergent.

False $\because$ Let $a_n = (-1)^n$

Then, $\lim\limits_{n\to\infty} a_{2n-1} = \lim\limits_{n\to\infty} (-1)^{2n-1} = -1 = a$,

$\lim\limits_{n\to\infty} a_{2n} = \lim\limits_{n\to\infty} (-1)^{2n} = 1 = b$, and $a < b$

But, $\lim\limits_{n\to\infty} a_n = \lim\limits_{n\to\infty} (-1)^n$; No limit exists.

(9) For two convergent sequences $\{a_n\}$ and $\{b_n\}$,

  if $a_n < c_n < b_n$ and $\lim\limits_{n\to\infty}(b_n - a_n) = 0$, then $\{c_n\}$ is convergent.

True

Let $\lim\limits_{n\to\infty} a_n = a$ and $\lim\limits_{n\to\infty} b_n = b$

Since $\lim\limits_{n\to\infty}(b_n - a_n) = \lim\limits_{n\to\infty} b_n - \lim\limits_{n\to\infty} a_n = b - a = 0$, $a = b$

Since $a_n < c_n < b_n$, $\lim\limits_{n\to\infty} c_n = \lim\limits_{n\to\infty} a_n = \lim\limits_{n\to\infty} b_n = a (= b)$;i.e., $\{c_n\}$ converges.

(10) For a sequence $\{a_n\}$ with $a_1 = 1$, $\dfrac{1}{a_{n+1}} = \dfrac{1}{a_n} + 1$ $(n = 1, 2, 3, \cdots\cdots)$, $\lim\limits_{n\to\infty} a_n = 0$

True

$a_1 = 1$, $\dfrac{1}{a_{n+1}} = \dfrac{1}{a_n} + 1$

$\Rightarrow \left\{\dfrac{1}{a_n}\right\}$ is an arithmetic sequence with first term $\dfrac{1}{a_1} = 1$ and common difference 1.

$$\therefore \ \frac{1}{a_n} = \frac{1}{a_1} + (n-1)\cdot 1 = n \qquad \therefore \ a_n = \frac{1}{n}$$

Therefore, $\lim\limits_{n\to\infty} a_n = \lim\limits_{n\to\infty} \frac{1}{n} = 0$

(11) If S_n is the finite sum of first n terms such that $S_n = ka_n + 1$ $(k\,(\neq 1)$; constant),

then i) $a_1 = \dfrac{1}{1-k}$

ii) $\{a_n\}$ is a geometric sequence.

iii) If $k = \dfrac{2}{3}$, then $\{a_n\}$ is convergent.

i)   True

Since $a_1 = S_1 = ka_1 + 1$, $a_1(1-k) = 1$ $\qquad \therefore \ a_1 = \dfrac{1}{1-k}$, $k \neq 1$

ii)   True

When $n \geq 2$, $a_n = S_n - S_{n-1} = (ka_n + 1) - (ka_{n-1} + 1) = ka_n - ka_{n-1}$

$\therefore \ (1-k)a_n = -ka_{n-1} \qquad \therefore \ a_n = \dfrac{k}{k-1} a_{n-1} \ (\because k \neq 1)$

$\therefore \ \{a_n\}$ is a geometric sequence with common ratio $\dfrac{k}{k-1}$.

iii)   False   Let $k = \dfrac{2}{3}$

Then the ratio is $\dfrac{k}{k-1} = \dfrac{\frac{2}{3}}{\frac{2}{3}-1} = \dfrac{\frac{2}{3}}{-\frac{1}{3}} = -2$

Since the ratio is less than -1, $\{a_n\}$ diverges.

#7 For a sequence $\{a_n\}$ with $a_1 = 2$ and $a_i > 0$ $(i = 1, 2, 3, \cdots\cdots)$,

a quadratic equation $x^2 - \sqrt{a_n}\, x + (a_{n+1} - 1) = 0$ has a double root for any positive integer

n. Find the value of $\lim\limits_{n\to\infty} a_n$.

Let D be the discriminant of the equation. Then $D = 0$ (to have a double root)

$D = (\sqrt{a_n})^2 - 4(a_{n+1} - 1) = a_n - 4a_{n+1} + 4 = 0$

$\therefore \ 4a_{n+1} = a_n + 4 \ ; \ a_{n+1} = \dfrac{1}{4} a_n + 1 \ ; \ a_{n+1} - \dfrac{4}{3} = \dfrac{1}{4}\left(a_n - \dfrac{4}{3}\right)$

$\therefore \ \left\{a_n - \dfrac{4}{3}\right\}$ is a geometric sequence with first term $a_1 - \dfrac{4}{3} = 2 - \dfrac{4}{3} = \dfrac{2}{3}$ and common ratio $\dfrac{1}{4}$.

$\therefore \ a_n - \dfrac{4}{3} = \dfrac{2}{3}\left(\dfrac{1}{4}\right)^{n-1} \ ; \ a_n = \dfrac{2}{3}\left(\dfrac{1}{4}\right)^{n-1} + \dfrac{4}{3}$

$\therefore \ \lim\limits_{n\to\infty} a_n = \lim\limits_{n\to\infty}\left\{\dfrac{2}{3}\left(\dfrac{1}{4}\right)^{n-1} + \dfrac{4}{3}\right\} = \dfrac{4}{3}$

#8 For a sequence $\{a_n\}$ with $a_n > 0$ $(n = 1, 2, 3, \cdots\cdots)$,

an inequality $x^2 - 4\sqrt{2a_{n+1}}\, x + a_n > 0$ is always true.

(1) Find the relationship between a_n and a_{n+1}.

Let D be the discriminant of the equation $x^2 - 4\sqrt{2a_{n+1}}\, x + a_n = 0$.

Since the inequality is always true, $D < 0$; $\dfrac{D}{4} < 0$

$\therefore \dfrac{D}{4} = (2\sqrt{2a_{n+1}})^2 - a_n = 8a_{n+1} - a_n < 0$

$\therefore a_{n+1} < \dfrac{1}{8} a_n$

(2) Find the value of $\lim\limits_{n \to \infty} a_n$.

$a_{n+1} < \dfrac{1}{8} a_n$

$\Rightarrow a_n < \dfrac{1}{8} a_{n-1} < \left(\dfrac{1}{8}\right)^2 a_{n-2} < \left(\dfrac{1}{8}\right)^3 a_{n-3} < \cdots\cdots < \left(\dfrac{1}{8}\right)^{n-1} a_1$

$\therefore 0 < a_n < \left(\dfrac{1}{8}\right)^{n-1} a_1$

Since $\lim\limits_{n \to \infty} 0 = 0$ and $\lim\limits_{n \to \infty}\left\{\left(\dfrac{1}{8}\right)^{n-1} a_1\right\} = 0$, $\lim\limits_{n \to \infty} a_n = 0$

(3) Find the value of $\lim\limits_{n \to \infty} \dfrac{3a_n + 2n - 1}{5a_n + 4n + 6}$.

$\lim\limits_{n \to \infty} \dfrac{3a_n + 2n - 1}{5a_n + 4n + 6} = \lim\limits_{n \to \infty} \dfrac{2n - 1}{4n + 6}$ $\quad(\because \lim\limits_{n \to \infty} a_n = 0)$

$= \lim\limits_{n \to \infty} \dfrac{\frac{2n}{n} - \frac{1}{n}}{\frac{4n}{n} + \frac{6}{n}} = \lim\limits_{n \to \infty} \dfrac{2 - \frac{1}{n}}{4 + \frac{6}{n}} = \dfrac{2}{4} = \dfrac{1}{2}$

#9 For a positive integer n, when a polynomial $f(x) = 2^n x^2 + 3^n x + 1$ is divided by $x - 1$

and $x - 2$, the remainders are a_n and b_n, respectively. Find the value of $\lim\limits_{n \to \infty} \dfrac{a_n}{b_n}$.

By the remainder theorem,

$a_n = f(1) = 2^n + 3^n + 1$ and $b_n = f(2) = 2^n \cdot 2^2 + 3^n \cdot 2 + 1$

$\therefore \lim\limits_{n \to \infty} \dfrac{a_n}{b_n} = \lim\limits_{n \to \infty} \dfrac{2^n + 3^n + 1}{4 \cdot 2^n + 2 \cdot 3^n + 1} = \lim\limits_{n \to \infty} \dfrac{\left(\frac{2}{3}\right)^n + 1 + \left(\frac{1}{3}\right)^n}{4 \cdot \left(\frac{2}{3}\right)^n + 2 + \left(\frac{1}{3}\right)^n} = \dfrac{1}{2}$

#10 Find the sum, when the sum exists.

(1) $\displaystyle\sum_{n=2}^{\infty} \log \dfrac{n^2}{n^2 - 1}$

Let $S_n = \displaystyle\sum_{k=2}^{n} \log \dfrac{k^2}{k^2 - 1}$

Then, $S_n = \log \frac{2^2}{2^2-1} + \log \frac{3^2}{3^2-1} + \log \frac{4^2}{4^2-1} + \cdots\cdots + \log \frac{n^2}{n^2-1}$

$= \log \frac{2\cdot2}{(2-1)(2+1)} + \log \frac{3\cdot3}{(3-1)(3+1)} + \log \frac{4\cdot4}{(4-1)(4+1)} + \cdots\cdots + \log \frac{n\cdot n}{(n-1)(n+1)}$

$= \log \left(\frac{2}{1}\cdot\frac{2}{3}\right) + \log \left(\frac{3}{2}\cdot\frac{3}{4}\right) + \log \left(\frac{4}{3}\cdot\frac{4}{5}\right) + \cdots\cdots + \log \left(\frac{n}{n-1}\cdot\frac{n}{n+1}\right)$

$= \log\left\{\left(\frac{2}{1}\cdot\frac{2}{3}\right)\cdot\left(\frac{3}{2}\cdot\frac{3}{4}\right)\cdot\left(\frac{4}{3}\cdot\frac{4}{5}\right)\cdots\cdots\left(\frac{n}{n-1}\cdot\frac{n}{n+1}\right)\right\} = \log\left(\frac{2}{1}\cdot\frac{n}{n+1}\right) = \log\frac{2n}{n+1}$

Since $\lim\limits_{n\to\infty} S_n = \lim\limits_{n\to\infty} \log\frac{2n}{n+1} = \log 2$; i.e., S_n converges to a number $\log 2$,

$$\sum_{n=2}^{\infty} \log \frac{n^2}{n^2-1} = \lim_{n\to\infty} S_n = \log 2$$

(2) $\sum\limits_{n=1}^{\infty}(\sqrt{n+2}-\sqrt{n+1})$

Let $S_n = \sum\limits_{k=1}^{n}(\sqrt{k+2}-\sqrt{k+1})$

Then, $S_n = (\sqrt{3}-\sqrt{2}) + (\sqrt{4}-\sqrt{3}) + (\sqrt{5}-\sqrt{4}) + \cdots\cdots + (\sqrt{n+2}-\sqrt{n+1}) = -\sqrt{2}+\sqrt{n+2}$

$\therefore \sum\limits_{n=1}^{\infty}(\sqrt{n+2}-\sqrt{n+1}) = \lim\limits_{n\to\infty} S_n = \lim\limits_{n\to\infty}(-\sqrt{2}+\sqrt{n+2}) = \infty$ (Diverge)

(3) $\left(\frac{3}{2}-\frac{4}{3}\right) + \left(\frac{4}{3}-\frac{5}{4}\right) + \left(\frac{5}{4}-\frac{6}{5}\right) + \cdots\cdots$

Let $S_n = \sum\limits_{k=1}^{n}\left(\frac{k+2}{k+1}-\frac{k+3}{k+2}\right) = \frac{3}{2}-\frac{n+3}{n+2}$ $\therefore \lim\limits_{n\to\infty} S_n = \lim\limits_{n\to\infty}\left(\frac{3}{2}-\frac{n+3}{n+2}\right) = \frac{3}{2}-1 = \frac{1}{2}$

(4) $\log_2\left(1-\frac{1}{2^2}\right) + \log_2\left(1-\frac{1}{3^2}\right) + \cdots\cdots + \log_2\left(1-\frac{1}{(n+1)^2}\right) + \cdots\cdots$

Let $a_n = \log_2\left(1-\frac{1}{(n+1)^2}\right) = \log_2\left\{\frac{(n+1)^2-1}{(n+1)^2}\right\} = \log_2\left\{\frac{n(n+2)}{(n+1)^2}\right\} = \log_2\left(\frac{n}{n+1}\cdot\frac{n+2}{n+1}\right)$

Then $S_n = \sum\limits_{k=1}^{n} a_k = \sum\limits_{k=1}^{n} \log_2\left(\frac{k}{k+1}\cdot\frac{k+2}{k+1}\right)$

$= \log_2\left(\frac{1}{2}\cdot\frac{3}{2}\right) + \log_2\left(\frac{2}{3}\cdot\frac{4}{3}\right) + \log_2\left(\frac{3}{4}\cdot\frac{5}{4}\right) + \cdots\cdots + \log_2\left(\frac{n}{n+1}\cdot\frac{n+2}{n+1}\right)$

$= \log_2\left\{\left(\frac{1}{2}\cdot\frac{3}{2}\right)\cdot\left(\frac{2}{3}\cdot\frac{4}{3}\right)\cdot\left(\frac{3}{4}\cdot\frac{5}{4}\right)\cdots\cdots\left(\frac{n}{n+1}\cdot\frac{n+2}{n+1}\right)\right\} = \log_2\left(\frac{1}{2}\cdot\frac{n+2}{n+1}\right)$

$\therefore \lim\limits_{n\to\infty} S_n = \lim\limits_{n\to\infty} \log_2\left(\frac{1}{2}\cdot\frac{n+2}{n+1}\right) = \log_2\left(\frac{1}{2}\cdot 1\right) = \log_2 2^{-1} = -1$

(5) $\sum\limits_{n=1}^{\infty}\left(\frac{2}{3}\right)^n \sin\left(n\pi + \frac{\pi}{6}\right)$

Substituting $n = 1, 2, 3, \cdots\cdots$,

we have $\frac{2}{3}\cdot\left(-\frac{1}{2}\right)$, $\left(\frac{2}{3}\right)^2\cdot\left(\frac{1}{2}\right)$, $\left(\frac{2}{3}\right)^3\cdot\left(-\frac{1}{2}\right)$, $\left(\frac{2}{3}\right)^4\cdot\left(\frac{1}{2}\right)$, $\cdots\cdots$

$$\therefore \ S_n = \frac{\left(-\frac{1}{3}\right)\left\{1-\left(-\frac{2}{3}\right)^n\right\}}{1-\left(-\frac{2}{3}\right)} = -\frac{1}{5}\left\{1-\left(-\frac{2}{3}\right)^n\right\}$$

Therefore, $\displaystyle\sum_{n=1}^{\infty}\left(\frac{2}{3}\right)^n \sin\left(n\pi + \frac{\pi}{6}\right) = \lim_{n\to\infty} S_n = \lim_{n\to\infty} -\frac{1}{5}\left\{1-\left(-\frac{2}{3}\right)^n\right\} = -\frac{1}{5}\lim_{n\to\infty}\left\{1-\left(-\frac{2}{3}\right)^n\right\}$

$$= -\frac{1}{5}\cdot 1 = -\frac{1}{5}$$

Note that $\displaystyle\lim_{n\to\infty} S_n = \frac{a}{1-r} = \frac{-\frac{1}{3}}{1-\left(-\frac{2}{3}\right)} = -\frac{1}{5}$

#11 Determine the convergence of the following series.

(1) $\displaystyle\sum_{n=1}^{\infty} 2n^{-\frac{3}{2}}$

$$\sum_{n=1}^{\infty} 2n^{-\frac{3}{2}} = 2\sum_{n=1}^{\infty} n^{-\frac{3}{2}} = 2\sum_{n=1}^{\infty}\left(\frac{1}{n^{\frac{3}{2}}}\right)$$

Since $\displaystyle\sum_{n=1}^{\infty}\left(\frac{1}{n^{\frac{3}{2}}}\right)$ is P-series with $P = \frac{3}{2}(> 1)$, the series is convergent.

(2) $\displaystyle\sum_{n=1}^{\infty} \frac{\sqrt[4]{n}}{3\sqrt[3]{n^2}}$

$$\sum_{n=1}^{\infty} \frac{\sqrt[4]{n}}{3\sqrt[3]{n^2}} = \frac{1}{3}\sum_{n=1}^{\infty} \frac{n^{\frac{1}{4}}}{n^{\frac{2}{3}}} = \frac{1}{3}\sum_{n=1}^{\infty} n^{\left(\frac{1}{4}-\frac{2}{3}\right)} = \frac{1}{3}\sum_{n=1}^{\infty} (n)^{\frac{-5}{12}} = \frac{1}{3}\sum_{n=1}^{\infty}\left(\frac{1}{n^{\frac{5}{12}}}\right)$$

Since $\displaystyle\frac{1}{3}\sum_{n=1}^{\infty}\left(\frac{1}{n^{\frac{5}{12}}}\right)$ is a P-series with $P = \frac{5}{12}$ $(0 < P \leq 1)$, the series is divergent.

#12 Determine the convergence of the following series. If the series converges, then calculate its sum.

(1) $1 + \frac{1}{2} + \frac{1}{2^2} + \frac{1}{2^3} + \cdots\cdots$

$a = 1, \ r = \frac{1}{2}$

Since $-1 < r < 1$, the sum is $S = \frac{a}{1-r} = \frac{1}{1-\frac{1}{2}} = 2$

(2) $\left(\sqrt{3}+1\right) + \left(\sqrt{3}-3\right) + \left(9\sqrt{3}-15\right) + \cdots\cdots$

$a = \sqrt{3}+1, \ r = \frac{\sqrt{3}-3}{\sqrt{3}+1} = \frac{(\sqrt{3}-3)(\sqrt{3}-1)}{(\sqrt{3}+1)(\sqrt{3}-1)} = \frac{3-3\sqrt{3}-\sqrt{3}+3}{3-1} = \frac{6-4\sqrt{3}}{2} = 3 - 2\sqrt{3}$

Since $-1 < r < 1$, the sum is

$$S = \frac{\sqrt{3}+1}{1-(3-2\sqrt{3})} = \frac{\sqrt{3}+1}{2\sqrt{3}-2} = \frac{(\sqrt{3}+1)(2\sqrt{3}+2)}{(2\sqrt{3}-2)(2\sqrt{3}+2)} = \frac{(\sqrt{3}+1)(2\sqrt{3}+2)}{12-4}$$

$$= \frac{6+4\sqrt{3}+2}{8} = \frac{8+4\sqrt{3}}{8} = \frac{2+\sqrt{3}}{2}$$

(3) $\sin 30° + \sin^2 30° + \sin^3 30° + \sin^4 30° + \cdots\cdots$

$a = \sin 30° = \dfrac{1}{2}$, $r = \sin 30° = \dfrac{1}{2}$

Since $-1 < r < 1$, the sum is $S = \dfrac{a}{1-r} = \dfrac{\frac{1}{2}}{1-\frac{1}{2}} = 1$

(4) $\displaystyle\sum_{n=1}^{\infty} 2^n \left(\dfrac{1}{3}\right)^{n-1}$

$\displaystyle\sum_{n=1}^{\infty} 2^n \left(\dfrac{1}{3}\right)^{n-1} = \sum_{n=1}^{\infty} 2 \cdot 2^{n-1} \left(\dfrac{1}{3}\right)^{n-1} = \sum_{n=1}^{\infty} 2 \left(\dfrac{2}{3}\right)^{n-1} = \dfrac{2}{1-\frac{2}{3}} = \dfrac{2}{\frac{1}{3}} = 6$

(5) $\displaystyle\sum_{n=0}^{\infty} 3\left(-\dfrac{2}{5}\right)^n$

It is a geometric series with $a = 3$ and $r = -\dfrac{2}{5}$.

Since $-1 < r < 1$, the series converges to the sum of $\dfrac{a}{1-r}$.

$\displaystyle\sum_{n=0}^{\infty} 3\left(-\dfrac{2}{5}\right)^n = \dfrac{a}{1-r} = \dfrac{3}{1-\left(-\frac{2}{5}\right)} = \dfrac{3}{\frac{7}{5}} = \dfrac{15}{7}$

(6) $\displaystyle\sum_{n=1}^{\infty} \dfrac{2^n + (-3)^n}{4^n}$

$\displaystyle\sum_{n=1}^{\infty} \dfrac{2^n + (-3)^n}{4^n} = \sum_{n=1}^{\infty}\left\{\left(\dfrac{1}{2}\right)^n + \left(-\dfrac{3}{4}\right)^n\right\} = \sum_{n=1}^{\infty}\left(\dfrac{1}{2}\right)^n + \sum_{n=1}^{\infty}\left(-\dfrac{3}{4}\right)^n = \dfrac{\frac{1}{2}}{1-\left(\frac{1}{2}\right)} + \dfrac{-\frac{3}{4}}{1-\left(-\frac{3}{4}\right)} = 1 + \dfrac{-\frac{3}{4}}{\frac{7}{4}} = \dfrac{4}{7}$

(7) $\displaystyle\sum_{n=0}^{\infty} \left(3^{n+1} - 1\right)\left(\dfrac{1}{4}\right)^n$

$\displaystyle\sum_{n=0}^{\infty} \left(3^{n+1} - 1\right)\left(\dfrac{1}{4}\right)^n = \sum_{n=0}^{\infty}\left\{3 \cdot 3^n \left(\dfrac{1}{4}\right)^n - \left(\dfrac{1}{4}\right)^n\right\} = 3\sum_{n=0}^{\infty}\left(\dfrac{3}{4}\right)^n - \sum_{n=0}^{\infty}\left(\dfrac{1}{4}\right)^n$

$= 3\dfrac{1}{1-\left(\frac{3}{4}\right)} - \dfrac{1}{1-\left(\frac{1}{4}\right)} = 3 \cdot 4 - \dfrac{4}{3} = \dfrac{32}{3}$

(8) $\displaystyle\sum_{n=1}^{\infty} \left(\dfrac{1}{2}\right)^n \cos n\pi$

$\displaystyle\sum_{n=1}^{\infty} \left(\dfrac{1}{2}\right)^n \cos n\pi = \dfrac{1}{2}\cos\pi + \left(\dfrac{1}{2}\right)^2 \cos 2\pi + \left(\dfrac{1}{2}\right)^3 \cos 3\pi + \cdots\cdots = -\dfrac{1}{2} + \left(\dfrac{1}{2}\right)^2 - \left(\dfrac{1}{2}\right)^3 + \cdots\cdots$

$= \dfrac{-\frac{1}{2}}{1-\left(-\frac{1}{2}\right)} = \dfrac{-\frac{1}{2}}{\frac{3}{2}} = -\dfrac{1}{3}$

(9) $\displaystyle\sum_{n=1}^{\infty} \left(-\dfrac{1}{2}\right)^n \sin\left(n\pi + \dfrac{\pi}{3}\right)$

$\displaystyle\sum_{n=1}^{\infty} \left(-\dfrac{1}{2}\right)^n \sin\left(n\pi + \dfrac{\pi}{3}\right) = \sum_{n=1}^{\infty}\left(-\dfrac{1}{2}\right)^n (-1)^n \sin\dfrac{\pi}{3} = \sum_{n=1}^{\infty}\left(\dfrac{1}{2}\right)^n \dfrac{\sqrt{3}}{2} = \dfrac{\sqrt{3}}{2}\sum_{n=1}^{\infty}\left(\dfrac{1}{2}\right)^n = \dfrac{\sqrt{3}}{2}\dfrac{\frac{1}{2}}{1-\left(\frac{1}{2}\right)} = \dfrac{\sqrt{3}}{2}$

(10) $\displaystyle \lim_{k \to \infty} \frac{1}{k} \sum_{n=1}^{k} \left(\sum_{j=1}^{n} \frac{1}{2^j} \right)$

$$\sum_{j=1}^{n} \frac{1}{2^j} = \frac{\left(\frac{1}{2}\right)\left\{1 - \left(\frac{1}{2}\right)^n\right\}}{1 - \left(\frac{1}{2}\right)} = 1 - \left(\frac{1}{2}\right)^n$$

$$\sum_{n=1}^{k} \left(\sum_{j=1}^{n} \frac{1}{2^j} \right) = \sum_{n=1}^{k} \left\{ 1 - \left(\frac{1}{2}\right)^n \right\} = \sum_{n=1}^{k} 1 - \sum_{n=1}^{k} \left(\frac{1}{2}\right)^n = k - \left\{ 1 - \left(\frac{1}{2}\right)^k \right\} = \left(\frac{1}{2}\right)^k + k - 1$$

$$\lim_{k \to \infty} \frac{1}{k} \sum_{n=1}^{k} \left(\sum_{j=1}^{n} \frac{1}{2^j} \right) = \lim_{k \to \infty} \frac{1}{k} \left\{ \left(\frac{1}{2}\right)^k + k - 1 \right\} = \lim_{k \to \infty} \left\{ \frac{1}{k} \left(\frac{1}{2}\right)^k + 1 - \frac{1}{k} \right\} = 1$$

#13 Find the value.

(1) When a_n is the nth term of a sequence 2, 4, 8, 16, $\cdots\cdots$ and S_n is the finite sum of first n terms, find the value of $\displaystyle \lim_{n \to \infty} \frac{a_n}{S_n}$.

$$a_n = 2 \cdot 2^{n-1} = 2^n$$

$$S_n = \frac{2\{1 - 2^n\}}{1 - 2} = 2(2^n - 1)$$

$$\lim_{n \to \infty} \frac{a_n}{S_n} = \lim_{n \to \infty} \frac{2^n}{2(2^n - 1)} = \lim_{n \to \infty} \frac{1}{2\left(1 - \frac{1}{2^n}\right)} = \lim_{n \to \infty} \frac{1}{2\left\{1 - \left(\frac{1}{2}\right)^n\right\}} = \frac{1}{2}$$

(2) For $f(x) = \dfrac{1}{x(x+1)}$,

1) Find the minimum value of the integer n such that $\displaystyle \sum_{k=1}^{n} f(k) > 0.9$

$$\sum_{k=1}^{n} f(k) = \sum_{k=1}^{n} \frac{1}{k(k+1)} = \sum_{k=1}^{n} \left(\frac{1}{k} - \frac{1}{k+1} \right) = \left(1 - \frac{1}{2} \right) + \left(\frac{1}{2} - \frac{1}{3} \right) + \cdots\cdots + \left(\frac{1}{n} - \frac{1}{n+1} \right)$$

$$= 1 - \frac{1}{n+1} > 0.9$$

$$\therefore \quad \frac{1}{n+1} < 1 - 0.9 = \frac{1}{10} \qquad \therefore \quad n + 1 > 10 \qquad \therefore \quad n > 9$$

Therefore, $n = 10$

2) Find the value of positive constant a such that $\displaystyle \lim_{n \to \infty} \sum_{k=1}^{n} f(a + k) = \frac{1}{2}$

$$\sum_{k=1}^{n} f(a + k) = \sum_{k=1}^{n} \frac{1}{(a+k)(a+k+1)} = \sum_{k=1}^{n} \left(\frac{1}{a+k} - \frac{1}{a+k+1} \right)$$

$$= \left(\frac{1}{a+1} - \frac{1}{a+2} \right) + \left(\frac{1}{a+2} - \frac{1}{a+3} \right) + \cdots\cdots + \left(\frac{1}{a+n} - \frac{1}{a+n+1} \right) = \frac{1}{a+1} - \frac{1}{a+n+1}$$

$$\lim_{n \to \infty} \sum_{k=1}^{n} f(a + k) = \lim_{n \to \infty} \left(\frac{1}{a+1} - \frac{1}{a+n+1} \right) = \frac{1}{a+1} = \frac{1}{2} \qquad \therefore \quad a + 1 = 2 \qquad \therefore \quad a = 1$$

(3) For a quadratic equation $x^2 + (n-1)x + n^2 = 0$,

let α_n and β_n be the two roots of the equation. Find the value of $\displaystyle\sum_{n=1}^{\infty} \frac{1}{(\alpha_n - 1)(\beta_n - 1)}$.

By the relationship between the roots and coefficients,

$\alpha_n + \beta_n = -(n-1)$ and $\alpha_n \beta_n = n^2$

$(\alpha_n - 1)(\beta_n - 1) = \alpha_n \beta_n - (\alpha_n + \beta_n) + 1 = n^2 + (n-1) + 1 = n^2 + n = n(n+1)$

$\displaystyle\sum_{n=1}^{\infty} \frac{1}{(\alpha_n - 1)(\beta_n - 1)} = \lim_{n\to\infty} \sum_{k=1}^{n} \frac{1}{k(k+1)} = \lim_{n\to\infty} \sum_{k=1}^{n} \left(\frac{1}{k} - \frac{1}{k+1}\right)$

$\displaystyle = \lim_{n\to\infty} \left\{ \left(1 - \frac{1}{2}\right) + \left(\frac{1}{2} - \frac{1}{3}\right) + \cdots\cdots + \left(\frac{1}{n} - \frac{1}{n+1}\right) \right\} = \lim_{n\to\infty} \left(1 - \frac{1}{n+1}\right) = 1$

(4) For a quadratic equation $(4n^2 - 1)x^2 - 4nx + 1 = 0$ (n; positive integer),

let α_n and β_n ($\alpha_n > \beta_n$) be two roots of the equation. Find the value of $\displaystyle\sum_{n=1}^{\infty} (\alpha_n - \beta_n)$.

$(4n^2 - 1)x^2 - 4nx + 1 = 0 \quad \Rightarrow \quad (2n-1)(2n+1)x^2 - 4nx + 1 = 0$

$\Rightarrow \quad \{(2n-1)x - 1\}\{(2n+1)x - 1\} = 0$

$\therefore \quad x = \frac{1}{2n-1} \quad \text{or} \quad x = \frac{1}{2n+1}$

Since $\alpha_n > \beta_n$, $\alpha_n = \frac{1}{2n-1}$ and $\beta_n = \frac{1}{2n+1}$

$\displaystyle\therefore \sum_{n=1}^{\infty} (\alpha_n - \beta_n) = \lim_{n\to\infty} \sum_{k=1}^{n} (\alpha_k - \beta_k) = \lim_{n\to\infty} \sum_{k=1}^{n} \left(\frac{1}{2k-1} - \frac{1}{2k+1}\right)$

$\displaystyle = \lim_{n\to\infty} \left\{ \left(1 - \frac{1}{3}\right) + \left(\frac{1}{3} - \frac{1}{5}\right) + \cdots\cdots + \left(\frac{1}{2n-1} - \frac{1}{2n+1}\right) \right\} = \lim_{n\to\infty} \left(1 - \frac{1}{2n+1}\right) = 1$

(5) For convergent series $(a_1 - 2) + (a_2 - \frac{3}{2}) + (a_3 - \frac{4}{3}) + (a_4 - \frac{5}{4}) + \cdots\cdots$,

find the value of $\displaystyle\lim_{n\to\infty} a_n$.

Let n^{th} term of the series be $a_n - \frac{n+1}{n}$.

Since $\displaystyle\sum_{n=1}^{\infty} \left(a_n - \frac{n+1}{n}\right)$ converges, $\displaystyle\lim_{n\to\infty} \left(a_n - \frac{n+1}{n}\right) = 0$.

Let $a_n - \frac{n+1}{n} = b_n$

Then, $\displaystyle\lim_{n\to\infty} b_n = 0$.

Since $a_n = b_n + \frac{n+1}{n}$, $\displaystyle\lim_{n\to\infty} a_n = \lim_{n\to\infty} \left(b_n + \frac{n+1}{n}\right) = \lim_{n\to\infty} b_n + 1 = 0 + 1 = 1$

(6) For a sequence $\{a_n\}$ such that $\displaystyle\sum_{n=1}^{\infty} \left(a_n - \frac{2^{n+1} + 3^n}{3^{n+1}}\right) = 5$, find the value of $\displaystyle\lim_{n\to\infty} \frac{-a_n + 1}{3a_n - 2}$.

Since the infinite series converges, $\displaystyle\lim_{n\to\infty} \left(a_n - \frac{2^{n+1} + 3^n}{3^{n+1}}\right) = 0$.

$\displaystyle\therefore \lim_{n\to\infty} a_n = \lim_{n\to\infty} \frac{2^{n+1} + 3^n}{3^{n+1}} = \lim_{n\to\infty} \frac{\left(\frac{2}{3}\right)^{n+1} + \frac{1}{3}}{1} = \frac{1}{3} \qquad\qquad \therefore \lim_{n\to\infty} \frac{-a_n + 1}{3a_n - 2} = \frac{-\frac{1}{3} + 1}{3\left(\frac{1}{3}\right) - 2} = \frac{\frac{2}{3}}{-1} = -\frac{2}{3}$

(7) When $a_n = \sum_{k=1}^{n} ak$, find the value of the constant a such that $\sum_{n=1}^{\infty} \dfrac{1}{a_n} = \dfrac{1}{2}$.

$$a_n = \sum_{k=1}^{n} ak = a \sum_{k=1}^{n} k = a \cdot \frac{n(n+1)}{2}$$

$$\therefore \sum_{n=1}^{\infty} \frac{1}{a_n} = \sum_{n=1}^{\infty} \frac{2}{an(n+1)} = \lim_{n \to \infty} \sum_{k=1}^{n} \frac{2}{ak(k+1)} = \lim_{n \to \infty} \frac{2}{a} \sum_{k=1}^{n} \frac{1}{k(k+1)} = \lim_{n \to \infty} \frac{2}{a} \sum_{k=1}^{n} \left(\frac{1}{k} - \frac{1}{k+1} \right)$$

$$= \lim_{n \to \infty} \frac{2}{a} \left\{ \left(1 - \frac{1}{2} \right) + \left(\frac{1}{2} - \frac{1}{3} \right) + \cdots\cdots + \left(\frac{1}{n} - \frac{1}{n+1} \right) \right\} = \lim_{n \to \infty} \frac{2}{a} \left(1 - \frac{1}{n+1} \right)$$

$$= \frac{2}{a} \lim_{n \to \infty} \left(1 - \frac{1}{n+1} \right) = \frac{2}{a} \cdot 1 = \frac{2}{a} = \frac{1}{2} \qquad \therefore \ a = 4$$

(8) For a geometric sequence $\{a_n\}$, $\sum_{n=1}^{\infty} a_n = 3$ and $\sum_{n=1}^{\infty} a_n{}^2 = 4$. Find the value of $\sum_{n=1}^{\infty} a_n{}^3$.

Let a be the first term and r be the common ratio of the geometric sequence $\{a_n\}$.

Then, a^2 and r^2 be the first term and common ratio of the sequence $\{a_n{}^2\}$, respectively.

Also, a^3 and r^3 be the first term and common ratio of the sequence $\{a_n{}^3\}$, respectively.

$$\therefore \ \sum_{n=1}^{\infty} a_n = 3 \quad \Rightarrow \quad \frac{a}{1-r} = 3 \ ; \quad a = 3 - 3r$$

$$\sum_{n=1}^{\infty} a_n{}^2 = 4 \quad \Rightarrow \quad \frac{a^2}{1-r^2} = 4 \ ; \quad a^2 = 4 - 4r^2$$

$$\therefore \ (3 - 3r)^2 = 4 - 4r^2 \ ; \quad 9 - 18r + 9r^2 = 4 - 4r^2 \ ; \quad 13r^2 - 18r + 5 = 0$$

$$\therefore \ (r-1)(13r-5) = 0 \ ; \quad r = 1 \ \text{ or } \ r = \frac{5}{13}$$

Since $\sum_{n=1}^{\infty} a_n = 3$ is converges, $-1 < r < 1$.

$$\therefore \ r = \frac{5}{13} , \quad a = 3 - 3r = 3 - 3 \cdot \frac{5}{13} = \frac{24}{13}$$

Therefore, $\displaystyle \sum_{n=1}^{\infty} a_n{}^3 = \frac{a^3}{1-r^3} = \frac{\left(\frac{24}{13} \right)^3}{1 - \left(\frac{5}{13} \right)^3} = \frac{(24)^3}{(13)^3 - (5)^3} = \frac{13824}{2197 - 125} \approx 6.672$

(9) For a sequence $\{a_n\}$, $a_1 = 1$, $a_2 = 2$, and $a_{n+2} = a_{n+1} + a_n$ $(n = 1, 2, 3, \cdots\cdots)$.

Find the value of $\displaystyle \sum_{n=1}^{\infty} \frac{a_n}{a_{n+1} a_{n+2}}$.

$\{a_n\}$; $1, 2, 3, 5, 8, 13, 21, 34, \cdots\cdots$

$$\therefore \ \sum_{n=1}^{\infty} \frac{a_n}{a_{n+1} a_{n+2}} = \sum_{n=1}^{\infty} \frac{a_{n+2} - a_{n+1}}{a_{n+1} a_{n+2}} = \sum_{n=1}^{\infty} \left(\frac{1}{a_{n+1}} - \frac{1}{a_{n+2}} \right) = \lim_{n \to \infty} \sum_{k=1}^{n} \left(\frac{1}{a_{k+1}} - \frac{1}{a_{k+2}} \right)$$

$$= \lim_{n \to \infty} \left\{ \left(\frac{1}{a_2} - \frac{1}{a_3} \right) + \left(\frac{1}{a_3} - \frac{1}{a_4} \right) + \cdots\cdots + \left(\frac{1}{a_{n+1}} - \frac{1}{a_{n+2}} \right) \right\} = \lim_{n \to \infty} \left(\frac{1}{a_2} - \frac{1}{a_{n+2}} \right)$$

$$= \lim_{n \to \infty} \frac{1}{a_2} - \lim_{n \to \infty} \frac{1}{a_{n+2}} = \frac{1}{a_2} - 0 = \frac{1}{a_2} = \frac{1}{2}$$

(10) **For an arithmetic sequence $\{a_n\}$ with first term -100 and common difference 4,**

a sequence $\{b_n\}$ is defined by $b_n = \sum_{k=1}^{n}(a_{k+1} - a_k)$. Find the value of $\lim_{n\to\infty} \sum_{k=1}^{n} \dfrac{1}{(2n+1)^2} b_k$.

Since the common difference of the sequence $\{a_n\}$ is 4, $a_{k+1} - a_k = 4$

$$\therefore \quad b_n = \sum_{k=1}^{n}(a_{k+1} - a_k) = \sum_{k=1}^{n} 4 = 4n$$

$$\therefore \quad \lim_{n\to\infty} \sum_{k=1}^{n} \frac{1}{(2n+1)^2} b_k = \lim_{n\to\infty} \sum_{k=1}^{n} \frac{4k}{(2n+1)^2} = \lim_{n\to\infty} \frac{1}{(2n+1)^2} \sum_{k=1}^{n} 4k = \lim_{n\to\infty} \frac{4}{(2n+1)^2} \cdot \frac{n(n+1)}{2}$$

$$= \lim_{n\to\infty} \frac{4n^2 + 4n}{8n^2 + 8n + 2} = \frac{4}{8} = \frac{1}{2}$$

(11) **When an infinite series $S = 1 + \dfrac{1}{4} + \dfrac{1}{9} + \dfrac{1}{16} + \cdots\cdots + \dfrac{1}{n^2} + \cdots\cdots$ converges,**

express the value of $1 + \dfrac{1}{9} + \dfrac{1}{25} + \dfrac{1}{49} + \cdots\cdots + \dfrac{1}{(2n-1)^2} + \cdots\cdots$ as S.

$$S = 1 + \frac{1}{4} + \frac{1}{9} + \frac{1}{16} + \cdots\cdots + \frac{1}{n^2} + \cdots\cdots \quad ①$$

$$\frac{1}{4}S = \frac{1}{4} + \frac{1}{16} + \frac{1}{36} + \frac{1}{64} + \cdots\cdots + \frac{1}{(2n)^2} + \cdots\cdots \quad ②$$

$$① - ②; \quad \frac{3}{4}S = 1 + \frac{1}{9} + \frac{1}{25} + \cdots\cdots + \frac{1}{(2n-1)^2} + \cdots\cdots \qquad \therefore \text{ We have } \frac{3}{4}S.$$

(12) **Find the range of x so that $\sum_{n=1}^{\infty}\left(\dfrac{x}{2}\right)^n$ and $\sum_{n=1}^{\infty}\left(\dfrac{1}{x}\right)^n$ are convergent.**

For the series $\sum_{n=1}^{\infty}\left(\dfrac{x}{2}\right)^n$, if $-1 < \dfrac{x}{2} < 1$, then it is convergent ;i.e., $-2 < x < 2$; $|x| < 2$

For the series $\sum_{n=1}^{\infty}\left(\dfrac{1}{x}\right)^n$, if $-1 < \dfrac{1}{x} < 1$, then it is convergent ;i.e., $x > 1$, $x < -1$; $|x| > 1$

$\therefore \ 1 < |x| < 2 \qquad$ Therefore, $1 < x < 2$ or $-2 < x < -1$

(13) **Find the range of x so that the infinite geometric sequence $\{(3x-1)^{n-1}\}$ and the**

infinite geometric series $\sum_{n=1}^{\infty}\left(\dfrac{\log_2 x - 1}{2}\right)^n$ are convergent.

Since $-1 < 3x - 1 \leq 1$, $0 < 3x \leq 2$; $0 < x \leq \dfrac{2}{3}$ $\quad\cdots\cdots ①$

Since $-1 < \dfrac{\log_2 x - 1}{2} < 1$, $\quad -2 < \log_2 x - 1 < 2$; $\quad -1 < \log_2 x < 3$;

$\log_2 2^{-1} < \log_2 x < \log_2 2^3$; $\dfrac{1}{2} < x < 8$ $\quad\cdots\cdots ②$

By ① and ②, $\dfrac{1}{2} < x \leq \dfrac{2}{3}$

(14) When $\displaystyle\sum_{n=1}^{\infty} \dfrac{a_n}{2^n} = 1$, find the value of $\displaystyle\lim_{n\to\infty} \dfrac{2^{n-1}+5a_n}{2^{n+3}-3a_n}$.

Since $\displaystyle\sum_{n=1}^{\infty} \dfrac{a_n}{2^n} = 1$, $\displaystyle\lim_{n\to\infty} \dfrac{a_n}{2^n} = 0$

$$\lim_{n\to\infty} \frac{2^{n-1}+5a_n}{2^{n+3}-3a_n} = \lim_{n\to\infty} \frac{2^{-1}+\frac{5a_n}{2^n}}{2^3-3\frac{a_n}{2^n}} = \frac{2^{-1}}{2^3} = \frac{\frac{1}{2}}{8} = \frac{1}{16}$$

(15) For a sequence $\{a_n\}$, let S_n be the finite partial sum of first n terms such that

$$\lim_{n\to\infty} S_n = 3. \text{ Find the value of } \lim_{n\to\infty} \frac{a_n+S_n}{S_n^{\,3}}.$$

Note that $\displaystyle\lim_{n\to\infty} S_n = \lim_{n\to\infty} S_{n-1} = 3$

Since $a_n = S_n - S_{n-1}$, $\displaystyle\lim_{n\to\infty} a_n = \lim_{n\to\infty} (S_n - S_{n-1}) = 3 - 3 = 0$

$\therefore \displaystyle\lim_{n\to\infty} \dfrac{a_n+S_n}{S_n^{\,3}} = \dfrac{0+3}{3^3} = \dfrac{1}{9}$

(16) For a sequence $a_1,\ 2a_2,\ 2^2 a_3, \cdots\cdots, 2^{n-1} a_n, \cdots\cdots$, the sum of first n terms is $10n$.

Find the value of $\displaystyle\sum_{n=1}^{\infty} a_n$.

Let $S_n = a_1 + 2a_2 + 2^2 a_3 + \cdots\cdots + 2^{n-1} a_n = 10n$

$S_{n-1} = a_1 + 2a_2 + 2^2 a_3 + \cdots\cdots + 2^{n-2} a_{n-1} = 10(n-1)$

When $n \geq 2$, $S_n - S_{n-1} = 2^{n-1} a_n = 10$ $\quad \therefore a_n = \dfrac{10}{2^{n-1}}$

Since $S_1 = a_1 = 10$, $a_n = \dfrac{10}{2^{n-1}}$ $(n \geq 1)$

$\therefore \{a_n\}$ is a geometric sequence with first term $a = 10$ and common ratio $r = \dfrac{1}{2}$

$\therefore \displaystyle\sum_{n=1}^{\infty} a_n = \dfrac{a}{1-r} = \dfrac{10}{1-\frac{1}{2}} = 20$

(17) When $\cos^2 \theta + \cos^2 \theta \sin \theta + \cos^2 \theta \sin^2 \theta + \cdots\cdots = \dfrac{18}{13}$ $\left(0 < \theta < \dfrac{\pi}{2}\right)$,

find the value of $\dfrac{10}{\tan \theta}$.

Since the series is an infinite geometric series with first term $\cos^2 \theta$ and common ratio $\sin \theta$,

the sum is $\dfrac{\cos^2 \theta}{1-\sin \theta} = \dfrac{18}{13}$

$\therefore \dfrac{\cos^2 \theta}{1-\sin \theta} = \dfrac{1-\sin^2 \theta}{1-\sin \theta} = \dfrac{(1-\sin \theta)(1+\sin \theta)}{1-\sin \theta} = 1 + \sin \theta = \dfrac{18}{13}$

$\therefore \sin \theta = \dfrac{18}{13} - 1 = \dfrac{5}{13}$

Since $0 < \theta < \dfrac{\pi}{2}$, $\tan \theta = \dfrac{5}{12}$ $\quad \therefore \dfrac{10}{\tan \theta} = 24$

(18) For the infinite sequence $\sqrt{3}, \ \sqrt{3\sqrt{3}}, \ \sqrt{3\sqrt{3\sqrt{3}}}, \ \cdots\cdots$ find the limit of the sequence.

$$\sqrt{3} = 3^{\frac{1}{2}}$$

$$\sqrt{3\sqrt{3}} = \left(3 \cdot 3^{\frac{1}{2}}\right)^{\frac{1}{2}} = \left(3^{1+\frac{1}{2}}\right)^{\frac{1}{2}} = 3^{\frac{1}{2}+\left(\frac{1}{2}\right)^2}$$

$$\sqrt{3\sqrt{3\sqrt{3}}} = \left(3^{1+\frac{1}{2}+\left(\frac{1}{2}\right)^2}\right)^{\frac{1}{2}} = 3^{\frac{1}{2}+\left(\frac{1}{2}\right)^2+\left(\frac{1}{2}\right)^3}$$

$\therefore$ The nth term is $a_n = 3^{\frac{1}{2}+\left(\frac{1}{2}\right)^2+\left(\frac{1}{2}\right)^3+\cdots+\left(\frac{1}{2}\right)^n}$

Since $\lim\limits_{n\to\infty} \left\{\frac{1}{2} + \left(\frac{1}{2}\right)^2 + \left(\frac{1}{2}\right)^3 + \cdots\cdots + \left(\frac{1}{2}\right)^n\right\} = \dfrac{\frac{1}{2}}{1-\frac{1}{2}} = 1$, $\lim\limits_{n\to\infty} a_n = 3^1 = 3$

(19) For a sequence $\{a_n\}$ with $a_1 = 4$ and $a_{n+1} = \sqrt[3]{a_n}$ $(n \geq 1)$,

find the value of $\lim\limits_{n\to\infty} (a_1 \cdot a_2 \cdot a_3 \cdots\cdots a_n)$.

$a_1 = 4$

$a_2 = \sqrt[3]{a_1} = a_1^{\frac{1}{3}}$

$a_3 = a_2^{\frac{1}{3}} = \left(a_1^{\frac{1}{3}}\right)^{\frac{1}{3}} = a_1^{\left(\frac{1}{3}\right)^2}$

$\vdots$

$a_n = a_1^{\left(\frac{1}{3}\right)^{n-1}}$

$\therefore \lim\limits_{n\to\infty} (a_1 \cdot a_2 \cdot a_3 \cdots\cdots a_n) = \lim\limits_{n\to\infty} a_1^{\left\{1+\frac{1}{3}+\left(\frac{1}{3}\right)^2+\cdots+\left(\frac{1}{3}\right)^{n-1}\right\}} = \lim\limits_{n\to\infty} a_1^{\left\{\frac{1\left\{1-\left(\frac{1}{3}\right)^n\right\}}{1-\left(\frac{1}{3}\right)}\right\}}$

$= \lim\limits_{n\to\infty} a_1^{\frac{3}{2}\left\{1-\left(\frac{1}{3}\right)^n\right\}} = a_1^{\frac{3}{2}} = 4^{\frac{3}{2}} = 2^3 = 8$

(20) For a geometric sequence $\{a_n\}$, the sum of the first term and second term is 4 and

the sum of the second term and third term is -2. Find the value of $\sum\limits_{n=1}^{\infty} a_n$.

Let a and r be the first term and common ratio of the sequence $\{a_n\}$, respectively.

Then, $a + ar = 4$, $ar + ar^2 = -2$

$\therefore \ a(1+r) = 4 \ \cdots\cdots ①$ $\qquad ar(1+r) = -2 \ \cdots\cdots ②$

$② \div ① ; \ r = -\dfrac{1}{2}$

By ①, $a\left(1 - \dfrac{1}{2}\right) = 4 \ ; \ a = 8$

$\therefore \ \sum\limits_{n=1}^{\infty} a_n = \dfrac{8}{1-\left(-\frac{1}{2}\right)} = \dfrac{8}{\frac{3}{2}} = \dfrac{16}{3}$

(21) When $S = 1 + \frac{1}{2} + \frac{1}{2^2} + \frac{1}{2^3} + \cdots\cdots$ and S_n is the finite sum of first n terms of the

series, find the minimum value of n such that $S - S_n \leq 0.01$ ($\log 2 = 0.3010$)

$$S_n = \frac{1\left\{1-\left(\frac{1}{2}\right)^n\right\}}{1-\left(\frac{1}{2}\right)} = 2\left\{1 - \left(\frac{1}{2}\right)^n\right\}$$

$$S = \frac{a}{1-r} = \frac{1}{1-\frac{1}{2}} = 2$$

$\therefore\ S - S_n = 2\left(\frac{1}{2}\right)^n \leq 0.01$; $\left(\frac{1}{2}\right)^{n-1} \leq \frac{1}{100}$; $2^{n-1} \geq 100$

Since $2^6 = 64$ and $2^7 = 128$, $n - 1 \geq 7$ $\therefore\ n \geq 8$

(22) For a sequence $\{a_n\}$ such that $\displaystyle\sum_{k=1}^{n} a_k = 3\left\{1 - \left(\frac{1}{3}\right)^n\right\}$, find the value of $\displaystyle\sum_{n=1}^{\infty} a_{2n}$.

Let $\displaystyle\sum_{k=1}^{n} a_k = S_n$

Then, $S_n = 3\left\{1 - \left(\frac{1}{3}\right)^n\right\}$

$$a_n = S_n - S_{n-1} = 3\left\{1 - \left(\frac{1}{3}\right)^n\right\} - 3\left\{1 - \left(\frac{1}{3}\right)^{n-1}\right\} = 3\left(\frac{1}{3}\right)^{n-1}\left\{1 - \left(\frac{1}{3}\right)\right\}$$

$$= 3\left(\frac{1}{3}\right)^{n-1} \cdot \frac{2}{3} = 2\left(\frac{1}{3}\right)^{n-1} \quad (n \geq 2)$$

Since $a_1 = S_1 = 2$, $a_n = 2\left(\frac{1}{3}\right)^{n-1}$ $(n \geq 1)$

$$\therefore\ \sum_{n=1}^{\infty} a_{2n} = \sum_{n=1}^{\infty} 2\left(\frac{1}{3}\right)^{2n-1} = 2 \cdot \frac{\frac{1}{3}}{1 - \frac{1}{9}} = 2 \cdot \frac{3}{8} = \frac{3}{4}$$

(23) For a sequence $\{a_n\}$ with $a_1 = 1$ and $a_n = \displaystyle\sum_{k=1}^{n-1} a_k$ $(n \geq 2)$, find the value of $\displaystyle\sum_{n=1}^{\infty} \frac{1}{a_n}$.

Let $\displaystyle\sum_{k=1}^{n} a_k = S_n$

Then, $\displaystyle\sum_{k=1}^{n-1} a_k = S_{n-1}$

Since $a_n = \displaystyle\sum_{k=1}^{n-1} a_k$ (by given), $a_n = S_{n-1}$

$\therefore\ a_{n+1} = S_n$

$\therefore\ a_n = S_n - S_{n-1} = a_{n+1} - a_n$; $2a_n = a_{n+1}$, $n \geq 2$

Since $a_n = \displaystyle\sum_{k=1}^{n-1} a_k$ $(n \geq 2)$, $a_2 = \displaystyle\sum_{k=1}^{1} a_k = a_1$

$$\therefore\ \sum_{n=1}^{\infty} \frac{1}{a_n} = 1 + \left(1 + \frac{1}{2} + \frac{1}{2^2} + \cdots\cdots\right) = 1 + \frac{1}{1 - \frac{1}{2}} = 1 + 2 = 3$$

(24) When the infinite series $\displaystyle\sum_{n=1}^{\infty} \dfrac{an^2 + 3}{n^2 + 2n}$ **is convergent, find the sum of the series.**

(a; constant)

Since the series converges, $\displaystyle\lim_{n\to\infty} \dfrac{an^2+3}{n^2+2n} = 0$.

Since $\displaystyle\lim_{n\to\infty} \dfrac{an^2+3}{n^2+2n} = a$, $\ a = 0$

$$\therefore \ \sum_{n=1}^{\infty} \frac{an^2+3}{n^2+2n} = \sum_{n=1}^{\infty} \frac{0+3}{n^2+2n} = \sum_{n=1}^{\infty} \frac{3}{n^2+2n} = 3\sum_{n=1}^{\infty} \frac{1}{n(n+2)} = 3\sum_{n=1}^{\infty} \frac{1}{2} \cdot \left(\frac{1}{n} - \frac{1}{n+2}\right)$$

$$= \frac{3}{2} \sum_{n=1}^{\infty} \left(\frac{1}{n} - \frac{1}{n+2}\right) = \frac{3}{2} \lim_{n\to\infty} \sum_{k=1}^{n} \left(\frac{1}{k} - \frac{1}{k+2}\right)$$

$$= \frac{3}{2} \lim_{n\to\infty} \left\{ \left(1 - \frac{1}{3}\right) + \left(\frac{1}{2} - \frac{1}{4}\right) + \left(\frac{1}{3} - \frac{1}{5}\right) + \cdots\cdots + \left(\frac{1}{n} - \frac{1}{n+2}\right) \right\}$$

$$= \frac{3}{2} \lim_{n\to\infty} \left(1 + \frac{1}{2} - \frac{1}{n+1} - \frac{1}{n+2}\right) = \frac{3}{2}\left(1 + \frac{1}{2}\right) = \frac{9}{4}$$

#14 For sequences $\{a_n\}$ and $\{b_n\}$, determine if the following statements are true or false.

(1) If $\displaystyle\sum_{n=1}^{\infty} 2a_n$ **converges, then** $\displaystyle\lim_{n\to\infty} a_n = 0$.

True   Since $\displaystyle\sum_{n=1}^{\infty} 2a_n$ converges, $\displaystyle\lim_{n\to\infty} 2a_n = 0$

$\therefore \ 2 \displaystyle\lim_{n\to\infty} a_n = 0$ $\therefore \ \displaystyle\lim_{n\to\infty} a_n = 0$

(2) If $\displaystyle\lim_{n\to\infty} a_n = 0$, **then** $\displaystyle\sum_{n=1}^{\infty} a_n{}^2$ **converges.**

False   Let $a_n = \dfrac{1}{\sqrt{n}}$

Then $\displaystyle\lim_{n\to\infty} a_n = \lim_{n\to\infty} \frac{1}{\sqrt{n}} = 0$ But, $\displaystyle\sum_{n=1}^{\infty} a_n{}^2 = \sum_{n=1}^{\infty} \frac{1}{n}$ diverges.

(3) If $\displaystyle\sum_{n=1}^{\infty} a_n b_n$ **converges, then** $\displaystyle\lim_{n\to\infty} a_n = 0$ **and** $\displaystyle\lim_{n\to\infty} b_n = 0$.

False   Let $\{a_n\} = 1,\ 0,\ 1,\ 0, \cdots\cdots$

$\{b_n\} = 0,\ 1,\ 0,\ 1, \cdots\cdots$

Then, $a_n b_n = 0$ $\therefore \displaystyle\sum_{n=1}^{\infty} a_n b_n = 0 \ ; \ \sum_{n=1}^{\infty} a_n b_n$ converges.

But, $\displaystyle\lim_{n\to\infty} a_n \neq 0$ and $\displaystyle\lim_{n\to\infty} b_n \neq 0$

(4) If $\displaystyle\sum_{n=1}^{\infty} a_n$ **and** $\displaystyle\sum_{n=1}^{\infty} (a_n + b_n)$ **converge, then** $\displaystyle\sum_{n=1}^{\infty} b_n$ **converges.**

True   Let $\displaystyle\sum_{n=1}^{\infty} a_n = a$ and $\displaystyle\sum_{n=1}^{\infty} (a_n + b_n) = c$

Then, $\displaystyle\sum_{n=1}^{\infty} b_n = \sum_{n=1}^{\infty} \{(a_n + b_n) - a_n\} = \sum_{n=1}^{\infty} (a_n + b_n) - \sum_{n=1}^{\infty} a_n = c - a$ $\therefore \displaystyle\sum_{n=1}^{\infty} b_n$ converges.

(5) If $\displaystyle\sum_{n=1}^{\infty} a_n b_n$ **converges and** $\displaystyle\lim_{n\to\infty} a_n \neq 0$, **then** $\displaystyle\lim_{n\to\infty} b_n = 0$.

    False   Let $\{a_n\} = 1,\ 0,\ 1,\ 0, \cdots\cdots$

                $\{b_n\} = 0,\ 1,\ 0,\ 1, \cdots\cdots$

            Then, $\displaystyle\lim_{n\to\infty} a_n \neq 0$

            Since $a_n b_n = 0,\ \displaystyle\sum_{n=1}^{\infty} a_n b_n = 0$ $\therefore \displaystyle\sum_{n=1}^{\infty} a_n b_n$ converges. But, $\displaystyle\lim_{n\to\infty} b_n \neq 0$

(6) If $\displaystyle\sum_{n=1}^{\infty} a_n = a$ **and** $\displaystyle\sum_{n=1}^{\infty} b_n = b$, **then** $\displaystyle\sum_{n=1}^{\infty} a_n b_n = ab$. **($a, b$; constants)**

    False   Let $a_n = \left(\frac{1}{2}\right)^{n-1}$, $b_n = \left(\frac{1}{2}\right)^{n-1}$

    Then, $a_n b_n = \left(\frac{1}{2}\right)^{n-1} \left(\frac{1}{2}\right)^{n-1} = \left(\frac{1}{4}\right)^{n-1}$

    $\therefore \displaystyle\sum_{n=1}^{\infty} a_n b_n = \displaystyle\sum_{n=1}^{\infty} \left(\frac{1}{4}\right)^{n-1} = \dfrac{1}{1-\frac{1}{4}} = \dfrac{4}{3}$

    But, $\displaystyle\sum_{n=1}^{\infty} a_n = \displaystyle\sum_{n=1}^{\infty} b_n = \displaystyle\sum_{n=1}^{\infty} \left(\frac{1}{2}\right)^{n-1} = \dfrac{1}{1-\frac{1}{2}} = 2$ $\therefore a = b = 2 ;\ ab = 4$

    $\therefore \displaystyle\sum_{n=1}^{\infty} a_n b_n \neq ab$

(7) If $\displaystyle\sum_{n=1}^{\infty}(2a_n + b_n)$ **and** $\displaystyle\sum_{n=1}^{\infty}(a_n - 2b_n)$ **converge, then** $\displaystyle\sum_{n=1}^{\infty} a_n$ **and** $\displaystyle\sum_{n=1}^{\infty} b_n$ **converge.**

    True   Let $2a_n + b_n = c_n ;\ a_n - 2b_n = d_n$

    Then, $\displaystyle\sum_{n=1}^{\infty}(2a_n + b_n) = \displaystyle\sum_{n=1}^{\infty} c_n \overset{let}{=} c$; $\displaystyle\sum_{n=1}^{\infty}(a_n - 2b_n) = \displaystyle\sum_{n=1}^{\infty} d_n \overset{let}{=} d$

$$\begin{array}{rl} 4a_n + 2b_n &= 2c_n \\ + \quad a_n - 2b_n &= d_n \\ \hline 5a_n \quad\quad &= 2c_n + d_n \end{array} \quad \therefore a_n = \frac{2c_n + d_n}{5}$$

$$\begin{array}{rl} 2a_n + b_n &= c_n \\ - \quad 2a_n - 4b_n &= 2d_n \\ \hline 5b_n &= c_n - 2d_n \end{array} \quad \therefore b_n = \frac{c_n - 2d_n}{5}$$

$\therefore \displaystyle\sum_{n=1}^{\infty} a_n = \displaystyle\sum_{n=1}^{\infty} \frac{2c_n + d_n}{5} = \frac{2}{5}\displaystyle\sum_{n=1}^{\infty} c_n + \frac{1}{5}\displaystyle\sum_{n=1}^{\infty} d_n = \frac{2}{5}c + \frac{1}{5}d$ (Converge)

$\therefore \displaystyle\sum_{n=1}^{\infty} b_n = \displaystyle\sum_{n=1}^{\infty} \frac{c_n - 2d_n}{5} = \frac{1}{5}\displaystyle\sum_{n=1}^{\infty} c_n - \frac{2}{5}\displaystyle\sum_{n=1}^{\infty} d_n = \frac{1}{5}c - \frac{2}{5}d$ (Converge)

(8) If $a_n > b_n,$ $\sum\limits_{n=1}^{\infty} a_n = a$ **, and** $\sum\limits_{n=1}^{\infty} b_n = b,$ **then** $a > b.$

$(a, b \text{ ; real numbers, } n; \text{ positive integer})$

True $\quad a - b = \sum\limits_{n=1}^{\infty} a_n - \sum\limits_{n=1}^{\infty} b_n = \sum\limits_{n=1}^{\infty} (a_n - b_n) = \lim\limits_{n \to \infty} \sum\limits_{k=1}^{n} (a_k - b_k) > 0 \quad \therefore \ a > b$

(9) If $\sum\limits_{n=1}^{\infty} a_{2n-1} = a$ **and** $\sum\limits_{n=1}^{\infty} a_{2n} = b,$ **then** $\sum\limits_{n=1}^{\infty} a_n = a + b.$

True $\quad \sum\limits_{n=1}^{\infty} a_n = \sum\limits_{n=1}^{\infty} (a_{2n-1} + a_{2n}) = \sum\limits_{n=1}^{\infty} a_{2n-1} + \sum\limits_{n=1}^{\infty} a_{2n} = a + b$

#15 For geometric sequences $\{a_n\}$ and $\{b_n\}$,

determine if the following statements are true or false.

(1) If $\sum\limits_{n=1}^{\infty} a_n$ **converges, then** $\sum\limits_{n=1}^{\infty} a_{2n}$ **converges.**

True $\quad$ Let $a_n = ar^{n-1}$

Then, $a_{2n} = ar^{2n-1} \quad \therefore$ The common ratio of $\{a_{2n}\}$ is $r^2.$

Since $\sum\limits_{n=1}^{\infty} a_n$ converges, $-1 < r < 1 \quad \therefore \ 0 \le r^2 < 1 \quad \therefore \ \sum\limits_{n=1}^{\infty} a_{2n}$ converges.

(2) If $\sum\limits_{n=1}^{\infty} a_n$ **diverges, then** $\sum\limits_{n=1}^{\infty} a_{2n}$ **diverges.**

True $\quad$ Let $a_n = ar^{n-1}$

If $\sum\limits_{n=1}^{\infty} a_n$ diverges, then $|r| \ge 1$

Since $\{a_{2n}\}$ has common ratio $r^2,$ $r^2 \ge 1 \quad \therefore \ \sum\limits_{n=1}^{\infty} a_{2n}$ diverges.

(3) If $\sum\limits_{n=1}^{\infty} a_n$ **converges, then** $\sum\limits_{n=1}^{\infty} \left(a_n + \dfrac{1}{2} \right)$ **converges.**

False $\quad$ If $\sum\limits_{n=1}^{\infty} a_n$ converges, then $\lim\limits_{n \to \infty} a_n = 0$

Since $\lim\limits_{n \to \infty} \left(a_n + \dfrac{1}{2} \right) = \lim\limits_{n \to \infty} a_n + \dfrac{1}{2} = 0 + \dfrac{1}{2} = \dfrac{1}{2} \ne 0,$ $\sum\limits_{n=1}^{\infty} \left(a_n + \dfrac{1}{2} \right)$ converges.

(4) If $\sum\limits_{n=1}^{\infty} a_n$ **and** $\sum\limits_{n=1}^{\infty} b_n$ **converge, then** $\sum\limits_{n=1}^{\infty} a_n b_n$ **converges.**

True $\quad$ Let $a_n = ar^{n-1},$ $b_n = bs^{n-1}$

Then $a_n b_n = ab(rs)^{n-1}$

$\therefore \ \{a_n b_n\}$ has common ratio $rs.$

Since $\sum\limits_{n-1}^{\infty} a_n$ and $\sum\limits_{n-1}^{\infty} b_n$ converge, $-1 < r < 1$ and $-1 < s < 1 \quad \therefore \ |r| < 1$ and $|s| < 1$

Therefore, $|rs| = |r||s| < 1$ and hence $\sum\limits_{n=1}^{\infty} a_n b_n$ converges.

(5) If $\displaystyle\sum_{n=1}^{\infty} a_n b_n$ **converges, then at least one of** $\displaystyle\sum_{n=1}^{\infty} a_n$ **and** $\displaystyle\sum_{n=1}^{\infty} b_n$ **is convergent.**

True   Let $a_n = ar^{n-1},\ \ b_n = bs^{n-1}$

Suppose $\displaystyle\sum_{n=1}^{\infty} a_n$ and $\displaystyle\sum_{n=1}^{\infty} b_n$ are divergent.

Then, $|r| \geq 1$ and $|s| \geq 1$ $\therefore |rs| = |r||s| \geq 1$ $\therefore \displaystyle\sum_{n=1}^{\infty} a_n b_n$ diverges.

(6) If $\displaystyle\sum_{n=1}^{\infty} a_n$ **and** $\displaystyle\sum_{n=1}^{\infty} b_n$ **are divergent, then** $\displaystyle\lim_{n\to\infty}(a_n + b_n) \neq 0.$

False   Let $a_n = (-1)^n,\ b_n = (-1)^{n-1}$

Then, $\displaystyle\lim_{n\to\infty} a_n \neq 0,\ \ \lim_{n\to\infty} b_n \neq 0$

$\therefore \displaystyle\sum_{n=1}^{\infty} a_n$ and $\displaystyle\sum_{n=1}^{\infty} b_n$ diverge.

But, $a_n + b_n = (-1)^n + (-1)^{n-1} = 0$ $\therefore \displaystyle\lim_{n\to\infty}(a_n + b_n) = 0$

(7) If $\displaystyle\sum_{n=1}^{\infty} a_n{}^3$ **and** $\displaystyle\sum_{n=1}^{\infty} b_n{}^3$ **are convergent, then** $\displaystyle\sum_{n=1}^{\infty} (a_n + b_n)$ **is convergent.**

False   Let $a_n = ar^{n-1},\ b_n = bs^{n-1}$

Then, $\{a_n{}^3\}$ has common ratio r^3, $\{b_n{}^3\}$ has common ratio s^3.

Since $\displaystyle\sum_{n=1}^{\infty} a_n{}^3$ and $\displaystyle\sum_{n=1}^{\infty} b_n{}^3$ converge, $|r^3| < 1$ and $|s^3| < 1$.

$\therefore |r| < 1$ and $|s| < 1$.

$\therefore \displaystyle\sum_{n=1}^{\infty} a_n$ and $\displaystyle\sum_{n=1}^{\infty} b_n$ converge.

Since $\displaystyle\sum_{n=1}^{\infty} (a_n + b_n) = \sum_{n=1}^{\infty} a_n + \sum_{n=1}^{\infty} b_n$, $\displaystyle\sum_{n=1}^{\infty} (a_n + b_n)$ converges.

Chapter 2. Limits of Functions and Continuity

#1 Evaluate the following limits:

(1) $\displaystyle\lim_{x\to2} 3x^2 = 3\lim_{x\to2} x^2 = 3\cdot2^2 = 12$

(2) $\displaystyle\lim_{x\to3} x(2x-1) = \lim_{x\to3} x \cdot \lim_{x\to3}(2x-1) = 3\cdot(2\cdot3-1) = 15$

(3) $\displaystyle\lim_{x\to2}\frac{x^2}{x+1} = \frac{\lim\limits_{x\to2} x^2}{\lim\limits_{x\to2}(x+1)} = \frac{2^2}{2+1} = \frac{4}{3}$

(4) $\displaystyle\lim_{x\to1}\sqrt{x+3} = \sqrt{1+3} = 2$

(5) $\displaystyle\lim_{x\to0}\cos x = \cos 0 = 1$

(6) $\displaystyle\lim_{x\to1}\sin\frac{\pi x}{2} = \sin\frac{\pi}{2} = 1$

(7) $\displaystyle\lim_{x\to0}\frac{2^{x+1}}{2^x+3^x} = \frac{2^{0+1}}{2^0+3^0} = \frac{2}{1+1} = 1$

(8) $\displaystyle\lim_{x\to10}\left(\log x^2 - \log\frac{1}{x}\right) = \log 10^2 - \log\frac{1}{10} = 2-(-1) = 3$

(9) $\displaystyle\lim_{x\to0}\frac{1}{\sqrt{x^2}}$ Since $\dfrac{1}{\sqrt{x^2}}\to 0^+$ as $x\to0$, $\displaystyle\lim_{x\to0}\frac{1}{\sqrt{x^2}} = \infty$

(10) $\displaystyle\lim_{x\to1}\frac{x}{(x-1)^2}$ Since $\dfrac{1}{(x-1)^2}\to 0^+$ as $x\to1$, $\displaystyle\lim_{x\to1}\frac{x}{(x-1)^2} = \infty$

(11) $\displaystyle\lim_{x\to2}\left(3-\frac{x}{|x-2|}\right)$ Since $|x-2|\to0^+$ as $x\to2$, $\displaystyle\lim_{x\to2}\frac{x}{|x-2|} = \infty$ $\therefore\ \displaystyle\lim_{x\to2}\left(3-\frac{x}{|x-2|}\right) = -\infty$

(12) $\displaystyle\lim_{x\to3}\frac{x^2+x-12}{x-3} = \lim_{x\to3}\frac{(x+4)(x-3)}{x-3} = \lim_{x\to3}(x+4) = 3+4 = 7$

(13) $\displaystyle\lim_{x\to1}\left(\frac{5x^2-7x+2}{x-1} + \frac{3-2x}{x+1}\right)$

$= \displaystyle\lim_{x\to1}\frac{(x-1)(5x-2)}{x-1} + \lim_{x\to1}\frac{3-2x}{x+1} = \lim_{x\to1}(5x-2) + \frac{3-2\cdot1}{1+1} = (5\cdot1-2) + \frac{1}{2} = 3\frac{1}{2}$

(14) $\displaystyle\lim_{x\to a}\frac{x^3-ax^2+a^2x-a^3}{x-a}$

$= \displaystyle\lim_{x\to a}\frac{(x^3-a^3)-ax(x-a)}{x-a} = \lim_{x\to a}\frac{(x-a)(x^2+ax+a^2)-ax(x-a)}{x-a} = \lim_{x\to a}\frac{(x-a)(x^2+ax+a^2-ax)}{x-a}$

$= \displaystyle\lim_{x\to a}(x^2+a^2) = a^2+a^2 = 2a^2$

(15) $\displaystyle\lim_{x\to1}\frac{\sqrt{x+8}-3}{x-1} = \lim_{x\to1}\frac{(\sqrt{x+8}-3)(\sqrt{x+8}+3)}{(x-1)(\sqrt{x+8}+3)} = \lim_{x\to1}\frac{(x+8)-9}{(x-1)(\sqrt{x+8}+3)} = \lim_{x\to1}\frac{1}{\sqrt{x+8}+3} = \frac{1}{\sqrt{1+8}+3} = \frac{1}{6}$

(16) $\displaystyle\lim_{x\to3}\frac{x-3}{\sqrt{x-3}} = \lim_{x\to3}\frac{x-3}{(x-3)^{\frac{1}{2}}} = \lim_{x\to3}(x-3)^{1-\frac{1}{2}} = \lim_{x\to3}(x-3)^{\frac{1}{2}} = \lim_{x\to3}\sqrt{x-3} = \sqrt{3-3} = 0$

(17) $\displaystyle\lim_{x\to-8}\frac{x+8}{\sqrt[3]{x}+2} = \lim_{x\to-8}\frac{(x+8)\left(\sqrt[3]{x^2}-2\sqrt[3]{x}+4\right)}{\left(\sqrt[3]{x}+2\right)\left(\sqrt[3]{x^2}-2\sqrt[3]{x}+4\right)} = \lim_{x\to-8}\frac{(x+8)\left(\sqrt[3]{x^2}-2\sqrt[3]{x}+4\right)}{x+8}$

$$= \lim_{x \to -8} \left(\sqrt[3]{x^2} - 2\sqrt[3]{x} + 4 \right) = \sqrt[3]{64} - 2\sqrt[3]{(-8)} + 4 = \sqrt[3]{4^3} - 2\sqrt[3]{(-2)^3} + 4$$

$$= 4 - 2(-2) + 4 = 12$$

(18) $\lim\limits_{x \to 1} \dfrac{\sqrt[3]{x}-1}{x-1} = \lim\limits_{x \to 1} \dfrac{\left(\sqrt[3]{x}-1\right)\left(\sqrt[3]{x^2}+\sqrt[3]{x}+1\right)}{(x-1)\left(\sqrt[3]{x^2}+\sqrt[3]{x}+1\right)} = \lim\limits_{x \to 1} \dfrac{x-1}{(x-1)\left(\sqrt[3]{x^2}+\sqrt[3]{x}+1\right)} = \lim\limits_{x \to 1} \dfrac{1}{\left(\sqrt[3]{x^2}+\sqrt[3]{x}+1\right)}$$

$$= \dfrac{1}{1+1+1} = \dfrac{1}{3}$$

(19) $\lim\limits_{x \to \infty} \dfrac{2x-3}{3x^2-2x+1} = \lim\limits_{x \to \infty} \dfrac{\frac{2}{x}-\frac{3}{x^2}}{3-\frac{2}{x}+\frac{1}{x^2}} = \dfrac{0}{3} = 0$

(20) $\lim\limits_{x \to \infty} \dfrac{5x^3+2x}{2x^3-4x^2+5x} = \lim\limits_{x \to \infty} \dfrac{5+\frac{2}{x^2}}{2-\frac{4}{x}+\frac{5}{x^2}} = \dfrac{5}{2}$

(21) $\lim\limits_{x \to \infty} \dfrac{3x^3-2x+4}{x^2+1} = \lim\limits_{x \to \infty} \dfrac{3x-\frac{2}{x}+\frac{4}{x^2}}{1+\frac{1}{x^2}} = \infty$

(22) $\lim\limits_{x \to \infty} \dfrac{\sqrt{x^2+1}-1}{x+1} = \lim\limits_{x \to \infty} \dfrac{\sqrt{1+\frac{1}{x^2}}-\frac{1}{x}}{1+\frac{1}{x}} = 1$

(23) $\lim\limits_{x \to -\infty} \dfrac{\sqrt{x^2+1}-1}{x+1} = \lim\limits_{x \to -\infty} \dfrac{-\sqrt{1+\frac{1}{x^2}}-\frac{1}{x}}{1+\frac{1}{x}} = -1$

(24) $\lim\limits_{x \to \infty} (2x^3 - 3x^2 + 4x - 1) = \lim\limits_{x \to \infty} x^3 \left(2 - \dfrac{3}{x} + \dfrac{4}{x^2} - \dfrac{1}{x^3} \right) = \infty$

(25) $\lim\limits_{x \to \infty} \left(\sqrt{x^2+2x+3} - x \right) = \lim\limits_{x \to \infty} \dfrac{\left(\sqrt{x^2+2x+3}-x\right)\left(\sqrt{x^2+2x+3}+x\right)}{\sqrt{x^2+2x+3}+x} = \lim\limits_{x \to \infty} \dfrac{(x^2+2x+3)-x^2}{\sqrt{x^2+2x+3}+x}$

$$= \lim\limits_{x \to \infty} \dfrac{2x+3}{\sqrt{x^2+2x+3}+x} \quad \left(\text{Form of } \dfrac{\infty}{\infty}\right) = \lim\limits_{x \to \infty} \dfrac{2+\frac{3}{x}}{\sqrt{1+\frac{2}{x}+\frac{3}{x^2}}+1} = \dfrac{2}{2} = 1$$

(26) $\lim\limits_{x \to 0} \dfrac{1}{x} \left(1 + \dfrac{1}{x-1} \right) = \lim\limits_{x \to 0} \dfrac{1}{x} \left(\dfrac{x-1+1}{x-1} \right) = \lim\limits_{x \to 0} \dfrac{1}{x} \left(\dfrac{x}{x-1} \right) = \lim\limits_{x \to 0} \dfrac{1}{x-1} = \dfrac{1}{0-1} = -1$

(27) $\lim\limits_{x \to 0} \dfrac{1}{x} \left(\dfrac{1}{\sqrt{x+1}} - 1 \right) = \lim\limits_{x \to 0} \dfrac{1}{x} \left(\dfrac{1-\sqrt{x+1}}{\sqrt{x+1}} \right) \left(\text{Form of } \dfrac{0}{0}\right) = \lim\limits_{x \to 0} \left\{ \dfrac{(1-\sqrt{x+1})(1+\sqrt{x+1})}{x(\sqrt{x+1})(1+\sqrt{x+1})} \right\}$

$$= \lim\limits_{x \to 0} \left\{ \dfrac{1-(x+1)}{x(\sqrt{x+1})(1+\sqrt{x+1})} \right\} = \lim\limits_{x \to 0} \left\{ \dfrac{-1}{(\sqrt{x+1})(1+\sqrt{x+1})} \right\}$$

$$= \dfrac{-1}{(\sqrt{1})(1+\sqrt{1})} = -\dfrac{1}{2}$$

(28) $\lim\limits_{x \to -\infty} \dfrac{x+1}{\sqrt{x^2+x}-x}$

Let $\quad -x = t$

Then, $x \to -\infty \quad \Rightarrow \quad t \to \infty$

$$\lim\limits_{x \to -\infty} \dfrac{x+1}{\sqrt{x^2+x}-x} = \lim\limits_{t \to \infty} \dfrac{-t+1}{\sqrt{t^2-t}+t} \left(\text{Form of } \dfrac{\infty}{\infty}\right) = \lim\limits_{t \to \infty} \dfrac{-1+\frac{1}{t}}{\sqrt{1-\frac{1}{t}}+1} = \dfrac{-1}{\sqrt{1}+1} = -\dfrac{1}{2}$$

#2 Find the value of the constant a such that the following expression:

(1) $\displaystyle\lim_{x\to 2} \frac{x^2-2x+2}{3x^2+ax+1} = 2$

Since $\displaystyle\lim_{x\to 2}(3x^2 + ax + 1) \neq 0$, $\displaystyle\lim_{x\to 2} \frac{x^2-2x+2}{3x^2+ax+1} = \frac{4-4+2}{12+2a+1} = \frac{2}{13+2a} = 2$

$\therefore 26 + 4a = 2$; $4a = -24$ $\therefore a = -6$

(2) $\displaystyle\lim_{x\to\infty} \frac{3x^3+4x-1}{ax^3+2x^2+3} = \frac{1}{2}$

Since $\displaystyle\lim_{x\to\infty} \frac{3x^3+4x-1}{ax^3+2x^2+3} = \lim_{x\to\infty} \frac{3+\frac{4}{x^2}-\frac{1}{x^3}}{a+\frac{2}{x}+\frac{3}{x^3}} = \frac{3}{a}$, $\frac{3}{a} = \frac{1}{2}$ $\therefore a = 6$

(3) $\displaystyle\lim_{x\to\infty} \frac{a4^x}{4^{x+1}-3^x} = 4$

Since $\displaystyle\lim_{x\to\infty} \frac{a4^x}{4^{x+1}-3^x} = \lim_{x\to\infty} \frac{a}{\frac{4\cdot 4^x}{4^x}-\frac{3^x}{4^x}} = \lim_{x\to\infty} \frac{a}{4-\left(\frac{3}{4}\right)^x} = \frac{a}{4}$, $\frac{a}{4} = 4$ $\therefore a = 16$

(4) $\displaystyle\lim_{x\to\infty} \frac{ax}{\sqrt{x^2+1}-1} = 2$

Since $\displaystyle\lim_{x\to\infty} \frac{ax}{\sqrt{x^2+1}-1} = \lim_{x\to\infty} \frac{a}{\sqrt{1+\frac{1}{x^2}}-\frac{1}{x}} = \frac{a}{1} = 2$, $a = 2$

(5) $\displaystyle\lim_{x\to\infty}\{\log(ax + 1) - \log x\} = 1$

Since $\displaystyle\lim_{x\to\infty}\{\log(ax + 1) - \log x\} = \lim_{x\to\infty}\left\{\log\frac{ax+1}{x}\right\} = \lim_{x\to\infty}\left\{\log\frac{a+\frac{1}{x}}{1}\right\} = \log a$, $\log a = 1$

$\therefore a = e$

#3 Find the values of the constants a and b such that the following expression:

(1) $\displaystyle\lim_{x\to 1} \frac{ax-3x^2}{x-1} = b$

Since the limit is a constant and the denominator approaches 0 as $x \to 1$, the numerator approaches 0.

$\therefore \displaystyle\lim_{x\to 1}(ax - 3x^2) = 0$ $\therefore a - 3 = 0$ $\therefore a = 3$

$\therefore b = \displaystyle\lim_{x\to 1}\frac{ax-3x^2}{x-1} = \lim_{x\to 1}\frac{3x-3x^2}{x-1} = \lim_{x\to 1}\frac{-3x(x-1)}{x-1} = \lim_{x\to 1}(-3x) = (-3)\cdot 1 = -3$

Therefore, $a = 3$, $b = -3$

(2) $\displaystyle\lim_{x\to 1} \frac{x^2+ax+b}{x-1} = 5$

Since the limit is a constant and (The denominator) $\to 0$ as $x \to 1$,

(The numerator) $\to 0$ as $x \to 1$

$\therefore \displaystyle\lim_{x\to 1}(x^2 + ax + b) = 0$ $\therefore 1 + a + b = 0$; $b = -a - 1$

$\therefore\ 5 = \lim_{x\to 1}\dfrac{x^2+ax+b}{x-1} = \lim_{x\to 1}\dfrac{x^2+ax-(a+1)}{x-1} = \lim_{x\to 1}\dfrac{(x-1)(x+a+1)}{x-1} = \lim_{x\to 1}(x+a+1)$

$= 1+a+1 = a+2 \qquad \therefore\ a = 3 \qquad$ Therefore, $a = 3,\ b = -4$

(3) $\displaystyle\lim_{x\to -3}\dfrac{\sqrt{x^2-x-3}+ax}{x+3} = b$

When $x \to -3,\ \displaystyle\lim_{x\to -3}(x+3) = 0$

Since $\displaystyle\lim_{x\to -3}\dfrac{\sqrt{x^2-x-3}+ax}{x+3} = b$; i. e., the limit exists, $\displaystyle\lim_{x\to -3}\left(\sqrt{x^2-x-3}+ax\right) = 0$

$\therefore\ \displaystyle\lim_{x\to -3}\left(\sqrt{x^2-x-3}+ax\right) = \sqrt{(-3)^2-(-3)-3}+a(-3) = 3 - 3a = 0$; $a = 1$

$\therefore\ b = \displaystyle\lim_{x\to -3}\dfrac{\sqrt{x^2-x-3}+ax}{x+3} = \lim_{x\to -3}\dfrac{\sqrt{x^2-x-3}+x}{x+3} = \lim_{x\to -3}\dfrac{\left(\sqrt{x^2-x-3}+x\right)\left(\sqrt{x^2-x-3}-x\right)}{(x+3)\left(\sqrt{x^2-x-3}-x\right)}$

$= \displaystyle\lim_{x\to -3}\dfrac{(x^2-x-3)-x^2}{(x+3)\left(\sqrt{x^2-x-3}-x\right)} = \lim_{x\to -3}\dfrac{-(x+3)}{(x+3)\left(\sqrt{x^2-x-3}-x\right)} = \lim_{x\to -3}\dfrac{-1}{\sqrt{x^2-x-3}-x}$

$= \dfrac{-1}{\sqrt{(-3)^2-(-3)-3}-(-3)} = \dfrac{-1}{6} = -\dfrac{1}{6} \qquad$ Therefore, $a = 1,\ b = -\dfrac{1}{6}$

#4 Find the expression of $f(x)$ such that $\displaystyle\lim_{x\to\infty}\dfrac{f(x)}{2x^2+x+1} = 1$ and $\displaystyle\lim_{x\to 2}\dfrac{f(x)}{x^2-x-2} = 1$.

Since $\displaystyle\lim_{x\to\infty}\dfrac{f(x)}{2x^2+x+1} = 1,\ f(x)$ is the form of: $f(x) = 2x^2 + bx + c$

Since $\displaystyle\lim_{x\to 2}\dfrac{f(x)}{x^2-x-2} = 1,\ \lim_{x\to 2}\dfrac{f(x)}{x^2-x-2} = \lim_{x\to 2}\dfrac{2x^2+bx+c}{(x-2)(x+1)} = 1$ (Limit exists.)

Since $\displaystyle\lim_{x\to 2}(x-2)(x+1) = 0,\ \lim_{x\to 2}(2x^2+bx+c) = 0$

$\therefore\ \displaystyle\lim_{x\to 2}(2x^2+bx+c) = 2(2)^2 + 2b + c = 8 + 2b + c = 0$; $c = -8 - 2b$

$\therefore\ \displaystyle\lim_{x\to 2}\dfrac{f(x)}{x^2-x-2} = \lim_{x\to 2}\dfrac{2x^2+bx-8-2b}{(x-2)(x+1)} = \lim_{x\to 2}\dfrac{(x-2)(2x+4+b)}{(x-2)(x+1)} = \lim_{x\to 2}\dfrac{2x+4+b}{x+1} = \dfrac{8+b}{3} = 1$

$\therefore\ b = -5$; $c = -8 + 10 = 2$

Therefore, $f(x) = 2x^2 - 5x + 2$

#5 Determine whether the limits exist.

(1) $\displaystyle\lim_{x\to 0}\dfrac{1}{x}$

Since $\displaystyle\lim_{x\to 0^-}\dfrac{1}{x} = -\infty$ and $\displaystyle\lim_{x\to 0^+}\dfrac{1}{x} = \infty,\ \lim_{x\to 0}\dfrac{1}{x}$ does not exist.

(2) $\displaystyle\lim_{x\to\infty}\sin\dfrac{1}{x}$

Since $\dfrac{1}{x}\to 0$ as $x\to\infty,\ \displaystyle\lim_{x\to\infty}\sin\dfrac{1}{x} = \sin 0 = 0 \qquad \therefore\ \lim_{x\to\infty}\sin\dfrac{1}{x}$ exists.

(3) $\lim\limits_{x\to 1}\dfrac{|x^2-1|}{|x-1|}$

$\lim\limits_{x\to 1}\dfrac{|x^2-1|}{|x-1|} = \lim\limits_{x\to 1}\dfrac{|(x-1)(x+1)|}{|x-1|} = \lim\limits_{x\to 1}\dfrac{|x-1||x+1|}{|x-1|} = \lim\limits_{x\to 1}|x+1| = |1+1| = 2$

$\therefore\ \lim\limits_{x\to 1}\dfrac{|x^2-1|}{|x-1|}$ exists.

(4) $\lim\limits_{x\to\infty}\left(\sqrt{x^2-x}-x\right)$

$\lim\limits_{x\to\infty}\left(\sqrt{x^2-x}-x\right) = \lim\limits_{x\to\infty}\dfrac{\left(\sqrt{x^2-x}-x\right)\left(\sqrt{x^2-x}+x\right)}{\sqrt{x^2-x}+x} = \lim\limits_{x\to\infty}\dfrac{(x^2-x)-x^2}{\sqrt{x^2-x}+x} = \lim\limits_{x\to\infty}\dfrac{-x}{\sqrt{x^2-x}+x}$

$\qquad\qquad\qquad\qquad = \lim\limits_{x\to\infty}\dfrac{-1}{\sqrt{1-\frac{1}{x}}+1} = \dfrac{-1}{\sqrt{1}+1} = -\dfrac{1}{2}$

$\therefore\ \lim\limits_{x\to\infty}\left(\sqrt{x^2-x}-x\right)$ exists.

#6 For functions $f(x), g(x)$, and $h(x)$ which are defined on all real numbers, determine if the following statements are true or false.

(1) If $\lim\limits_{x\to 0} f(x) = 1$, then $f(0) = 1$.

False   Let $f(x) = \begin{cases} 1, & x \neq 0 \\ 0, & x = 0 \end{cases}$

Then, $\lim\limits_{x\to 0} f(x) = 1$ But, $f(0) = 0$

(2) If $\lim\limits_{x\to 1} f(x) = 1$, then $\lim\limits_{x\to\infty} f\left(1+\dfrac{1}{x}\right) = 1$.

True   Let $1+\dfrac{1}{x} = t$

Then, $t \to 1$ as $x \to \infty$

$\therefore\ \lim\limits_{x\to\infty} f\left(1+\dfrac{1}{x}\right) = \lim\limits_{t\to 1} f(t) = \lim\limits_{x\to 1} f(x) = 1$

(3) If $f(x) < g(x) < h(x)$, $\lim\limits_{x\to 0} f(x) = 0$, and $\lim\limits_{x\to 0} h(x) = 0$, then $\lim\limits_{x\to 0} g(x) = 0$.

True $f(x) < g(x) < h(x)\ \Rightarrow\ \lim\limits_{x\to 0} f(x) \leq \lim\limits_{x\to 0} g(x) \leq \lim\limits_{x\to 0} h(x)$

Since $\lim\limits_{x\to 0} f(x) = \lim\limits_{x\to 0} h(x) = 0$, $\lim\limits_{x\to 0} g(x) = 0$

(4) If $\lim\limits_{x\to\infty} xf(x)$ exists, then $\lim\limits_{x\to\infty} f(x)$ exists.

True   Let $xf(x) = g(x)$, and let $\lim\limits_{x\to\infty} xf(x) = \lim\limits_{x\to\infty} g(x) = k$ $(k;\text{constant})$

Then, $\lim\limits_{x\to\infty} f(x) = \lim\limits_{x\to\infty}\dfrac{xf(x)}{x} = \lim\limits_{x\to\infty}\dfrac{g(x)}{x} = \lim\limits_{x\to\infty}\dfrac{k}{x} = 0$

$\therefore\ \lim\limits_{x\to\infty} f(x)$ exists.

(5) If $\lim\limits_{x\to\infty} \dfrac{1}{f(x)}$ **exists, then** $\lim\limits_{x\to\infty} f(x)$ **exists.**

False   Let $f(x) = x$

Then, $\lim\limits_{x\to\infty} \dfrac{1}{f(x)} = \lim\limits_{x\to\infty} \dfrac{1}{x} = 0$;i.e., $\lim\limits_{x\to\infty} \dfrac{1}{f(x)}$ exists.

But, $\lim\limits_{x\to\infty} f(x) = \lim\limits_{x\to\infty} x = \infty$ $\therefore$ $\lim\limits_{x\to\infty} f(x)$ does not exist.

(6) If $\lim\limits_{x\to a} f(x) = \infty$ **and** $\lim\limits_{x\to a} f(x)\,g(x)$ **exists, then** $\lim\limits_{x\to a} g(x) = 0.$

True   Let $f(x)g(x) = h(x)$

Then, $g(x) = \dfrac{h(x)}{f(x)}$

Since $\lim\limits_{x\to a} f(x) = \infty,$ $\lim\limits_{x\to a} \dfrac{1}{f(x)} = 0$

Since $\lim\limits_{x\to a} h(x)$ exists, $\lim\limits_{x\to a} h(x) = k$ $(k;\text{constant})$

$\therefore$ $\lim\limits_{x\to a} g(x) = \lim\limits_{x\to a} \dfrac{h(x)}{f(x)} = k \cdot 0 = 0$

(7) If $\lim\limits_{x\to a} f(x)$ **and** $\lim\limits_{x\to a} f(x)\,g(x)$ **exist, then** $\lim\limits_{x\to a} g(x)$ **exists.**

False   Let $f(x) = x,\ g(x) = [x]$ $([x];\text{the greatest integer in } x)$

Then, $\lim\limits_{x\to 0} f(x) = \lim\limits_{x\to 0} x = 0$;i. e., $\lim\limits_{x\to a} f(x)$ exists.

$\lim\limits_{x\to 0} f(x)\,g(x) = \lim\limits_{x\to 0} x[x] = 0$;i. e., $\lim\limits_{x\to a} f(x)\,g(x)$ exists.

But, $\lim\limits_{x\to 0^+} g(x) \neq \lim\limits_{x\to 0^-} g(x)$ $\therefore$ $\lim\limits_{x\to 0} g(x)$ does not exist.

(8) If $\lim\limits_{x\to a} g(x)$ **and** $\lim\limits_{x\to\infty} \dfrac{f(x)}{g(x)}$ **exist, then** $\lim\limits_{x\to a} f(x)$ **exists.**

True   Let $\lim\limits_{x\to a} g(x) = a$ and $\lim\limits_{x\to\infty} \dfrac{f(x)}{g(x)} = b$ $(a,b;\text{constants})$

Then, $\lim\limits_{x\to a} f(x) = \lim\limits_{x\to a} g(x)\dfrac{f(x)}{g(x)} = \lim\limits_{x\to a} g(x) \cdot \lim\limits_{x\to a} \dfrac{f(x)}{g(x)} = ab$ $(ab;\text{constant})$

$\therefore$ $\lim\limits_{x\to a} f(x)$ exists.

(9) If $\lim\limits_{x\to a} g(x)$ **exists,** $\lim\limits_{x\to a} f(g(x))$ **exists.**

False   Let $f(x) = [x]$ $([x];\text{the greatest integer in } x)$ and $g(x) = x$

Then, $\lim\limits_{x\to 0} g(x) = \lim\limits_{x\to 0} x = 0$

$\lim\limits_{x\to 0} f(g(x)) = \lim\limits_{x\to 0} f(x) = \lim\limits_{x\to 0} [x]$

Since $\lim\limits_{x\to 0^+} [x] \neq \lim\limits_{x\to 0^-} [x],$ $\lim\limits_{x\to 0} [x]$ does not exist.

$\therefore$ $\lim\limits_{x\to a} f(g(x))$ does not exist.

#7 Find the value. ($[x]$; the greatest integer in x)

(1) Find the value of the limit: $\lim\limits_{x \to 2^+} \left(\dfrac{[x]^2 - 3[x] + 2}{[x] - 1} \right) \cdot \lim\limits_{x \to 2^-} \left(\dfrac{|x-2|}{x-2} \right)$

Since $[x] = 2$ when $2 \le x < 3$, $\lim\limits_{x \to 2^+} \left(\dfrac{[x]^2 - 3[x] + 2}{[x] - 1} \right) = \dfrac{2^2 - 3 \cdot 2 + 2}{2 - 1} = 0$

Note that $\lim\limits_{x \to 2^-} \left(\dfrac{|x-2|}{x-2} \right) = \lim\limits_{x \to 2^-} \left\{ \dfrac{-(x-2)}{x-2} \right\} = -1$

$\therefore \lim\limits_{x \to 2^+} \left(\dfrac{[x]^2 - 3[x] + 2}{[x] - 1} \right) \cdot \lim\limits_{x \to 2^-} \left(\dfrac{|x-2|}{x-2} \right) = 0 \cdot -1 = 0$

(2) Find the value of the limit: $\lim\limits_{x \to \infty} \dfrac{3}{x} \left[\dfrac{x}{2} \right]$

Let $\dfrac{x}{2} = n + \alpha$ (n; integer, $0 \le \alpha < 1$)

Then, $\left[\dfrac{x}{2} \right] = n = \dfrac{x}{2} - \alpha$

$\therefore \lim\limits_{x \to \infty} \dfrac{3}{x} \left[\dfrac{x}{2} \right] = \lim\limits_{x \to \infty} \dfrac{3}{x} \left(\dfrac{x}{2} - \alpha \right) = \lim\limits_{x \to \infty} \left(\dfrac{3}{2} - \dfrac{3\alpha}{x} \right) = \dfrac{3}{2} - 0 = \dfrac{3}{2}$

(3) Find the value of the limit: $\lim\limits_{x \to -\infty} (2x + \sqrt{[4x^2 + x]})$

Let $x = -t$

Then $t \to \infty$ as $x \to -\infty$

Since $4x^2 + x = [4x^2 + x] + \alpha$ $(0 \le \alpha < 1)$, $[4x^2 + x] = [4t^2 - t] = 4t^2 - t - \alpha$

$\therefore \lim\limits_{x \to -\infty} (2x + \sqrt{[4x^2 + x]}) = \lim\limits_{t \to \infty} (-2t + \sqrt{4t^2 - t - \alpha})$

$= \lim\limits_{t \to \infty} \dfrac{(-2t + \sqrt{4t^2 - t - \alpha})(-2t - \sqrt{4t^2 - t - \alpha})}{(-2t - \sqrt{4t^2 - t - \alpha})} = \lim\limits_{t \to \infty} \dfrac{4t^2 - (4t^2 - t - \alpha)}{(-2t - \sqrt{4t^2 - t - \alpha})}$

$= \lim\limits_{t \to \infty} \dfrac{t + \alpha}{-2t - \sqrt{4t^2 - t - \alpha}} = \lim\limits_{t \to \infty} \dfrac{1 + \frac{\alpha}{t}}{-2 - \sqrt{4 - \frac{1}{t} - \frac{\alpha}{t^2}}} = \dfrac{1}{-2 - \sqrt{4}} = -\dfrac{1}{4}$

(4) For a function $f(x) = [x]^2 - a[x]$,

    find the value of the real number a so that $\lim\limits_{x \to 2} f(x)$ exists.

When $1 \le x < 2$, $[x] = 1$

When $2 \le x < 3$, $[x] = 2$

$\therefore \lim\limits_{x \to 2^-} f(x) = \lim\limits_{x \to 2^-} ([x]^2 - a[x]) = 1^2 - a \cdot 1 = 1 - a$

$\lim\limits_{x \to 2^+} f(x) = \lim\limits_{x \to 2^+} ([x]^2 - a[x]) = 2^2 - a \cdot 2 = 4 - 2a$

Since $\lim\limits_{x \to 2} f(x)$ exists, $\lim\limits_{x \to 2^-} f(x) = \lim\limits_{x \to 2^+} f(x)$

$\therefore 1 - a = 4 - 2a \qquad \therefore a = 3$

(5) Find the value of a positive integer a so that $\lim\limits_{x\to a}\dfrac{[x]^2+2x}{[x]}$ exists.

Since $\lim\limits_{x\to a^-}[x]=a-1$ and $\lim\limits_{x\to a^+}[x]=a,$

$$\lim_{x\to a^-}\frac{[x]^2+2x}{[x]}=\frac{(a-1)^2+2a}{a-1}=\frac{a^2+1}{a-1}\text{ and }\lim_{x\to a^+}\frac{[x]^2+2x}{[x]}=\frac{a^2+2a}{a}=\frac{a(a+2)}{a}=a+2$$

Since $\lim\limits_{x\to a}\dfrac{[x]^2+2x}{[x]}$ exists, $\lim\limits_{x\to a^-}\dfrac{[x]^2+2x}{[x]}=\lim\limits_{x\to a^+}\dfrac{[x]^2+2x}{[x]}$

$\therefore \dfrac{a^2+1}{a-1}=a+2$; $a^2+1=(a-1)(a+2)$; $a-2=1$ $\quad\therefore a=3$

(6) For constants $a, b,$ and $c,$ find the value of $a+b+c$ such that $\lim\limits_{x\to1}\dfrac{x^3+ax+b}{(x-1)^2}=c$.

Since $\lim\limits_{x\to1}\dfrac{x^3+ax+b}{(x-1)^2}$ exists and $\lim\limits_{x\to1}(x-1)^2=0,$ $\lim\limits_{x\to1}(x^3+ax+b)=0$

$\therefore 1^3+a\cdot1+b=a+b+1=0$; $b=-(a+1)$

$\therefore \lim\limits_{x\to1}\dfrac{x^3+ax+b}{(x-1)^2}=\lim\limits_{x\to1}\dfrac{x^3+ax-(a+1)}{(x-1)^2}=\lim\limits_{x\to1}\dfrac{(x-1)(x^2+x+a+1)}{(x-1)^2}$

$$=\lim_{x\to1}\frac{x^2+x+a+1}{x-1}=c$$

Since $\lim\limits_{x\to1}\dfrac{x^2+x+a+1}{x-1}$ exists and $\lim\limits_{x\to1}(x-1)=0,$ $\lim\limits_{x\to1}(x^2+x+a+1)=0$

$\therefore 1^2+1+a+1=a+3=0$; $a=-3,$ $b=-(-3+1)=2$

$c=\lim\limits_{x\to1}\dfrac{x^2+x+a+1}{x-1}=\lim\limits_{x\to1}\dfrac{x^2+x-3+1}{x-1}=\lim\limits_{x\to1}\dfrac{x^2+x-2}{x-1}=\lim\limits_{x\to1}\dfrac{(x+2)(x-1)}{x-1}=\lim\limits_{x\to1}(x+2)=3$

$\therefore a+b+c=(-3)+2+3=2$

(7) For a cubic function $f(x)$ such that $\lim\limits_{x\to0}\dfrac{f(x)}{x}=\lim\limits_{x\to1}\dfrac{f(x)}{x-1}=1,$ find the value of $f(2)$.

Note that $\lim\limits_{x\to0}\dfrac{f(x)}{x}$ and $\lim\limits_{x\to1}\dfrac{f(x)}{x-1}$ exist.

Since $\lim\limits_{x\to0}x=0$ and $\lim\limits_{x\to1}(x-1)=0,$ $\lim\limits_{x\to0}f(x)=0$ and $\lim\limits_{x\to1}f(x)=0$

$\therefore f(0)=0$ and $f(1)=0$

Since $f(x)$ is a cubic, $f(x)=(x-0)(x-1)(ax+b)=x(x-1)(ax+b)$

$\therefore \lim\limits_{x\to0}\dfrac{f(x)}{x}=\lim\limits_{x\to0}\dfrac{x(x-1)(ax+b)}{x}=\lim\limits_{x\to0}(x-1)(ax+b)=-b$

$\lim\limits_{x\to1}\dfrac{f(x)}{x-1}=\lim\limits_{x\to1}\dfrac{x(x-1)(ax+b)}{x-1}=\lim\limits_{x\to1}x(ax+b)=a+b$

Since $\lim\limits_{x\to0}\dfrac{f(x)}{x}=\lim\limits_{x\to1}\dfrac{f(x)}{x-1}=1,$ $-b=1$ and $a+b=1$

$\therefore a=2, b=-1$

$\therefore f(x)=x(x-1)(ax+b)=x(x-1)(2x-1)$

Therefore, $f(2)=2(2-1)(4-1)=6$

(8) For two functions f and g such that $\lim\limits_{x\to\infty} f(x) = \infty$ and $\lim\limits_{x\to\infty}\{f(x) - 2g(x)\} = a$,

find the value of $\lim\limits_{x\to\infty} \dfrac{f(x)+2g(x)+3}{2f(x)-3g(x)-4}$. ($a$; constant)

Since $\lim\limits_{x\to\infty}\{f(x) - 2g(x)\} = \lim\limits_{x\to\infty} f(x)\left\{1 - \dfrac{2g(x)}{f(x)}\right\} = \lim\limits_{x\to\infty} f(x)\cdot \lim\limits_{x\to\infty}\left\{1 - \dfrac{2g(x)}{f(x)}\right\} = a$ (exist)

and $\lim\limits_{x\to\infty} f(x) = \infty$, $\lim\limits_{x\to\infty}\left\{1 - \dfrac{2g(x)}{f(x)}\right\} = 0$

$\therefore\ \lim\limits_{x\to\infty}\dfrac{2g(x)}{f(x)} = 1$; $\lim\limits_{x\to\infty}\dfrac{g(x)}{f(x)} = \dfrac{1}{2}$

$\therefore\ \lim\limits_{x\to\infty}\dfrac{f(x)+2g(x)+3}{2f(x)-3g(x)-4} = \lim\limits_{x\to\infty}\dfrac{1+2\frac{g(x)}{f(x)}+\frac{3}{f(x)}}{2-3\frac{g(x)}{f(x)}-\frac{4}{f(x)}} = \dfrac{1+2\frac{1}{2}+0}{2-3\frac{1}{2}-0} = \dfrac{2}{\frac{1}{2}} = 4$

(9) For a function $f(x) = x^3 + x^2 + x$, let $f^{-1}(x)$ be the inverse of $f(x)$.

Find the value of $\lim\limits_{x\to 0}\dfrac{f^{-1}(2x)}{x}$.

Let $f^{-1}(2x) = y$

Then, $f(y) = 2x$

$\therefore\ f(y) = y^3 + y^2 + y = 2x$　　$\therefore\ x = \dfrac{y^3+y^2+y}{2}$

Substituting $x = 0$ and $y = 0$ into $f(y) = 2x$, we have $f(0) = 0$

$\therefore\ y$ approaches 0 as x approaches 0.

$\therefore\ \lim\limits_{x\to 0}\dfrac{f^{-1}(2x)}{x} = \lim\limits_{x\to 0}\dfrac{y}{x} = \lim\limits_{y\to 0}\dfrac{y}{\frac{y^3+y^2+y}{2}} = \lim\limits_{y\to 0}\dfrac{2y}{y^3+y^2+y} = \lim\limits_{y\to 0}\dfrac{2}{y^2+y+1} = \dfrac{2}{1} = 2$

(10) For polynomial functions f and g such that $\lim\limits_{x\to 1}\dfrac{f(x)-1}{x-1} = 1$ and $\lim\limits_{x\to\infty}\dfrac{f(x)-x^3}{x^2+1} = 1$,

find the value of $f(2)$.

Since $\lim\limits_{x\to 1}\dfrac{f(x)-1}{x-1}$ exists and $\lim\limits_{x\to 1}(x-1) = 0$, $\lim\limits_{x\to 1}(f(x)-1) = 0$　$\therefore\ f(1) = 1$ $\cdots\cdots$ ①

Since $\lim\limits_{x\to\infty}\dfrac{f(x)-x^3}{x^2+1} = 1$, $f(x) = x^3 + x^2 + ax + b$

$\therefore\ f(1) = 1^3 + 1^2 + a + b = a + b + 2 = 1$ by ①　　　$\therefore\ b = -a - 1$

$\lim\limits_{x\to 1}\dfrac{f(x)-1}{x-1} = \lim\limits_{x\to 1}\dfrac{(x^3+x^2+ax+b)-1}{x-1} = \lim\limits_{x\to 1}\dfrac{(x^3+x^2+ax-a-1)-1}{x-1} = \lim\limits_{x\to 1}\dfrac{x^3+x^2+ax-a-2}{x-1}$

$\qquad = \lim\limits_{x\to 1}\dfrac{(x-1)(x^2+2x+2+a)}{x-1} = \lim\limits_{x\to 1}(x^2+2x+2+a) = 5 + a$

$\therefore\ 5 + a = 1$; $a = -4$, $b = 3$

$\therefore\ f(x) = x^3 + x^2 - 4x + 3$

$\therefore\ f(2) = 8 + 4 - 8 + 3 = 7$

(11) For polynomial function $f(x)$ such that $\displaystyle\lim_{x\to0^+}\dfrac{x^3f\left(\frac{1}{x}\right)-1}{x^3+x}=5$ and $\displaystyle\lim_{x\to1}\dfrac{f(x)}{x^2+x-2}=\dfrac{1}{3}$,

find the value of $f(1)$.

Let $\dfrac{1}{x}=t$

Since $t\to\infty$ as $x\to0^+$, $\displaystyle\lim_{x\to0^+}\dfrac{x^3f\left(\frac{1}{x}\right)-1}{x^3+x}=\lim_{t\to\infty}\dfrac{\frac{1}{t^3}f(t)-1}{\frac{1}{t^3}+\frac{1}{t}}=\lim_{t\to\infty}\dfrac{f(t)-t^3}{1+t^2}=5$

$\therefore f(t)=t^3+5t^2+at+b$

Since $\displaystyle\lim_{x\to1}\dfrac{f(x)}{x^2+x-2}$ exists and $\displaystyle\lim_{x\to1}(x^2+x-2)=0$, $\displaystyle\lim_{x\to1}f(x)=0$

$\therefore f(1)=0$; $f(1)=1^3+5\cdot1^2+a\cdot1+b=a+b+6=0$; $b=-a-6$

$\displaystyle\lim_{x\to1}\dfrac{f(x)}{x^2+x-2}=\lim_{x\to1}\dfrac{x^3+5x^2+ax-a-6}{(x+2)(x-1)}=\lim_{x\to1}\dfrac{(x-1)(x^2+6x+a+6)}{(x+2)(x-1)}=\lim_{x\to1}\dfrac{x^2+6x+a+6}{x+2}$

$\qquad=\dfrac{1+6+a+6}{1+2}=\dfrac{a+13}{3}=\dfrac{1}{3}$

$\therefore\ a+13=1$; $a=-12,\ b=6$

$\therefore\ f(x)=x^3+5x^2-12x+6$

$\therefore\ f(1)=1^3+5\cdot1^2-12\cdot1+6=0$

(12) For polynomial function $g(x)$, $\displaystyle\lim_{x\to1}\dfrac{g(x)-2x}{x-1}$ exists. For a polynomial function $f(x)$

such that $f(x)+x-1=(x-1)g(x)$, find the value of $\displaystyle\lim_{x\to1}\dfrac{f(x)g(x)}{x^2-1}$.

Since $\displaystyle\lim_{x\to1}\dfrac{g(x)-2x}{x-1}$ exists and $\displaystyle\lim_{x\to1}(x-1)=0$, $\displaystyle\lim_{x\to1}\{g(x)-2x\}=g(1)-2=0$

$\therefore\quad g(1)=2$

Since $f(x)+x-1=(x-1)g(x)$, $f(x)=(x-1)\{g(x)-1\}$

$\therefore\ \displaystyle\lim_{x\to1}\dfrac{f(x)g(x)}{x^2-1}=\lim_{x\to1}\dfrac{(x-1)\{g(x)-1\}g(x)}{(x-1)(x+1)}=\lim_{x\to1}\dfrac{\{g(x)-1\}g(x)}{x+1}=\dfrac{\{g(1)-1\}g(1)}{1+1}=\dfrac{(2-1)\cdot2}{2}=1$

(13) When $\displaystyle\lim_{x\to\infty}f(x)=\infty$ and $\displaystyle\lim_{x\to a}\dfrac{\sqrt{g(x)}}{f(x)}=2$, find the value of $\displaystyle\lim_{x\to a}\dfrac{\log g(x)}{\log f(x)}$.

Note that $\log\sqrt{g(x)}=\log g(x)^{\frac{1}{2}}=\dfrac{1}{2}\log g(x)$; $2\log\sqrt{g(x)}=\log g(x)$

$\therefore\ \displaystyle\lim_{x\to a}\dfrac{\log g(x)}{\log f(x)}=\lim_{x\to a}\dfrac{2\log\sqrt{g(x)}}{\log f(x)}=\lim_{x\to a}\dfrac{2\log\left\{f(x)\frac{\sqrt{g(x)}}{f(x)}\right\}}{\log f(x)}=\lim_{x\to a}\dfrac{2\left\{\log f(x)+\log\frac{\sqrt{g(x)}}{f(x)}\right\}}{\log f(x)}$

$\qquad=2\displaystyle\lim_{x\to a}\left\{1+\dfrac{\log\frac{\sqrt{g(x)}}{f(x)}}{\log f(x)}\right\}=2\left\{1+\dfrac{\displaystyle\lim_{x\to a}\log\frac{\sqrt{g(x)}}{f(x)}}{\displaystyle\lim_{x\to a}\log f(x)}\right\}$

$\qquad=2\left\{1+\dfrac{\log2}{\displaystyle\lim_{x\to a}\log f(x)}\right\}=2(1+0)=2$

(14) For two polynomials f and g, such that $\lim\limits_{x \to a} \dfrac{f(x)}{x-a} = 3$ and $\lim\limits_{x \to a} \dfrac{g(x)}{x-a} = 2$,

find the value of $\lim\limits_{x \to a} \dfrac{2f(x)+3g(x)}{f(x)-g(x)}$.

$$\lim_{x \to a} \frac{2f(x)+3g(x)}{f(x)-g(x)} = \lim_{x \to a} \frac{\frac{2f(x)+3g(x)}{x-a}}{\frac{f(x)-g(x)}{x-a}} = \lim_{x \to a} \frac{2\frac{f(x)}{x-a}+3\frac{g(x)}{x-a}}{\frac{f(x)}{x-a}-\frac{g(x)}{x-a}} = \frac{2\lim\limits_{x \to a}\frac{f(x)}{x-a}+3\lim\limits_{x \to a}\frac{g(x)}{x-a}}{\lim\limits_{x \to a}\frac{f(x)}{x-a}-\lim\limits_{x \to a}\frac{g(x)}{x-a}} = \frac{2\cdot3+3\cdot2}{3-2} = 12$$

(15) For two real numbers x and y, a function $f(x)$ satisfies:

i) $f(x + y) = f(x) + f(y) + a$ and

ii) $\lim\limits_{x \to 2} \dfrac{f(x-2)}{x-2} = 1$

When $\lim\limits_{x \to 0} f(x) = f(0)$, find the value of the constant a.

Substituting $x = 0$ and $y = 0$ into i) , $f(0) = f(0 + 0) = f(0) + f(0) + a$

$\therefore f(0) = -a$

Let $x - 2 = t$

Then, $t \to 0$ as $x \to 2$

$\therefore \lim\limits_{x \to 2} \dfrac{f(x-2)}{x-2} = \lim\limits_{t \to 0} \dfrac{f(t)}{t} = 1$; $\lim\limits_{t \to 0} \dfrac{f(t)}{t}$ exists.

Since $\lim\limits_{t \to 0} t = 0$, $\lim\limits_{t \to 0} f(t) = 0$

$\therefore f(0) = \lim\limits_{t \to 0} f(t) = 0$ and $f(0) = -a$

Therefore, $a = 0$

(16) Find the value of a constant a at which $\lim\limits_{x \to 1} \dfrac{2x^2+a^2x-3a}{3x^2+a^2x-4a}$ does not exist.

Since $\lim\limits_{x \to 1} \dfrac{2x^2+a^2x-3a}{3x^2+a^2x-4a}$ does not exist,

$\lim\limits_{x \to 1}(3x^2 + a^2x - 4a) = 0$ and $\lim\limits_{x \to 1}(2x^2 + a^2x - 3a) = k$ $(k \neq 0,\ k;$ constant$)$

$\therefore \lim\limits_{x \to 1}(2x^2 + a^2x - 3a) = 2 + a^2 - 3a \neq 0$; $(a - 1)(a - 2) \neq 0$

$\therefore a \neq 1$ and $a \neq 2$ $\cdots\cdots$①

Since $\lim\limits_{x \to 1}(3x^2 + a^2x - 4a) = 3 + a^2 - 4a = (a - 1)(a - 3) = 0$,

$a = 1$ or $a = 3$ $\cdots\cdots$②

Therefore, by ① and ②, $a = 3$

#8 For two functions $f(x)$ and $g(x)$ such that

i) $x + f(x) = g(x)\{x - f(x)\}$ and

ii) $\lim\limits_{x \to 0} g(x) = 2$,

determine if the limit exists.

(1) $\lim\limits_{x\to 0}\dfrac{f(x)}{x}$

$x + f(x) = g(x)\{x - f(x)\} \;\Rightarrow\; f(x)\{1 + g(x)\} = x\{g(x) - 1\}$

$\therefore\; \dfrac{f(x)}{x} = \dfrac{g(x)-1}{1+g(x)} \;\; (x \neq 0,\; g(x) \neq -1)$

$\therefore\; \lim\limits_{x\to 0}\dfrac{f(x)}{x} = \lim\limits_{x\to 0}\dfrac{g(x)-1}{1+g(x)} = \dfrac{\lim\limits_{x\to 0} g(x)-1}{1+ \lim\limits_{x\to 0} g(x)} = \dfrac{2-1}{1+2} = \dfrac{1}{3}$ 　　$\therefore$ The limit exists.

(2) $\lim\limits_{x\to 0} f(x)$

$\lim\limits_{x\to 0} f(x) = \lim\limits_{x\to 0}\left\{ x \cdot \dfrac{f(x)}{x}\right\} = \lim\limits_{x\to 0} x \cdot \lim\limits_{x\to 0}\dfrac{f(x)}{x} = 0 \cdot \dfrac{1}{3} = 0$ 　　$\therefore$ The limit exists.

(3) $\lim\limits_{x\to 0}\dfrac{x^2+f(x)}{x^2-f(x)}$

$\lim\limits_{x\to 0}\dfrac{x^2+f(x)}{x^2-f(x)} = \lim\limits_{x\to 0}\dfrac{x+\frac{f(x)}{x}}{x-\frac{f(x)}{x}} = \dfrac{\lim\limits_{x\to 0} x+ \lim\limits_{x\to 0}\frac{f(x)}{x}}{\lim\limits_{x\to 0} x- \lim\limits_{x\to 0}\frac{f(x)}{x}} = \dfrac{0+\frac{1}{3}}{0-\frac{1}{3}} = -1$ 　　$\therefore$ The limit exists.

#9 Find the limit.

(1) $\lim\limits_{x\to 0}\dfrac{\sin 3x}{\sin 2x} = \lim\limits_{x\to 0}\left(\dfrac{\sin 3x}{3x} \cdot \dfrac{2x}{\sin 2x} \cdot \dfrac{3}{2}\right) = 1 \cdot 1 \cdot \dfrac{3}{2} = \dfrac{3}{2}$

(2) $\lim\limits_{x\to 0}\dfrac{\sin x^\circ}{x}$

Since $180^\circ = \pi,\; 1^\circ = \dfrac{\pi}{180}$ 　　$\therefore\; x^\circ = \dfrac{\pi}{180}x$

$\therefore\; \lim\limits_{x\to 0}\dfrac{\sin x^\circ}{x} = \lim\limits_{x\to 0}\dfrac{\sin\frac{\pi}{180}x}{x} = \lim\limits_{x\to 0}\left(\dfrac{\sin\frac{\pi}{180}x}{\frac{\pi}{180}x} \cdot \dfrac{\pi}{180}\right) = 1 \cdot \dfrac{\pi}{180} = \dfrac{\pi}{180}$

(3) $\lim\limits_{x\to 0}\dfrac{\sin x-2\sin 2x}{x\cos x} = \lim\limits_{x\to 0}\dfrac{\sin x-4\sin x\cos x}{x\cos x} = \lim\limits_{x\to 0}\left(\dfrac{\sin x}{x} \cdot \dfrac{1}{\cos x} - \dfrac{4\sin x}{x}\right) = 1 \cdot 1 - 4 \cdot 1 = -3$

(4) $\lim\limits_{x\to 0}\dfrac{\tan 4x}{\tan 3x} = \lim\limits_{x\to 0}\left(\dfrac{\tan 4x}{4x} \cdot \dfrac{3x}{\tan 3x} \cdot \dfrac{4}{3}\right) = 1 \cdot 1 \cdot \dfrac{4}{3} = \dfrac{4}{3}$

(5) $\lim\limits_{x\to 0}\dfrac{1-\cos 4x}{x^2} = \lim\limits_{x\to 0}\dfrac{(1-\cos 4x)(1+\cos 4x)}{x^2(1+\cos 4x)} = \lim\limits_{x\to 0}\dfrac{1-\cos^2 4x}{x^2(1+\cos 4x)} = \lim\limits_{x\to 0}\left(\dfrac{\sin^2 4x}{x^2} \cdot \dfrac{1}{1+\cos 4x}\right)$

$= \lim\limits_{x\to 0}\left\{\left(\dfrac{\sin 4x}{4x}\right)^2 \cdot 4^2 \cdot \dfrac{1}{1+\cos 4x}\right\} = 1^2 \cdot 4^2 \cdot \dfrac{1}{2} = 8$

(6) $\lim\limits_{x\to 3}\dfrac{x-3}{\sin \pi x}$

Let $x - 3 = t$ Then, $x = t + 3$

Since $t \to 0$ as $x \to 3$,

$\lim\limits_{x\to 3}\dfrac{x-3}{\sin \pi x} = \lim\limits_{t\to 0}\dfrac{t}{\sin \pi(t+3)} = \lim\limits_{t\to 0}\dfrac{t}{\sin(t\pi+3\pi)} = \lim\limits_{t\to 0}\dfrac{t}{-\sin t\pi} = \lim\limits_{t\to 0}\left(\dfrac{t\pi}{-\sin t\pi} \cdot \dfrac{1}{\pi}\right)$

$= -1 \cdot \dfrac{1}{\pi} = -\dfrac{1}{\pi}$

(7) $\lim\limits_{x \to 2\pi} \dfrac{\sin x}{x^2 - 4\pi^2}$

Let $x - 2\pi = t$ Then, $x = 2\pi + t$

Since $t \to 0$ as $x \to 2\pi$,

$$\lim_{x \to 2\pi} \frac{\sin x}{x^2 - 4\pi^2} = \lim_{t \to 0} \frac{\sin(2\pi + t)}{(2\pi + t)^2 - 4\pi^2} = \lim_{t \to 0} \frac{\sin t}{4\pi t + t^2} = \lim_{t \to 0} \frac{\sin t}{t(4\pi + t)} = \lim_{t \to 0} \left(\frac{\sin t}{t} \cdot \frac{1}{4\pi + t} \right)$$

$$= 1 \cdot \frac{1}{4\pi + 0} = \frac{1}{4\pi}$$

(8) $\lim\limits_{x \to 0} \dfrac{\cos 6x - 1}{2x} = \lim\limits_{x \to 0} \dfrac{\cos 6x - 1}{6x} \cdot 3 = 3 \lim\limits_{x \to 0} \left(\dfrac{\cos 6x - 1}{6x} \right) = 3 \cdot 1 = 3$

(9) $\lim\limits_{x \to \infty} \dfrac{3^x}{3^x - 2^x} = \lim\limits_{x \to \infty} \dfrac{1}{1 - \left(\frac{2}{3}\right)^x} = 1$

(10) $\lim\limits_{x \to \infty} (3^x + 2^x)^{\frac{1}{x}} = \lim\limits_{x \to \infty} \left[3^x \left\{ 1 + \left(\frac{2}{3}\right)^x \right\} \right]^{\frac{1}{x}} = \lim\limits_{x \to \infty} 3 \left\{ 1 + \left(\frac{2}{3}\right)^x \right\}^{\frac{1}{x}} = 3(1 + 0) = 3$

(11) $\lim\limits_{x \to \infty} \{\log(2 + 3x) - \log x\} = \lim\limits_{x \to \infty} \log \left(\dfrac{2 + 3x}{x} \right) = \lim\limits_{x \to \infty} \log \left(\dfrac{2}{x} + 3 \right) = \log 3$

(12) $\lim\limits_{x \to 3} \{\log|x^2 - 9| - \log|x - 3|\} = \lim\limits_{x \to 3} \log \left| \dfrac{x^2 - 9}{x - 3} \right| = \lim\limits_{x \to 3} \log \left| \dfrac{(x-3)(x+3)}{x-3} \right|$

$$= \lim_{x \to 3} \log|x + 3| = \log 6$$

(13) $\lim\limits_{x \to 0} \dfrac{\log(1+x)}{x} = \lim\limits_{x \to 0} \dfrac{1}{x} \log(1 + x) = \lim\limits_{x \to 0} \log(1 + x)^{\frac{1}{x}} = \log e = 1$

(14) $\lim\limits_{x \to 0} \dfrac{a^x - 1}{x}$ $(a > 0)$

Let $a^x - 1 = t$ Then, $a^x = t + 1$ $\therefore x = \log_a(t + 1)$

Since $t \to 0$ as $x \to 0$,

$$\lim_{x \to 0} \frac{a^x - 1}{x} = \lim_{t \to 0} \frac{t}{\log_a(t+1)} = \lim_{t \to 0} \frac{1}{\frac{1}{t}\log_a(t+1)} = \lim_{t \to 0} \frac{1}{\log_a(t+1)^{\frac{1}{t}}} = \frac{1}{\log_a e} = \log_e a = \ln a$$

(15) $\lim\limits_{x \to 0} \dfrac{\log(1+x)}{\tan x} = \lim\limits_{x \to 0} \left\{ \dfrac{\log(1+x)}{x} \cdot \dfrac{x}{\tan x} \right\} = 1 \cdot 1 = 1$

(16) $\lim\limits_{x \to 0} \dfrac{e^{2x} - 1}{\sin x} = \lim\limits_{x \to 0} \left\{ \dfrac{e^{2x} - 1}{2x} \cdot \dfrac{x}{\sin x} \cdot 2 \right\} = 1 \cdot 1 \cdot 2 = 2$

(17) $\lim\limits_{x \to 1} \dfrac{\sin\left(\cos\frac{\pi}{2}x\right)}{x - 1}$

Let $x - 1 = t$

Since $t \to 0$ as $x \to 1$ and $\cos\dfrac{\pi}{2}x = \cos\dfrac{\pi}{2}(t + 1) = \cos\left(\dfrac{\pi}{2}t + \dfrac{\pi}{2}\right) = -\sin\dfrac{\pi}{2}t$,

$$\lim_{x \to 1} \frac{\sin\left(\cos\frac{\pi}{2}x\right)}{x - 1} = \lim_{t \to 0} \frac{\sin\left(-\sin\frac{\pi}{2}t\right)}{t} = \lim_{t \to 0} \left\{ \frac{\sin\left(-\sin\frac{\pi}{2}t\right)}{-\sin\frac{\pi}{2}t} \cdot \frac{\sin\frac{\pi}{2}t}{-t} \right\}$$

$$= -\frac{\pi}{2} \lim_{t \to 0} \left\{ \frac{\sin\left(-\sin\frac{\pi}{2}t\right)}{-\sin\frac{\pi}{2}t} \cdot \frac{\sin\frac{\pi}{2}t}{\frac{\pi}{2}t} \right\} = -\frac{\pi}{2} \lim_{t \to 0} \left\{ \frac{\sin\left(-\sin\frac{\pi}{2}t\right)}{-\sin\frac{\pi}{2}t} \right\} \cdot \lim_{t \to 0} \left\{ \frac{\sin\frac{\pi}{2}t}{\frac{\pi}{2}t} \right\} = -\frac{\pi}{2}(1 \cdot 1) = -\frac{\pi}{2}$$

(18) $\displaystyle\lim_{x\to\frac{\pi}{2}}\left(\frac{1}{\cos x}-\tan x\right)$

Note that $\displaystyle\frac{1}{\cos x}-\tan x=\frac{1-\sin x}{\cos x}=\frac{(1-\sin x)(1+\sin x)}{\cos x(1+\sin x)}=\frac{1-\sin^2 x}{\cos x(1+\sin x)}=\frac{\cos^2 x}{\cos x(1+\sin x)}$

$$=\frac{\cos x}{1+\sin x}$$

$$\lim_{x\to\frac{\pi}{2}}\left(\frac{1}{\cos x}-\tan x\right)=\lim_{x\to\frac{\pi}{2}}\frac{\cos x}{1+\sin x}=\frac{0}{1+1}=0$$

(19) $\displaystyle\lim_{x\to\infty}\left(1+\sin\frac{1}{x}\right)^x$

Let $\dfrac{1}{x}=t$

Since $t\to 0$ as $x\to\infty$,

$$\lim_{x\to\infty}\left(1+\sin\frac{1}{x}\right)^x=\lim_{t\to 0}(1+\sin t)^{\frac{1}{t}}=\lim_{t\to 0}(1+\sin t)^{\frac{1}{\sin t}\cdot\frac{\sin t}{t}}$$

$$=\lim_{t\to 0}\left\{(1+\sin t)^{\frac{1}{\sin t}}\right\}^{\frac{\sin t}{t}}=e^1=e$$

(20) $\displaystyle\lim_{x\to 2}\{\log_4|x^2-4|-\log_4\left|x-\sqrt{x^2+x-2}\right|\}$

$$=\lim_{x\to 2}\left\{\log_4\frac{|x^2-4|}{\left|x-\sqrt{x^2+x-2}\right|}\right\}=\lim_{x\to 2}\left\{\log_4\frac{|x^2-4|\left|x+\sqrt{x^2+x-2}\right|}{\left|x-\sqrt{x^2+x-2}\right|\left|x+\sqrt{x^2+x-2}\right|}\right\}$$

$$=\lim_{x\to 2}\left\{\log_4\frac{|x^2-4|\left|x+\sqrt{x^2+x-2}\right|}{\left|x^2-(x^2+x-2)\right|}\right\}=\lim_{x\to 2}\left\{\log_4\frac{|(x-2)(x+2)|\left|x+\sqrt{x^2+x-2}\right|}{|-x+2|}\right\}$$

$$=\lim_{x\to 2}\{\log_4|(x+2)(x+\sqrt{x^2+x-2})|\}=\log_4\left\{\lim_{x\to 2}|(x+2)(x+\sqrt{x^2+x-2})|\right\}$$

$$=\log_4\{|4(2+\sqrt{2^2+2-2})|\}=\log_4 16=\log_4 4^2=2$$

(21) $\displaystyle\lim_{x\to 0}\left\{\frac{\sin x\tan x}{x^2}+\frac{\sin 2x\tan 2x}{x^2}+\cdots\cdots+\frac{\sin 10x\tan 10x}{x^2}\right\}$

Since $\tan x=\dfrac{\sin x}{\cos x}$,

$$\lim_{x\to 0}\left\{\frac{\sin x\tan x}{x^2}+\frac{\sin 2x\tan 2x}{x^2}+\cdots\cdots+\frac{\sin 10x\tan 10x}{x^2}\right\}$$

$$=\lim_{x\to 0}\left\{\frac{\sin^2 x}{x^2\cos x}+\frac{\sin^2 2x}{x^2\cos 2x}+\cdots\cdots+\frac{\sin^2 10x}{x^2\cos 10x}\right\}$$

$$=\lim_{x\to 0}\left\{\frac{\sin^2 x}{x^2}\cdot\frac{1}{\cos x}+\frac{\sin^2 2x}{(2x)^2}\cdot\frac{2^2}{\cos 2x}+\cdots\cdots+\frac{\sin^2 10x}{(10x)^2}\cdot\frac{10^2}{\cos 10x}\right\}$$

$$=1^2\cdot 1+1^2\cdot 2^2+\cdots\cdots+1^2\cdot 10^2=1^2+2^2+3^2\cdots\cdots+10^2=\sum_{k=1}^{10}k^2$$

$$=\frac{10(10+1)(2\cdot 10+1)}{6}=\frac{10\cdot 11\cdot 21}{6}=385$$

Alternative Approach:

Since $\lim\limits_{x\to 0}\dfrac{\sin x}{x}=1$ and $\lim\limits_{x\to 0}\dfrac{\tan x}{x}=1$,

$$\lim_{x\to 0}\left\{\frac{\sin x\tan x}{x^2}+\frac{\sin 2x\tan 2x}{x^2}+\cdots\cdots+\frac{\sin 10x\tan 10x}{x^2}\right\}$$

$$=\lim_{x\to 0}\left\{\frac{\sin x}{x}\cdot\frac{\tan x}{x}+2^2\cdot\frac{\sin 2x}{2x}\cdot\frac{\tan 2x}{2x}+\cdots\cdots+10^2\cdot\frac{\sin 10x}{10x}\cdot\frac{\tan 10x}{10x}\right\}$$

$$=1^2+2^2+3^2\cdots\cdots+10^2=\sum_{k=1}^{10}k^2=385$$

(22) $\lim\limits_{x\to\infty}\sin\left(\tan\dfrac{1}{x}\right)\cot\dfrac{1}{x}$

Let $\tan\dfrac{1}{x}=t$

Since $t\to 0$ as $x\to\infty$,

$$\lim_{x\to\infty}\sin\left(\tan\frac{1}{x}\right)\cot\frac{1}{x}=\lim_{x\to\infty}\frac{\sin\left(\tan\frac{1}{x}\right)}{\tan\frac{1}{x}}\tan\frac{1}{x}\cot\frac{1}{x}=\lim_{x\to\infty}\frac{\sin\left(\tan\frac{1}{x}\right)}{\tan\frac{1}{x}}=\lim_{t\to 0}\frac{\sin t}{t}=1$$

(23) $\lim\limits_{n\to\infty}\left\{\dfrac{1}{2}\left(1+\dfrac{1}{n}\right)\left(1+\dfrac{1}{n+1}\right)\cdots\cdots\left(1+\dfrac{1}{n+n}\right)\right\}^n$

$$=\lim_{n\to\infty}\left\{\frac{1}{2}\cdot\frac{n+1}{n}\cdot\frac{n+2}{n+1}\cdots\cdots\frac{2n+1}{n+n}\right\}^n=\lim_{n\to\infty}\left(\frac{1}{2}\cdot\frac{2n+1}{n}\right)^n=\lim_{n\to\infty}\left(\frac{2n+1}{2n}\right)^n$$

$$=\lim_{n\to\infty}\left(1+\frac{1}{2n}\right)^n=\lim_{n\to\infty}\left\{\left(1+\frac{1}{2n}\right)^{2n}\right\}^{\frac{1}{2}}=e^{\frac{1}{2}}=\sqrt{e}$$

(24) $\lim\limits_{x\to\infty}\left(\dfrac{x}{x-1}\right)^x$ $(x>1)$

Let $x-1=t$

Since $t\to\infty$ as $x\to\infty$,

$$\lim_{x\to\infty}\left(\frac{x}{x-1}\right)^x=\lim_{t\to\infty}\left(\frac{t+1}{t}\right)^{t+1}=\lim_{t\to\infty}\left(1+\frac{1}{t}\right)^{t+1}=\lim_{t\to\infty}\left\{\left(1+\frac{1}{t}\right)^t\left(1+\frac{1}{t}\right)\right\}=e(1+0)=e$$

(25) $\lim\limits_{x\to 0}\dfrac{1-\cos 2x}{2\sin^2 x}=\lim\limits_{x\to 0}\dfrac{1-\cos 2x}{\frac{2\sin^2 x}{x^2}\cdot x^2}=\lim\limits_{x\to 0}\left(\dfrac{x^2}{\sin^2 x}\cdot\dfrac{1-\cos 2x}{2x^2}\right)=\lim\limits_{x\to 0}\left(\dfrac{x^2}{\sin^2 x}\right)\cdot\lim\limits_{x\to 0}\left(\dfrac{1-\cos 2x}{2x^2}\right)$

$$=\lim_{x\to 0}\left(\frac{1-\cos 2x}{2x^2}\right)=\lim_{x\to 0}\frac{(1-\cos 2x)(1+\cos 2x)}{2x^2(1+\cos 2x)}=\lim_{x\to 0}\frac{1-\cos^2 2x}{2x^2(1+\cos 2x)}$$

$$=\lim_{x\to 0}\frac{\sin^2 2x}{2x^2(1+\cos 2x)}=\lim_{x\to 0}\frac{\sin^2 2x}{2x^2}\lim_{x\to 0}\frac{1}{1+\cos 2x}=\lim_{x\to 0}\frac{\sin^2 2x}{2x^2}\cdot\frac{1}{2}=\lim_{x\to 0}\frac{\sin^2 2x}{4x^2}=\lim_{x\to 0}\frac{\sin^2 2x}{(2x)^2}$$

$$=\lim_{x\to 0}\left(\frac{\sin 2x}{2x}\right)^2=\left(\lim_{t\to 0}\frac{\sin t}{t}\right)^2=1^2=1$$

(26) $\lim\limits_{x\to -\infty}\dfrac{e^x+x^3-1}{1+2x^3}$

Let $-x=t$

Since $t\to\infty$ as $x\to-\infty$,

$$\lim_{x\to -\infty}\frac{e^x+x^3-1}{1+2x^3}=\lim_{t\to\infty}\frac{e^{-t}-t^3-1}{1-2t^3}=\lim_{t\to\infty}\left(\frac{e^{-t}}{1-2t^3}-\frac{t^3+1}{1-2t^3}\right)=\lim_{t\to\infty}\left\{\frac{1}{e^t(1-2t^3)}+\frac{t^3+1}{2t^3-1}\right\}$$

Since $\displaystyle\lim_{t\to\infty} e^t(1-2t^3) = \lim_{t\to\infty} t^3 e^t\left(\frac{1}{t^3}-2\right) = -\infty,$

$$\lim_{x\to-\infty}\frac{e^x+x^3-1}{1+2x^3} = \lim_{t\to\infty}\left\{\frac{1}{e^t(1-2t^3)}+\frac{t^3+1}{2t^3-1}\right\} = \lim_{t\to\infty}\left\{\frac{1}{e^t(1-2t^3)}\right\} + \lim_{t\to\infty}\left\{\frac{t^3+1}{2t^3-1}\right\} = 0+\frac{1}{2} = \frac{1}{2}$$

(27) $\displaystyle\lim_{x\to0}\frac{\sin\pi(1-x)}{\ln(1+x)} = \lim_{x\to0}\frac{\sin\pi x}{\ln(1+x)} = \lim_{x\to0}\left\{\frac{\sin\pi x}{\pi x}\cdot\frac{x}{\ln(1+x)}\cdot\pi\right\} = \lim_{x\to0}\left\{\frac{\sin\pi x}{\pi x}\cdot\frac{1}{\frac{\ln(1+x)}{x}}\cdot\pi\right\}$

$$= \pi\lim_{x\to0}\left\{\frac{\sin\pi x}{\pi x}\cdot\frac{1}{\frac{\ln(1+x)}{x}}\right\} = \pi\cdot1\cdot1 = \pi$$

(28) $\displaystyle\lim_{x\to0}\left(\frac{\sin x+\cos x}{\cos x}\right)^{\cos x}$

$$= \lim_{x\to0}\left(1+\frac{\sin x}{\cos x}\right)^{\cos x} = \lim_{x\to0}\left(1+\frac{\sin x}{\cos x}\right)^{\frac{\cos x}{\sin x}\cdot\sin x} = \lim_{x\to0}\left\{\left(1+\frac{\sin x}{\cos x}\right)^{\frac{\cos x}{\sin x}}\right\}^{\sin x}$$

Let $\dfrac{\sin x}{\cos x} = t$

Since $t\to0$ as $x\to0$,

$$\lim_{x\to0}\left(\frac{\sin x+\cos x}{\cos x}\right)^{\cos x} = \lim_{x\to0}\left\{\left(1+\frac{\sin x}{\cos x}\right)^{\frac{\cos x}{\sin x}}\right\}^{\sin x} = \lim_{t\to0}\left\{(1+t)^{\frac{1}{t}}\right\}^{\lim\limits_{x\to0}\sin x} = e^0 = 1$$

(29) $\displaystyle\lim_{x\to0}\frac{x}{\sqrt[3]{e^x}-1} = \lim_{x\to0}\frac{x}{e^{\frac{x}{3}}-1} = \lim_{x\to0}\frac{x}{\frac{e^{\frac{x}{3}}-1}{\frac{x}{3}}\cdot\left(\frac{x}{3}\right)} = \lim_{x\to0}\frac{1}{\frac{e^{\frac{x}{3}}-1}{\frac{x}{3}}\cdot\left(\frac{x}{3}\right)} = \frac{1}{1\cdot\left(\frac{x}{3}\right)} = 3$

(30) $\displaystyle\lim_{x\to2}\left(\frac{x}{2}\right)^{\frac{1}{x-2}}$

Let $x-2 = t$

Since $t\to0$ as $x\to2$,

$$\lim_{x\to2}\left(\frac{x}{2}\right)^{\frac{1}{x-2}} = \lim_{t\to0}\left(\frac{t+2}{2}\right)^{\frac{1}{t}} = \lim_{t\to0}\left(1+\frac{t}{2}\right)^{\frac{2}{t}\cdot\frac{1}{2}} = \lim_{t\to0}\left\{\left(1+\frac{t}{2}\right)^{\frac{2}{t}}\right\}^{\frac{1}{2}} = e^{\frac{1}{2}} = \sqrt{e}$$

(31) $\displaystyle\lim_{x\to1}x^{\frac{1}{1-x}}$

Let $1-x = t$

Since $t\to0$ as $x\to1$,

$$\lim_{x\to1}x^{\frac{1}{1-x}} = \lim_{t\to0}(1-t)^{\frac{1}{t}} = \lim_{t\to0}\left\{(1-t)^{-\frac{1}{t}}\right\}^{-1} = e^{-1} = \frac{1}{e}$$

#10 **Find the values of constants a and b for the following limit.**

(1) $\displaystyle\lim_{x\to0}\frac{ax\sin x+b}{1-\cos x} = 1$

Since the limit is a constant (; i.e., the limit exists) and $\displaystyle\lim_{x\to0}(1-\cos x) = 0,$

$$\lim_{x \to 0}(ax \sin x + b) = 0. \qquad \therefore \ 0 + b = 0 \ ; \ b = 0$$

$$\lim_{x \to 0}\frac{ax \sin x + b}{1 - \cos x} = \lim_{x \to 0}\frac{ax \sin x}{1 - \cos x} = \lim_{x \to 0}\frac{ax \sin x(1 + \cos x)}{(1 - \cos x)(1 + \cos x)} = \lim_{x \to 0}\frac{ax \sin x(1 + \cos x)}{1 - \cos^2 x}$$

$$= \lim_{x \to 0}\frac{ax \sin x(1 + \cos x)}{\sin^2 x} = \lim_{x \to 0}\frac{ax(1 + \cos x)}{\sin x} = \lim_{x \to 0}\left\{a \cdot \frac{x}{\sin x} \cdot (1 + \cos x)\right\} = a \cdot 1 \cdot 2 = 2a = 1$$

$$\therefore \ a = \frac{1}{2}$$

Therefore, $a = \frac{1}{2}, \ b = 0$

(2) $\lim\limits_{x \to 0}\dfrac{\sin 2x}{\sqrt{ax + b} - 1} = 2$

Since the limit is a constant (; i.e., the limit exists) and $\lim\limits_{x \to 0}\sin 2x = 0$,

$$\lim_{x \to 0}(\sqrt{ax + b} - 1) = 0. \qquad \therefore \ \sqrt{b} - 1 = 0 \ ; \ b = 1$$

$$\lim_{x \to 0}\frac{\sin 2x}{\sqrt{ax + b} - 1} = \lim_{x \to 0}\frac{\sin 2x(\sqrt{ax + 1} + 1)}{(\sqrt{ax + 1} - 1)(\sqrt{ax + 1} + 1)} = \lim_{x \to 0}\frac{\sin 2x(\sqrt{ax + 1} + 1)}{(ax + 1) - 1}$$

$$= \lim_{x \to 0}\frac{\sin 2x(\sqrt{ax + 1} + 1)}{ax} = \lim_{x \to 0}\left\{\frac{\sin 2x}{2x} \cdot \frac{2(\sqrt{ax + 1} + 1)}{a}\right\} = \lim_{x \to 0}\left\{\frac{\sin 2x}{2x} \cdot \frac{2\left(\sqrt{\frac{x}{a} + \frac{1}{a^2}} + \frac{1}{a}\right)}{1}\right\}$$

$$= 1 \cdot 2\left(\frac{1}{a} + \frac{1}{a}\right) = \frac{4}{a} = 2 \ ; \ a = 2$$

Therefore, $a = 2, \ b = 1$

(3) $\lim\limits_{x \to 0}\dfrac{\tan x}{\sin(ax + b)} = \dfrac{1}{2} \ (0 \le b < \dfrac{\pi}{2})$

Since the limit is a constant (; i.e., the limit exists) and $\lim\limits_{x \to 0}\tan x = 0$,

$$\lim_{x \to 0}\sin(ax + b) = 0. \qquad \therefore \ \sin b = 0$$

Since $0 \le b < \dfrac{\pi}{2}$, $b = 0$

$$\lim_{x \to 0}\frac{\tan x}{\sin(ax + b)} = \lim_{x \to 0}\frac{\tan x}{\sin ax} = \lim_{x \to 0}\left(\frac{ax}{\sin ax} \cdot \frac{\tan x}{ax}\right) = \lim_{x \to 0}\left(\frac{ax}{\sin ax} \cdot \frac{\tan x}{x} \cdot \frac{1}{a}\right) = 1 \cdot 1 \cdot \frac{1}{a} = \frac{1}{2}$$

Therefore, $a = 2, \ b = 0$

(4) $\lim\limits_{x \to \infty}\dfrac{2^{x+1} - 3^{x+1}}{2^x - 3^x} = a \ ; \ \lim\limits_{x \to \infty}(4^x + 5^x)^{\frac{1}{x}} = b$

$$\lim_{x \to \infty}\frac{2^{x+1} - 3^{x+1}}{2^x - 3^x} = \lim_{x \to \infty}\frac{2\left(\frac{2}{3}\right)^x - 3}{\left(\frac{2}{3}\right)^x - 1} = \frac{2 \cdot 0 - 3}{0 - 1} = 3 = a$$

$$\lim_{x \to \infty}(4^x + 5^x)^{\frac{1}{x}} = \lim_{x \to \infty}\left[5^x\left\{\left(\frac{4}{5}\right)^x + 1\right\}\right]^{\frac{1}{x}} = \lim_{x \to \infty}5\left\{1 + \left(\frac{4}{5}\right)^x\right\}^{\frac{1}{x}} = 5(1 + 0)^0 = 5 = b$$

Therefore, $a = 3, \ b = 5$

(5) $\displaystyle\lim_{x\to 0}\frac{e^x\ln(x+a)}{b\sin x}=\frac{1}{2}$

Since the limit is a constant (; i.e., the limit exists) and $\displaystyle\lim_{x\to 0}b\sin x=0$,

$\displaystyle\lim_{x\to 0}e^x\ln(x+a)=0.$ $\quad\therefore\ \lim_{x\to 0}e^x\ln(x+a)=e^0\ln(0+a)=\ln a=0\ ;\ a=1$

$\displaystyle\lim_{x\to 0}\frac{e^x\ln(x+a)}{b\sin x}=\lim_{x\to 0}\frac{e^x\ln(x+1)}{b\sin x}=\lim_{x\to 0}\left\{\frac{1}{b}\cdot\frac{x}{\sin x}\cdot e^x\cdot\frac{\ln(x+1)}{x}\right\}=\frac{1}{b}\cdot 1\cdot e^0\cdot 1=\frac{1}{2}\ ;\ b=2$

Therefore, $a=1,\ \ b=2$

(6) $\displaystyle\lim_{x\to 1}\frac{\sin(x-1)}{\sqrt{ax-1}-b}=1$

Since the limit is a constant (; i.e., the limit exists) and $\displaystyle\lim_{x\to 1}\sin(x-1)=0$,

$\displaystyle\lim_{x\to 1}(\sqrt{ax-1}-b)=0.$ $\therefore\ \sqrt{a-1}-b=0\ ;\ b=\sqrt{a-1}$

$\displaystyle\lim_{x\to 1}\frac{\sin(x-1)}{\sqrt{ax-1}-b}=\lim_{x\to 1}\frac{\sin(x-1)}{\sqrt{ax-1}-\sqrt{a-1}}=\lim_{x\to 1}\frac{\sin(x-1)(\sqrt{ax-1}+\sqrt{a-1})}{(\sqrt{ax-1}-\sqrt{a-1})(\sqrt{ax-1}+\sqrt{a-1})}$

$\displaystyle=\lim_{x\to 1}\frac{\sin(x-1)(\sqrt{ax-1}+\sqrt{a-1})}{(ax-1)-(a-1)}=\lim_{x\to 1}\frac{\sin(x-1)(\sqrt{ax-1}+\sqrt{a-1})}{a(x-1)}$

$\displaystyle=\lim_{x\to 1}\left\{\frac{\sin(x-1)}{(x-1)}\cdot\frac{(\sqrt{ax-1}+\sqrt{a-1})}{a}\right\}=1\cdot\frac{2\sqrt{a-1}}{a}=1$

$\therefore\ a=2\sqrt{a-1}\ ;\ a^2=4(a-1)\ ;\ a^2-4a+4=(a-2)^2=0\ ;\ a=2$

Therefore, $a=2,\ \ b=\sqrt{2-1}=1$

(7) $\displaystyle\lim_{x\to\frac{\pi}{2}}\frac{ax^2-\pi^2}{\cos x}=b\ \ (b\neq 0)$

Since the limit is a constant (; i.e., the limit exists) and $\displaystyle\lim_{x\to\frac{\pi}{2}}\cos x=0$,

$\displaystyle\lim_{x\to\frac{\pi}{2}}(ax^2-\pi^2)=0.$ $\therefore\ a\left(\frac{\pi}{2}\right)^2-\pi^2=0\ ;\ a=4$

Let $\ x-\dfrac{\pi}{2}=t$

Then, $t\to 0$ as $x\to\dfrac{\pi}{2}$

$\displaystyle\lim_{x\to\frac{\pi}{2}}\frac{ax^2-\pi^2}{\cos x}=\lim_{t\to 0}\frac{4\left(t+\frac{\pi}{2}\right)^2-\pi^2}{\cos\left(t+\frac{\pi}{2}\right)}=\lim_{t\to 0}\frac{4t^2+4t\pi}{-\sin t}=\lim_{t\to 0}\frac{4t(t+\pi)}{-\sin t}=-4\lim_{t\to 0}\frac{t}{\sin t}(t+\pi)$

$\displaystyle=-4\cdot 1\cdot\pi=b\ ;\ b=-4\pi$

Therefore, $a=4,\ \ b=-4\pi$

#11 Find the value.

(1) For a continuous function $f(x)$ such that $(e^{4x} - 1)f(x) = \sin 2\pi x$, find the value of $f(0)$.

$(e^{4x} - 1)f(x) = \sin 2\pi x \implies f(x) = \dfrac{\sin 2\pi x}{e^{4x}-1}, \quad x \neq 0$

Since $f(x)$ is continuous, $f(0) = \lim\limits_{x \to 0} \dfrac{\sin 2\pi x}{e^{4x}-1}$

$\lim\limits_{x \to 0} \dfrac{\sin 2\pi x}{e^{4x}-1} = \lim\limits_{x \to 0} \left(\dfrac{\sin 2\pi x}{2\pi x} \cdot \dfrac{2\pi x}{e^{4x}-1} \right) = \lim\limits_{x \to 0} \left(\dfrac{\sin 2\pi x}{2\pi x} \cdot \dfrac{4x}{e^{4x}-1} \cdot \dfrac{\pi}{2} \right) = 1 \cdot 1 \cdot \dfrac{\pi}{2} = \dfrac{\pi}{2}$

$\therefore \ f(0) = \dfrac{\pi}{2}$

(2) $f(x)$ is a cubic function with $f(1) = 3$ and $f(2) = 12$.

When $f(-x) = -f(x)$ for any real number x, find the value of $\lim\limits_{x \to 0} \dfrac{f(\sin x)}{\sin f(x)}$.

Since $f(x)$ is a cubic function and $f(-x) = -f(x)$, $f(x) = ax^3 + bx$

Since $f(1) = a + b = 3$ and $f(2) = 8a + 2b = 12$, we have

$\begin{cases} a + b = 3 \\ 8a + 2b = 12 \end{cases} \implies \begin{cases} 2a + 2b = 6 \\ 8a + 2b = 12 \end{cases} \implies 6a = 6 \ ; \ a = 1, \ b = 2$

$\therefore \ f(x) = x^3 + 2x$

$\lim\limits_{x \to 0} \dfrac{f(\sin x)}{\sin f(x)} = \lim\limits_{x \to 0} \dfrac{\sin^3 x + 2(\sin x)}{\sin(x^3+2x)} = \lim\limits_{x \to 0} \dfrac{\sin x(\sin^2 x+2)}{\sin(x^3+2x)} = \lim\limits_{x \to 0} \left\{ \dfrac{x^3+2x}{\sin(x^3+2x)} \cdot \dfrac{\sin x(\sin^2 x+2)}{x^3+2x} \right\}$

$= \lim\limits_{x \to 0} \left\{ \dfrac{x^3+2x}{\sin(x^3+2x)} \cdot \dfrac{\sin x}{x} \cdot \dfrac{(\sin^2 x+2)}{x^2+2} \right\} = \lim\limits_{x \to 0} \left\{ \dfrac{x^3+2x}{\sin(x^3+2x)} \right\} \cdot \lim\limits_{x \to 0} \left\{ \dfrac{\sin x}{x} \right\} \cdot \lim\limits_{x \to 0} \left\{ \dfrac{\sin^2 x+2}{x^2+2} \right\}$

$= \lim\limits_{x \to 0} \left\{ \dfrac{x^3+2x}{\sin(x^3+2x)} \right\} \cdot \lim\limits_{x \to 0} \left\{ \dfrac{\sin x}{x} \right\} \cdot \lim\limits_{x \to 0} \left\{ \left(\dfrac{\sin^2 x}{x^2} \cdot \dfrac{x^2}{x^2+2} \right) + \left(\dfrac{2}{x^2+2} \right) \right\}$

$= 1 \cdot 1 \cdot \left(1 \cdot \dfrac{0}{2} + \dfrac{2}{2} \right) = 1 \cdot (0 + 1) = 1$

(3) For a sequence $\{a_n\}$ with $\lim\limits_{n \to \infty} \left(1 + \dfrac{3}{n} \right)^{a_n} = \dfrac{1}{e}$, find the value of $\lim\limits_{n \to \infty} \dfrac{a_n}{n}$.

Let $\dfrac{3}{n} = t$

Since $t \to 0$ as $n \to \infty$,

$\lim\limits_{n \to \infty} \left(1 + \dfrac{3}{n} \right)^{a_n} = \lim\limits_{t \to 0}(1 + t)^{\frac{1}{t}t \lim\limits_{n \to \infty} a_n} = \lim\limits_{t \to 0}(1 + t)^{\frac{1}{t} \lim\limits_{n \to \infty} \frac{3a_n}{n}} = e^{\lim\limits_{n \to \infty} \frac{3a_n}{n}} = e^{3 \lim\limits_{n \to \infty} \frac{a_n}{n}}$

Since $\lim\limits_{n \to \infty} \left(1 + \dfrac{3}{n} \right)^{a_n} = \dfrac{1}{e}$, $e^{3 \lim\limits_{n \to \infty} \frac{a_n}{n}} = \dfrac{1}{e} = e^{-1}$ $\quad \therefore \ 3 \lim\limits_{n \to \infty} \dfrac{a_n}{n} = -1$

$\therefore \ \lim\limits_{n \to \infty} \dfrac{a_n}{n} = -\dfrac{1}{3}$

(4) For a function $f(x)$ such that $\lim\limits_{x\to 0}\{f(x)\ln(1+3x)\} = 4$, find the value of $\lim\limits_{x\to 0} xf(x)$.

Since $\lim\limits_{x\to 0}\{f(x)\ln(1+3x)\} = 4$, $\lim\limits_{x\to 0}\left\{f(x)\cdot 3x\cdot \dfrac{\ln(1+3x)}{3x}\right\} = 4$

Since $\lim\limits_{x\to 0}\dfrac{\ln(1+3x)}{3x} = 1$, $\lim\limits_{x\to 0}\{f(x)\cdot 3x\} = 4$ $\qquad \therefore\ 3\lim\limits_{x\to 0} xf(x) = 4$

$\therefore\ \lim\limits_{x\to 0} xf(x) = \dfrac{4}{3}$

(5) For a function $f(x)$ such that $\lim\limits_{x\to 0}\dfrac{e^{2x}-1}{f(x)} = 5$, find the value of $\lim\limits_{x\to 0}\dfrac{f(x)}{x}$.

$\lim\limits_{x\to 0}\dfrac{f(x)}{x} = \lim\limits_{x\to 0}\left\{2\cdot \dfrac{e^{2x}-1}{2x}\cdot \dfrac{f(x)}{e^{2x}-1}\right\} = 2\lim\limits_{x\to 0}\left\{\dfrac{e^{2x}-1}{2x}\cdot \dfrac{1}{\frac{e^{2x}-1}{f(x)}}\right\} = 2\cdot 1\cdot \dfrac{1}{5} = \dfrac{2}{5}$

(6) For a function $f(x)$ such that $\lim\limits_{x\to\infty} x^2 f(x) = 2$, find the value of $\lim\limits_{x\to\infty} x^2 \ln\{1+2f(x)\}$.

Note that $\lim\limits_{x\to\infty} x^2 f(x) = \lim\limits_{x\to\infty}\dfrac{f(x)}{\frac{1}{x^2}} = 2$

Since $\lim\limits_{x\to\infty}\dfrac{f(x)}{\frac{1}{x^2}}$ exists and $\lim\limits_{x\to\infty}\dfrac{1}{x^2} = 0$, $\lim\limits_{x\to\infty} f(x) = 0$

$\lim\limits_{x\to\infty} x^2\ln\{1+2f(x)\} = \lim\limits_{x\to\infty}\left[x^2 2f(x)\dfrac{\ln\{1+2f(x)\}}{2f(x)}\right] = 2\lim\limits_{x\to\infty}\left[x^2 f(x)\dfrac{\ln\{1+2f(x)\}}{2f(x)}\right]$

$= 2\lim\limits_{x\to\infty} x^2 f(x)\cdot \lim\limits_{x\to\infty}\left[\dfrac{\ln\{1+2f(x)\}}{2f(x)}\right] = 2\cdot 2\cdot \lim\limits_{x\to\infty}\left[\dfrac{\ln\{1+2f(x)\}}{2f(x)}\right]$

Let $f(x) = t$

Since $\lim\limits_{x\to\infty} f(x) = 0$, $t \to 0$ as $x \to \infty$

$\therefore\ \lim\limits_{x\to\infty} x^2\ln\{1+2f(x)\} = 4\lim\limits_{x\to\infty}\left[\dfrac{\ln\{1+2f(x)\}}{2f(x)}\right] = 4\lim\limits_{t\to 0}\left[\dfrac{\ln(1+2t)}{2t}\right] = 4\cdot 1 = 4$

(7) For a positive constant a such that $\lim\limits_{x\to 0}\dfrac{(a+2)^x - a^x}{x} = \ln 3$, find the value of a.

$\lim\limits_{x\to 0}\dfrac{(a+2)^x - a^x}{x} = \lim\limits_{x\to 0}\dfrac{(a+2)^x - 1 + 1 - a^x}{x} = \lim\limits_{x\to 0}\left\{\dfrac{(a+2)^x - 1}{x} - \dfrac{a^x - 1}{x}\right\}$

$= \lim\limits_{x\to 0}\left\{\dfrac{(a+2)^x - 1}{x}\right\} - \lim\limits_{x\to 0}\left\{\dfrac{a^x - 1}{x}\right\} = \ln(a+2) - \ln a = \ln\dfrac{a+2}{a} = \ln 3$

$\therefore\ \dfrac{a+2}{a} = 3$ $\qquad \therefore\ a+2 = 3a$ $\qquad \therefore\ a = 1$

(8) For a function $f(x) = 2\ln(x+1) + 1$, let the inverse function of $f(x)$ be $g(x)$.

Find the value of $\lim\limits_{x\to 1}\dfrac{f(x-1)-f(0)}{g(x)-g(1)}$.

$y = 2\ln(x+1) + 1 \Rightarrow 2\ln(x+1) = y-1$ $\qquad \therefore\ \ln(x+1) = \dfrac{y-1}{2}$

$\therefore\ x+1 = e^{\frac{y-1}{2}}$ $\qquad \therefore\ x = e^{\frac{y-1}{2}} - 1$

Changing x and y, $y = e^{\frac{x-1}{2}} - 1$ $\qquad \therefore\ g(x) = e^{\frac{x-1}{2}} - 1$

Let $x - 1 = t$

Since $t \to 0$ as $x \to 1$,

$$\lim_{x \to 1} \frac{f(x-1)-f(0)}{g(x)-g(1)} = \lim_{x \to 1} \frac{2\ln x + 1 - 1}{\left(e^{\frac{x-1}{2}}-1\right)-0} = \lim_{x \to 1} \frac{2\ln x}{e^{\frac{x-1}{2}}-1} = \lim_{t \to 0} \frac{2\ln(1+t)}{e^{\frac{t}{2}}-1}$$

$$= 2\lim_{t \to 0}\left\{\frac{\ln(1+t)}{t} \cdot \frac{t}{e^{\frac{t}{2}}-1}\right\} = 2\lim_{t \to 0}\left\{\frac{\ln(1+t)}{t} \cdot \frac{\frac{t}{2}\cdot 2}{e^{\frac{t}{2}}-1}\right\} = 4\lim_{t \to 0}\left\{\frac{\ln(1+t)}{t} \cdot \frac{\frac{t}{2}}{e^{\frac{t}{2}}-1}\right\} = 4\cdot 1 \cdot 1 = 4$$

#12 For a function $f(x)$ such that $\displaystyle\lim_{x \to 0} \frac{f(x)}{x} = 2$,

determine whether the following statements are true or false.

(1) $\displaystyle\lim_{x \to 0} \frac{\tan x}{f(x)} = 1$

False $\quad \displaystyle\lim_{x \to 0} \frac{\tan x}{f(x)} = \lim_{x \to 0}\left\{\frac{x}{f(x)} \cdot \frac{\tan x}{x}\right\} = \lim_{x \to 0}\frac{x}{f(x)} \cdot \lim_{x \to 0}\frac{\tan x}{x} = \frac{1}{2} \cdot 1 = \frac{1}{2}$

(2) $\displaystyle\lim_{x \to 0} \frac{\tan f(x)}{f(x)} = 1$

True $\quad$ Since $\displaystyle\lim_{x \to 0} \frac{f(x)}{x} = 2$ and $\displaystyle\lim_{x \to 0} x = 0, \quad \lim_{x \to 0} f(x) = 0$

Let $f(x) = t$

Since $t \to 0$ as $x \to 0, \quad \displaystyle\lim_{x \to 0}\frac{\tan f(x)}{f(x)} = \lim_{t \to 0}\frac{\tan t}{t} = 1$

(3) For a function $g(x)$, if $\displaystyle\lim_{x \to 0} \frac{f(x)}{g(x)} = 1$, then $\displaystyle\lim_{x \to 0} \frac{\tan f(x)}{\tan g(x)} = 1$.

True $\quad$ Since $\displaystyle\lim_{x \to 0} \frac{f(x)}{g(x)} = 1$ and $\displaystyle\lim_{x \to 0} f(x) = 0$ (by (2)), $\displaystyle\lim_{x \to 0} g(x) = 0$

$$\lim_{x \to 0}\frac{\tan f(x)}{\tan g(x)} = \lim_{x \to 0}\left\{\frac{\tan f(x)}{f(x)} \cdot \frac{g(x)}{\tan g(x)} \cdot \frac{f(x)}{g(x)}\right\} = 1 \cdot 1 \cdot 1 = 1$$

#13 For a function $f(x) = \dfrac{b^x + \log_a x}{a^x + \log_b x}$ $(a > 0, \ b > 0, \ a \neq 1, \ b \neq 1)$, determine whether the

following statements are true or false.

(1) If $1 < a < b$, then $f(x) > 1$ for all x. $(x > 1)$

$1 < a < b, \ x > 1 \ \Rightarrow \ 1 < a^x < b^x$

Since $1 < \log_b x < \log_a x$, $a^x + \log_b x < b^x + \log_a x$

$\therefore \ \dfrac{b^x + \log_a x}{a^x + \log_b x} > 1 \qquad \therefore \ f(x) > 1$

$\therefore \ $ True

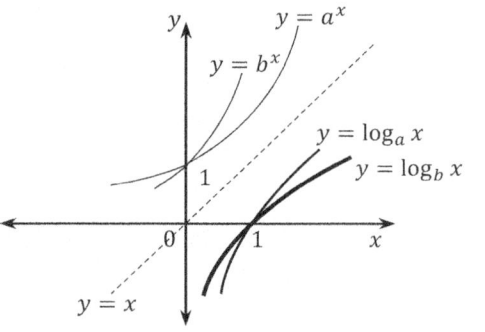

(2) If $b < a < 1$, then $\lim\limits_{x \to \infty} f(x) = 0$.

$$\lim_{x \to \infty} a^x = \lim_{x \to \infty} b^x = 0$$

$$\lim_{x \to \infty} \log_a x = \lim_{x \to \infty} \log_b x = -\infty$$

$$\therefore \ \lim_{x \to \infty} \frac{b^x}{\log_b x} = \lim_{x \to \infty} \frac{a^x}{\log_b x} = 0$$

$$\therefore \ \lim_{x \to \infty} f(x) = \lim_{x \to \infty} \frac{b^x + \log_a x}{a^x + \log_b x} = \lim_{x \to \infty} \frac{\frac{b^x}{\log_b x} + \frac{\log_a x}{\log_b x}}{\frac{a^x}{\log_b x} + 1} = \lim_{x \to \infty} \frac{\log_a x}{\log_b x} = \log_a b \neq 0$$

$\therefore$ False

(3) $\lim\limits_{x \to 0^+} f(x) = \log_a b$

When $x \to 0^+$, $\lim\limits_{x \to 0^+} a^x = \lim\limits_{x \to 0^+} b^x = 1$ and

$\log_a x$, $\log_b x$ are equal to ∞ or $-\infty$. (Diverge)

$$\therefore \ \lim_{x \to 0^+} \frac{b^x}{\log_b x} = \lim_{x \to 0^+} \frac{a^x}{\log_b x} = 0$$

$$\therefore \ \lim_{x \to 0^+} f(x) = \lim_{x \to 0^+} \left\{ \frac{b^x + \log_a x}{a^x + \log_b x} \right\} = \lim_{x \to 0^+} \frac{\frac{b^x}{\log_b x} + \frac{\log_a x}{\log_b x}}{\frac{a^x}{\log_b x} + 1} = \lim_{x \to 0^+} \frac{\log_a x}{\log_b x} = \log_a b$$

$\therefore$ True

#14 For two functions $f(x)$ and $g(x)$ such that $\lim\limits_{x \to \infty} f(x) = \infty$ and $\lim\limits_{x \to \infty} g(x) = \infty$,

determine whether the following statements are true or false.

(1) If $\lim\limits_{x \to \infty} \dfrac{g(x)}{f(x)} = 0$, then $\lim\limits_{x \to \infty} \dfrac{e^{g(x)}}{e^{f(x)}} = 1$.

$$\lim_{x \to \infty} \frac{e^{g(x)}}{e^{f(x)}} = \lim_{x \to \infty} e^{\{g(x) - f(x)\}} = \lim_{x \to \infty} e^{f(x)\left\{\frac{g(x)}{f(x)} - 1\right\}} = \lim_{x \to \infty} e^{-f(x)} = \lim_{x \to \infty} \frac{1}{e^{f(x)}} = 0 \ \text{(False)}$$

(2) If $\lim\limits_{x \to \infty} \dfrac{g(x)}{f(x)} = 1$, then $\lim\limits_{x \to \infty} \dfrac{\ln g(x)}{\ln f(x)} = 0$.

$$\lim_{x \to \infty} \frac{\ln g(x)}{\ln f(x)} = \lim_{x \to \infty} \frac{\ln\left\{f(x) \cdot \frac{g(x)}{f(x)}\right\}}{\ln f(x)} = \lim_{x \to \infty} \left\{ \frac{\ln f(x) + \ln \frac{g(x)}{f(x)}}{\ln f(x)} \right\} = \lim_{x \to \infty} \left\{ 1 + \frac{\ln \frac{g(x)}{f(x)}}{\ln f(x)} \right\} = 1 + 0 = 1 \ \text{(False)}$$

(3) If $\lim\limits_{x \to \infty} \dfrac{g(x)}{f(x)} = 1$, then $\lim\limits_{x \to \infty} \dfrac{\ln\left\{1 + \frac{1}{g(x)}\right\}}{\ln\left\{1 + \frac{1}{f(x)}\right\}} = 1$.

$$\lim_{x \to \infty} \frac{\ln\left\{1 + \frac{1}{g(x)}\right\}}{\ln\left\{1 + \frac{1}{f(x)}\right\}} = \lim_{x \to \infty} \frac{\ln\left\{1 + \frac{1}{g(x)}\right\}}{\ln\left\{1 + \frac{1}{f(x)}\right\}} \cdot 1 = \lim_{x \to \infty} \frac{\ln\left\{1 + \frac{1}{g(x)}\right\}}{\ln\left\{1 + \frac{1}{f(x)}\right\}} \cdot \lim_{x \to \infty} \frac{g(x)}{f(x)}$$

$$= \lim_{x \to \infty} \frac{\ln\left\{1 + \frac{1}{g(x)}\right\}}{\ln\left\{1 + \frac{1}{f(x)}\right\}} \cdot \lim_{x \to \infty} \frac{\frac{1}{f(x)}}{\frac{1}{g(x)}} = \lim_{x \to \infty} \frac{\ln\left\{1 + \frac{1}{g(x)}\right\}}{\frac{1}{g(x)}} \cdot \lim_{x \to \infty} \frac{\frac{1}{f(x)}}{\ln\left\{1 + \frac{1}{f(x)}\right\}} = 1 \cdot 1 = 1 \ \text{(True)}$$

#15 For a function $f(x)$, determine whether the following statements are true or false.

(1) If $f(x) = x^2$, then $\lim\limits_{x \to 0} \dfrac{e^{f(x)}-1}{x} = 0$.

$$\lim_{x \to 0} \frac{e^{f(x)}-1}{x} = \lim_{x \to 0} \frac{e^{x^2}-1}{x^2} \cdot x = 1 \cdot 0 = 0 \quad \text{(True)}$$

(2) If $\lim\limits_{x \to 0} \dfrac{e^x-1}{f(x)} = 1$, then $\lim\limits_{x \to 0} \dfrac{2^x-1}{f(x)} = \ln 2$

Note that $\lim\limits_{x \to 0} \dfrac{e^x-1}{f(x)} = \lim\limits_{x \to 0} \left\{ \dfrac{e^x-1}{x} \cdot \dfrac{x}{f(x)} \right\} = 1$

Since $\lim\limits_{x \to 0} \dfrac{e^x-1}{f(x)} = 1$, $\lim\limits_{x \to 0} \dfrac{x}{f(x)} = 1$

$\therefore \quad \lim\limits_{x \to 0} \dfrac{2^x-1}{f(x)} = \lim\limits_{x \to 0} \left\{ \dfrac{2^x-1}{x} \cdot \dfrac{x}{f(x)} \right\} = \ln 2 \cdot 1 = \ln 2 \quad \text{(True)}$

(3) If $\lim\limits_{x \to 0} f(x) = 0$, then $\lim\limits_{x \to 0} \dfrac{e^{f(x)}-1}{x}$ exists.

Let $f(x) = |x|$

Then, $\lim\limits_{x \to 0} f(x) = 0$

$$\lim_{x \to 0^+} \frac{e^{f(x)}-1}{x} = \lim_{x \to 0^+} \frac{e^{|x|}-1}{x} = \lim_{x \to 0^+} \frac{e^{|x|}-1}{|x|} \cdot \frac{|x|}{x} = 1 \cdot 1 = 1$$

$$\lim_{x \to 0^-} \frac{e^{f(x)}-1}{x} = \lim_{x \to 0^-} \frac{e^{|x|}-1}{x} = \lim_{x \to 0^-} \frac{e^{|x|}-1}{|x|} \cdot \frac{|x|}{x} = 1 \cdot (-1) = -1$$

$\therefore \quad \lim\limits_{x \to 0^+} \dfrac{e^{f(x)}-1}{x} \neq \lim\limits_{x \to 0^-} \dfrac{e^{f(x)}-1}{x} \; ; \quad \lim\limits_{x \to 0} \dfrac{e^{f(x)}-1}{x}$ does not exist. (False)

#16 For a function $f(x)$ such that $\lim\limits_{x \to 0} \dfrac{f(x)}{\ln(1+x)} = 1$, find the limit.

(1) $\lim\limits_{x \to 0} \dfrac{\sin x}{f(x)}$

Since $\lim\limits_{x \to 0} \dfrac{f(x)}{\ln(1+x)} = \lim\limits_{x \to 0} \left\{ \dfrac{x}{\ln(1+x)} \cdot \dfrac{f(x)}{x} \right\} = 1 \cdot \lim\limits_{x \to 0} \dfrac{f(x)}{x} = 1$, $\lim\limits_{x \to 0} \dfrac{f(x)}{x} = 1$

$\therefore \quad \lim\limits_{x \to 0} \dfrac{\sin x}{f(x)} = \lim\limits_{x \to 0} \dfrac{\sin x}{x} \cdot \dfrac{x}{f(x)} = 1 \cdot 1 = 1$

(2) $\lim\limits_{x \to 0} \dfrac{f(x)+x}{\ln(1+x)}$

$$\lim_{x \to 0} \frac{f(x)+x}{\ln(1+x)} = \lim_{x \to 0} \frac{f(x)+x}{x} \cdot \frac{x}{\ln(1+x)} = \lim_{x \to 0} \left\{ \frac{f(x)}{x} + 1 \right\} \cdot 1 = (1+1) \cdot 1 = 2$$

(3) $\lim\limits_{x \to 0} \dfrac{\{f(x)\}^2}{\ln(1+x)}$

Since $\lim\limits_{x \to 0} \dfrac{f(x)}{\ln(1+x)} = 1$ and $\lim\limits_{x \to 0} \ln(1+x) = 0$, $\lim\limits_{x \to 0} \{f(x)\}^2 = 0$. $\therefore \lim\limits_{x \to 0} f(x) = 0$

$\therefore \quad \lim\limits_{x \to 0} \dfrac{\{f(x)\}^2}{\ln(1+x)} = \lim\limits_{x \to 0} \dfrac{f(x)}{\ln(1+x)} \cdot f(x) = \lim\limits_{x \to 0} \dfrac{f(x)}{\ln(1+x)} \cdot \lim\limits_{x \to 0} f(x) = 1 \cdot 0 = 0$

#17 At what points is the function discontinuous?

(1) $f(x) = \dfrac{x-1}{|x-1|}x^2$

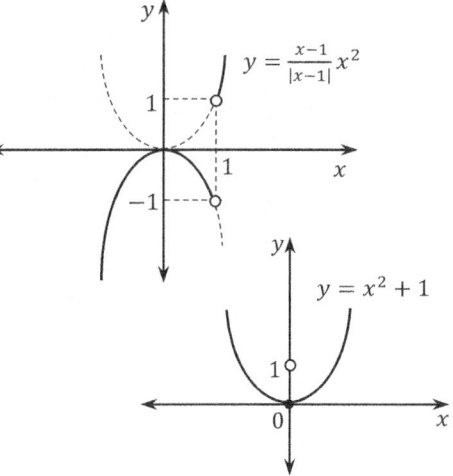

i)   When $x - 1 > 0$, $f(x) = x^2$

ii) When $x - 1 < 0$, $f(x) = -x^2$

iii) When $x - 1 = 0$, $f(x)$ does not exist.

Therefore, $f(x)$ is discontinuous at $x = 1$.

(2) $f(x) = x^2 + \dfrac{x^2}{1+x^2} + \dfrac{x^2}{(1+x^2)^2} + \cdots\cdots$

i)   When $x = 0$, $f(0) = 0 + 0 + \cdots\cdots = 0$

ii) When $x \neq 0$,

$f(x)$ is an infinite geometric series with first term x^2 and common ratio $\dfrac{1}{1+x^2}$.

Since $0 < \dfrac{1}{1+x^2} < 1$, the series converges.

$\therefore f(x) = \dfrac{a}{1-r} = \dfrac{x^2}{1-\frac{1}{1+x^2}} = \dfrac{x^2}{\frac{1+x^2-1}{1+x^2}} = \dfrac{x^2(1+x^2)}{x^2} = 1 + x^2$

$\therefore \lim_{x \to 0} f(x) = \lim_{x \to 0} (1 + x^2) = 1$

$\therefore \lim_{x \to 0} f(x) \neq f(0)$

Therefore, $f(x)$ is discontinuous at $x = 0$.

#18 Given the piecewise-defined function $f(x)$ defined below, identify any value(s) of x at which $f(x)$ is discontinuous and describe the discontinuity exhibited.

(1) $f(x) = \begin{cases} x^2 - 2x + 4, & x \leq 2 \\ x^3 - 5 & , x > 2 \end{cases}$

Both polynomials of $f(x)$ are continuous over their entire domains.

At $x = 2$, $f(2) = 2^2 - 2 \cdot 2 + 4 = 4$ $\therefore \lim_{x \to 2^-} f(x) = 4$

Similarly, at $x = 2$, the remaining function is $f(x) = x^3 - 5$.

$f(2) = 2^3 - 5 = 3$ $\therefore \lim_{x \to 2^+} f(x) = 3$

Since the left- and right- hand limits are unequal, $\lim_{x \to 2} f(x)$ does not exist.

Therefore, $f(x)$ is discontinuous at $x = 2$.

(2) $f(x) = \begin{cases} x & , \ x \text{ is a rational number in } [0, 1] \\ 1 - x & , \ x \text{ is an irrational number in } [0, 1] \end{cases}$

Suppose $f(x)$ is a rational number in $[0, 1]$.

i) When x is a rational number, $\displaystyle\lim_{x \to a} f(x) = \lim_{x \to a} x = a$

ii) When x is an irrational number, $\displaystyle\lim_{x \to a} f(x) = \lim_{x \to a}(1 - x) = 1 - a$

$\therefore \ a = 1 - a \ ; \ 2a = 1$ (To have continuity at $x = a$) $\qquad \therefore \ a = \dfrac{1}{2}$

Therefore, $f(x)$ is continuous at all real numbers in $[0, 1]$ except $x = \dfrac{1}{2}$.

(3) $f(x) = \begin{cases} \dfrac{x}{\sqrt{1+x}-1}, & x \neq 0 \\ 1 & , \ x = 0 \end{cases}$

$\displaystyle\lim_{x \to 0} f(x) = \lim_{x \to 0} \frac{x}{\sqrt{1+x}-1} = \lim_{x \to 0}\left(\frac{x}{\sqrt{1+x}-1} \cdot \frac{\sqrt{1+x}+1}{\sqrt{1+x}+1}\right) = \lim_{x \to 0}\left\{\frac{x(\sqrt{1+x}+1)}{(1+x-1)}\right\}$

$= \displaystyle\lim_{x \to 0}\left(\sqrt{1+x}+1\right) = 1 + 1 = 2$

Since $f(0) = 1$ and $\displaystyle\lim_{x \to 0} f(x) \neq f(0)$, $f(x)$ is discontinuous at $x = 0$.

(4) $f(x) = \begin{cases} \dfrac{[x]}{x}, & x \neq 0 \\ 1, & x = 0 \end{cases}$ ($[x]$ is the greatest integer in x)

Since $\displaystyle\lim_{x \to 0^-} [x] = -1$, $\displaystyle\lim_{x \to 0^-} \frac{[x]}{x} = \lim_{x \to 0^-} \frac{-1}{x} = \infty$

$\therefore \ \displaystyle\lim_{x \to 0} \frac{[x]}{x}$ does not exist. $\qquad \therefore \ f(x)$ is discontinuous at $x = 0$.

#19 Examine for continuity of the function.

(1) $f(x) = \begin{cases} x \sin\dfrac{1}{x}, & x \neq 0 \\ 0 & , \ x = 0 \end{cases}$

Since the rational function $y = \dfrac{1}{x}$ is continuous at $x = a$ $(a \neq 0)$ and sine function is

continuous at $x = a$, the composite function $y = \sin\dfrac{1}{x}$ is continuous at $x = a$ $(a \neq 0)$.

Since polynomial function $y = x$ is continuous at $x = a$, $y = x\sin\dfrac{1}{x}$ is continuous at $x = a$ $(a \neq 0)$.

Since $\displaystyle\lim_{x \to 0} x\sin\dfrac{1}{x} = 0$ and $f(0) = 0$, $y = x\sin\dfrac{1}{x}$ is continuous at $x = 0$.

Therefore, $y = f(x)$ is continuous at all x in $(-\infty, \infty)$.

(2) $f(x) = \lim\limits_{n\to\infty} \dfrac{1}{1+(-x^2+5)^{2n}}$

i) When $|-x^2+5| < 1$, $f(x) = \dfrac{1}{1+0} = 1$

ii) When $|-x^2+5| = 1$, $f(x) = \dfrac{1}{1+1} = \dfrac{1}{2}$

iii) When $|-x^2+5| > 1$,

$$f(x) = \lim_{n\to\infty} \frac{1}{1+(-x^2+5)^{2n}} = \lim_{n\to\infty} \frac{\frac{1}{(-x^2+5)^{2n}}}{\frac{1+(-x^2+5)^{2n}}{(-x^2+5)^{2n}}} = \lim_{n\to\infty} \frac{\frac{1}{(-x^2+5)^{2n}}}{\frac{1}{(-x^2+5)^{2n}}+1} = \frac{0}{0+1} = 0$$

Therefore, $f(x)$ is discontinuous at $-x^2+5 = 1$, $-x^2+5 = -1$.

(3) $f(x) = \lim\limits_{n\to\infty} \dfrac{1-\{\log_2(1+|x|)\}^n}{1+\{\log_2(1+|x|)\}^n}$

i) When $|x| < 1$,

Since $0 \le \log_2(1+|x|) < 1$, $\lim\limits_{n\to\infty} \{\log_2(1+|x|)\}^n = 0$

$\therefore f(x) = \dfrac{1-0}{1+0} = 1$ $\qquad \therefore \lim\limits_{x\to 1^-} f(x) = 1$

ii) When $x > 1$,

Since $\log_2(1+|x|) > 1$, $\lim\limits_{n\to\infty} \{\log_2(1+|x|)\}^n = \infty$

$$\therefore f(x) = \lim_{n\to\infty} \frac{1-\{\log_2(1+|x|)\}^n}{1+\{\log_2(1+|x|)\}^n} = \lim_{n\to\infty} \frac{\frac{1-\{\log_2(1+|x|)\}^n}{\{\log_2(1+|x|)\}^n}}{\frac{1+\{\log_2(1+|x|)\}^n}{\{\log_2(1+|x|)\}^n}} = \lim_{n\to\infty} \frac{\frac{1}{\{\log_2(1+|x|)\}^n}-1}{\frac{1}{\{\log_2(1+|x|)\}^n}+1} = \frac{0-1}{0+1} = -1$$

$\therefore \lim\limits_{x\to 1^+} f(x) = -1$

iii) By i) and ii), $f(x)$ is discontinuous at $x = 1$.

(4) $f(x) = \lim\limits_{n\to\infty} \dfrac{x^n+2x+1}{x^{n-1}+1}$ (*n*; **positive integer**)

Note that $|x| < 1 \Rightarrow \lim\limits_{n\to\infty} x^n = 0$

$\qquad\qquad |x| > 1 \Rightarrow \lim\limits_{n\to\infty} |x^n| = \infty$

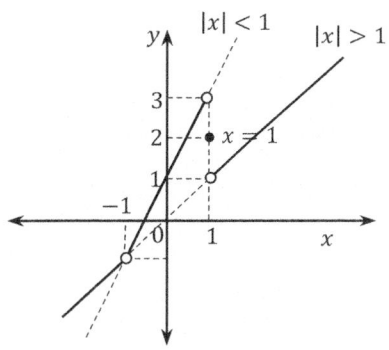

i) When $|x| < 1$, $f(x) = \lim\limits_{n\to\infty} \dfrac{x^n+2x+1}{x^{n-1}+1} = \dfrac{2x+1}{1} = 2x+1$

ii) When $x = 1$, $f(x) = \lim\limits_{n\to\infty} \dfrac{x^n+2x+1}{x^{n-1}+1} = \dfrac{1+2+1}{1+1} = \dfrac{4}{2} = 2$

iii) When $x = -1$, $f(x) = \lim\limits_{n\to\infty} \dfrac{(-1)^n-1}{(-1)^{n-1}+1}$; $f(-1)$ does not exist.

iv) When $|x| > 1$, $f(x) = \lim\limits_{n\to\infty} \dfrac{x^n+2x+1}{x^{n-1}+1} = \lim\limits_{n\to\infty} \dfrac{\frac{x^n+2x+1}{x^{n-1}}}{\frac{x^{n-1}+1}{x^{n-1}}} = \lim\limits_{n\to\infty} \dfrac{x+\frac{2}{x^{n-2}}+\frac{1}{x^{n-1}}}{1+\frac{1}{x^{n-1}}} = \dfrac{x}{1} = x$

Therefore, $f(x)$ is discontinuous at $x = \pm 1$.

#20 Show the indicated function has at least one real number solution in the interval.

(1) $x^4 + x^3 - 5x + 1 = 0$, $(1, 3)$

Note that: If $f(x)$ is continuous in $[a, b]$ and $f(a) \cdot f(b) < 0$,

then there is at least one real number solution in (a, b) such that $f(x) = 0$.

Polynomial $f(x) = x^4 + x^3 - 5x + 1$ is continuous at all real number x.

$f(1) = 1 + 1 - 5 + 1 = -2 < 0$; $f(3) = 3^4 + 3^3 - 15 + 1 = 94 > 0$

By the intermediate-value theorem, there is at least one real number solution in $(1, 3)$ such that $f(x) = 0$.

(2) $\sin x = x \cos x$, $\left(\pi, \dfrac{3}{2}\pi\right)$

Let $f(x) = \sin x - x \cos x$

Since $\sin x$ and $x \cos x$ are continuous at all real numbers, $f(x)$ is continuous in the interval $[\pi, \dfrac{3}{2}\pi]$.

Since $f(\pi) = \sin \pi - \pi \cos \pi = 0 - \pi(-1) = \pi > 0$ and

$$f\left(\frac{3}{2}\pi\right) = \sin\frac{3}{2}\pi - \frac{3}{2}\pi(\cos\frac{3}{2}\pi) = -1 < 0, \ \ f(\pi) \cdot f\left(\frac{3}{2}\pi\right) < 0$$

Therefore, by the intermediate-value theorem, there is at least one real number solution in the interval $\left(\pi, \dfrac{3}{2}\pi\right)$ such that $f(x) = 0$.

#21 Find the values of constants a and b:

(1) When a function $f(x) = \begin{cases} \dfrac{a\sqrt{x+1}-b}{x-1}, & x > 1 \\ 2x - 1 & , & x \le 1 \end{cases}$ is continuous at $x = 1$.

Since $f(x)$ is continuous at $x = 1$, $\displaystyle\lim_{x \to 1^-} f(x) = \lim_{x \to 1^+} f(x) = f(1)$.

i) When $x \to 1^-$, $f(x) = 2x - 1$

$\therefore \ \displaystyle\lim_{x \to 1^-} f(x) = \lim_{x \to 1^-} (2x - 1) = 2 \cdot 1 - 1 = 1$

ii) When $x \to 1^+$,

Since $\displaystyle\lim_{x \to 1^+} f(x) = \lim_{x \to 1^+} \frac{a\sqrt{x+1}-b}{x-1} = f(1)$ and $\displaystyle\lim_{x \to 1^+}(x - 1) = 0$, $\displaystyle\lim_{x \to 1^+}(a\sqrt{x+1} - b) = 0$

$\therefore \ a\sqrt{1+1} - b = a\sqrt{2} - b = 0 \quad \therefore \ b = a\sqrt{2}$

$\therefore \ \displaystyle\lim_{x \to 1^+} \frac{a\sqrt{x+1}-b}{x-1} = \lim_{x \to 1^+} \frac{a\sqrt{x+1}-a\sqrt{2}}{x-1} = \lim_{x \to 1^+} \frac{a(\sqrt{x+1}-\sqrt{2})}{x-1}$

$\displaystyle = \lim_{x \to 1^+} \frac{a(\sqrt{x+1}-\sqrt{2})(\sqrt{x+1}+\sqrt{2})}{(x-1)(\sqrt{x+1}+\sqrt{2})} = \lim_{x \to 1^+} \frac{a(x+1-2)}{(x-1)(\sqrt{x+1}+\sqrt{2})} = \lim_{x \to 1^+} \frac{a}{\sqrt{x+1}+\sqrt{2}} = \frac{a}{2\sqrt{2}}$

By i) and ii), $\dfrac{a}{2\sqrt{2}} = 1 \quad \therefore \ a = 2\sqrt{2} \quad$ Therefore, $a = 2\sqrt{2}, \ b(= a\sqrt{2}) = 4$

(2) When a function $f(x) = \lim\limits_{n \to \infty} \dfrac{x^{2n+1}+ax+b}{x^{2n}+1}$ is continuous at all real number x.

i) When $|x| < 1$, $f(x) = \lim\limits_{n \to \infty} \dfrac{x^{2n+1}+ax+b}{x^{2n}+1} = \dfrac{ax+b}{1} = ax + b$

ii) When $x = 1$, $f(x) = \lim\limits_{n \to \infty} \dfrac{x^{2n+1}+ax+b}{x^{2n}+1} = \dfrac{1+a+b}{2}$

iii) When $x = -1$, $f(x) = \lim\limits_{n \to \infty} \dfrac{x^{2n+1}+ax+b}{x^{2n}+1} = \dfrac{-1-a+b}{2}$

iv) When $|x| > 1$, $f(x) = \lim\limits_{n \to \infty} \dfrac{x^{2n+1}+ax+b}{x^{2n}+1} = \lim\limits_{n \to \infty} \dfrac{\frac{x^{2n+1}+ax+b}{x^{2n}}}{\frac{x^{2n}+1}{x^{2n}}} = \lim\limits_{n \to \infty} \dfrac{x+\frac{a}{x^{2n-1}}+\frac{b}{x^{2n}}}{1+\frac{1}{x^{2n}}} = x$

$\therefore \ f(x) = \begin{cases} ax + b, & |x| < 1 \\ \dfrac{1+a+b}{2}, & x = 1 \\ \dfrac{-1-a+b}{2}, & x = -1 \\ x, & |x| > 1 \end{cases}$

Since $f(x)$ is continuous at all real number x,

$\lim\limits_{x \to 1^-} f(x) = \lim\limits_{x \to 1^+} f(x) = f(1)$ and $\lim\limits_{x \to (-1)^-} f(x) = \lim\limits_{x \to (-1)^+} f(x) = f(-1)$.

$\therefore \ a + b = \dfrac{1+a+b}{2}, \quad -1 = \dfrac{-1-a+b}{2}$

$\therefore \ a + b = 1, \ a - b = 1$

Therefore, $a = 1, \ b = 0$

(3) When a function $f(x) = [x]^2 + (ax + b)[x]$ is continuous at all real number x.

($[x]$ is the greatest integer in x.)

Note that $x = n + \alpha$ (n; integer, $0 \le \alpha < 1$)

Then $[x] = n$

$\lim\limits_{x \to n^+} [x] = n, \quad \lim\limits_{x \to n^-} [x] = n - 1$

If x is an integer, then $y = [x]$ is discontinuous. Otherwise, $y = [x]$ is continuous.

Since $f(x)$ is continuous at all real number x, $f(x)$ is continuous at $x = n$ (n; integer).

i) Since $\lim\limits_{x \to n^-} [x] = n - 1$,

$\lim\limits_{x \to n^-} f(x) = \lim\limits_{x \to n^-} \{[x]^2 + (ax + b)[x]\} = (n - 1)^2 + (an + b)(n - 1)$

ii) Since $\lim\limits_{x \to n^+} [x] = n$,

$\lim\limits_{x \to n^+} f(x) = \lim\limits_{x \to n^+} \{[x]^2 + (ax + b)[x]\} = n^2 + (an + b)n$

By i) and ii), $(n - 1)^2 + (an + b)(n - 1) = n^2 + (an + b)n$

$\therefore \ (n - 1)^2 - an - b = n^2 \ ; \ -2n + 1 - an - b = 0 \ ; \ -n(2 + a) + 1 - b = 0$

Since the equality is always true for any integer n, $2 + a = 0$ and $1 - b = 0$.

Therefore, $a = -2, \ b = 1$

(4) When a function $f(x) = \begin{cases} x(x-1) & , \ |x| > 1 \\ -x^2 + ax + b & , \ |x| \leq 1 \end{cases}$ is continuous at all real number x.

Since $y = x(x-1)$ and $y = -x^2 + ax + b$ are polynomials, both are continuous functions.

Since $f(x)$ is continuous at all real numbers, $f(x)$ is continuous at $x = -1$ and $x = 1$.

i) Since $\displaystyle\lim_{x \to (-1)^+} f(x) = \lim_{x \to (-1)^-} f(x)$, $\displaystyle\lim_{x \to (-1)^+}(-x^2 + ax + b) = \lim_{x \to (-1)^-}\{x(x-1)\}$

$\therefore \ -1 - a + b = (-1)(-2); \ -a + b = 3 \ \cdots\cdots ①$

ii) Since $\displaystyle\lim_{x \to 1^+} f(x) = \lim_{x \to 1^-} f(x)$, $\displaystyle\lim_{x \to 1^+}\{x(x-1)\} = \lim_{x \to 1^-}(-x^2 + ax + b)$

$\therefore \ 1 \cdot 0 = -1 + a + b; \ a + b = 1 \ \cdots\cdots ②$

By ① and ②, $a = -1$, $b = 2$

(5) For a function $f(x) = x^2 - 4x + a$ and a function $g(x) = \displaystyle\lim_{n \to \infty} \frac{2|x-b|^n + 1}{|x-b|^n + 1}$,

$\quad h(x) = f(x)g(x).$

When the function $h(x)$ is continuous at all real number x, find the values of a and b.

Note that $f(x) = x^2 - 4x + a = (x-2)^2 - 4 + a$ (Continuous function).

i) If $|x - b| > 1$, then $g(x) = \displaystyle\lim_{n \to \infty} \frac{2|x-b|^n + 1}{|x-b|^n + 1} = \lim_{n \to \infty} \frac{\frac{2|x-b|^n + 1}{|x-b|^n}}{\frac{|x-b|^n + 1}{|x-b|^n}} = \frac{2+0}{1+0} = 2$

ii) If $|x - b| = 1$, then $g(x) = \displaystyle\lim_{n \to \infty} \frac{2|x-b|^n + 1}{|x-b|^n + 1} = \lim_{n \to \infty} \frac{2+1}{1+1} = \frac{3}{2}$

iii) If $|x - b| < 1$, then $g(x) = \displaystyle\lim_{n \to \infty} \frac{2|x-b|^n + 1}{|x-b|^n + 1} = \frac{1}{1} = 1$

Note that $|x - b| = 1 \iff f(x) = \begin{cases} x - b = 1, & x - b \geq 0 \\ -x + b = 1, & x - b < 0 \end{cases}$

$\qquad\qquad\qquad \iff f(x) = \begin{cases} x = b + 1, & x \geq b \\ x = b - 1, & x < b \end{cases}$

The graph of $y = g(x)$ is discontinuous at $x = b + 1$, $x = b - 1$.

Since $h(x) = f(x)g(x)$ is continuous at all real numbers, $f(b + 1) = f(b - 1)$.

Since $x = 2$ is the axis of symmetry of $y = f(x)$, $b = 2$

Since $f(b + 1) = 0$, $f(3) = 0$

$\therefore \ f(3) = 3^2 - 4 \cdot 3 + a = -3 + a = 0 \ ; \ a = 3$

Therefore, $a = 3$, $b = 2$

#22 Find the value of the constant a.

(1) For real number , $f(x) = \begin{cases} \frac{\sin 2(x-1)}{x-1} &, |x| \neq 1 \\ a &, |x| = 1 \end{cases}$.

   When $f(x)$ is continuous at $x = 1$, find the value of a.

Since $f(x)$ is continuous at $x = 1$, $\lim_{x \to 1} f(x) = f(1) = a$.

$\therefore \lim_{x \to 1} f(x) = \lim_{x \to 1} \frac{\sin 2(x-1)}{x-1} = \lim_{t \to 0} \frac{\sin 2t}{t} = \lim_{t \to 0} \frac{\sin 2t}{2t} \cdot 2 = 1 \cdot 2 = 2$

$\therefore a = 2$

(2) When $f(x) = \begin{cases} \frac{e^{3x}-1}{\sin 2x} &, -\frac{\pi}{2} \leq x < 0, \ 0 < x \leq \frac{\pi}{2} \\ a &, x = 0 \end{cases}$ is continuous at $x = 0$,

   find the value of a.

Since $f(x)$ is continuous at $x = 0$, $\lim_{x \to 0} f(x) = f(0) = a$.

$\therefore \lim_{x \to 0} f(x) = \lim_{x \to 0} \frac{e^{3x}-1}{\sin 2x} = \lim_{x \to 0} \frac{e^{3x}-1}{3x} \cdot \frac{2x}{\sin 2x} \cdot \frac{3x}{2x} = 1 \cdot 1 \cdot \frac{3}{2} = \frac{3}{2}$

$\therefore a = \frac{3}{2}$

(3) For two functions $f(x) = \begin{cases} x^2 + 2x + 2, & x \geq 1 \\ x + 1 &, x < 1 \end{cases}$ and $g(x) = |x - 2a|$,

   a function $h(x) = (g \circ f)(x)$ is continuous at all real number x. Find the value of a.

Since $h(x) = (g \circ f)(x)$ is continuous at all real number x, $h(x)$ is continuous at $x = 1$.

$\therefore \lim_{x \to 1^-} (g \circ f)(x) = \lim_{x \to 1^+} (g \circ f)(x)$

$\lim_{x \to 1^-} (g \circ f)(x) = \lim_{x \to 1^-} g(x + 1) = \lim_{x \to 1^-} |x + 1 - 2a| = |1 + 1 - 2a| = |2 - 2a|$

$\lim_{x \to 1^+} (g \circ f)(x) = \lim_{x \to 1^+} g(x^2 + 2x + 2) = \lim_{x \to 1^+} |x^2 + 2x + 2 - 2a|$

$\qquad = |1 + 2 + 2 - 2a| = |5 - 2a|$

$\therefore |2 - 2a| = |5 - 2a|$

$\therefore (2 - 2a)^2 = (5 - 2a)^2 \ ; \ 4 - 8a + 4a^2 = 25 - 20a + 4a^2 \ ; \ 12a = 21$

$\therefore a = \frac{7}{4}$

(4) When a function defined by $f(x) = \begin{cases} x^2 + \frac{x^2}{1+\tan^2 x} + \frac{x^2}{(1+\tan^2 x)^2} + \cdots\cdots &, x \neq 0 \\ a &, x = 0 \end{cases}$ is

   continuous at $x = 0$ in the interval $\left(-\frac{\pi}{2}, \frac{\pi}{2}\right)$, find the value of the constant a.

In the interval $\left(-\frac{\pi}{2}, \frac{\pi}{2}\right)$, $\tan^2 x > 0 \ (x \neq 0)$ $\qquad \therefore 0 < \frac{1}{1+\tan^2 x} < 1$

$\therefore \ x^2 + \dfrac{x^2}{1+\tan^2 x} + \dfrac{x^2}{(1+\tan^2 x)^2} + \cdots\cdots$ is an infinite geometric series with first term x^2 and

common ratio $\dfrac{1}{1+\tan^2 x}$.

$\therefore \ f(x) = \dfrac{x^2}{1-\frac{1}{1+\tan^2 x}} = \dfrac{x^2}{\frac{\tan^2 x}{1+\tan^2 x}} = \dfrac{x^2(1+\tan^2 x)}{\tan^2 x}, \ x \neq 0$

Since $f(x)$ is continuous at $x = 0$, $\displaystyle\lim_{x\to0} f(x) = f(0)$

$\therefore \ \displaystyle\lim_{x\to0} f(x) = \lim_{x\to0} \dfrac{x^2(1+\tan^2 x)}{\tan^2 x} = \lim_{x\to0} \left(\dfrac{x}{\tan x}\right)^2 (1 + \tan^2 x) = 1^2(1 + 0) = 1$

$\therefore \ a = 1$

#23 Find the value.

(1) For any real number x, the continuous function $f(x)$ such that $f(x + 4) = f(x)$ is

defined by $f(x) = \begin{cases} 3x & , \ 0 \le x < 1 \\ x^2 + ax + b & , \ 1 \le x \le 4 \end{cases}$. Find the value of $f(10)$.

Since $f(x + 4) = f(x)$, $f(0 + 4) = f(0)$ when $x = 0$.

$\therefore \ f(4) = f(0)$

Since $f(4) = 4^2 + 4a + b$ and $f(0) = 3 \cdot 0 = 0$, $16 + 4a + b = 0 \ \cdots\cdots①$

Since $f(x)$ is continuous at all real number x, $f(x)$ is continuous at $x = 1$.

$\therefore \ \displaystyle\lim_{x\to1^-} f(x) = \lim_{x\to1^+} f(x) = f(1)$

$\therefore \ \displaystyle\lim_{x\to1^-} 3x = \lim_{x\to1^+} (x^2 + ax + b) \ ; \ 3\cdot 1 = 1 + a + b \ ; \ a + b = 2 \ \cdots\cdots②$

By ① and ②, $16 + 4a + (2 - a) = 0 \ ; \ 3a + 18 = 0 \quad \therefore \ a = -6, \ b = 8$

$\therefore \ f(x) = x^2 - 6x + 8 \ , \ 1 \le x \le 4$

Therefore, $f(10) = f(6) = f(2) = 2^2 - 6 \cdot 2 + 8 = 0$

(2) When a quadratic function $f(x)$ has the leading coefficient 1, and two other functions

$g(x)$ and $h(x)$ are defined by $g(x) = \displaystyle\lim_{n\to\infty} \dfrac{x^{2n-1}-1}{x^{2n}+1}$, $h(x) = \begin{cases} \dfrac{|x|}{x} & , \ x \neq 0 \\ 0 & , \ x = 0 \end{cases}$,

$f(x)g(x)$ and $f(x)h(x)$ are continuous. Find the value of $f(10)$.

For $g(x)$, i) When $|x| > 1$, $g(x) = \displaystyle\lim_{n\to\infty} \dfrac{x^{2n-1}-1}{x^{2n}+1} = \lim_{n\to\infty} \dfrac{\frac{x^{2n-1}-1}{x^{2n}}}{\frac{x^{2n}+1}{x^{2n}}} = \lim_{n\to\infty} \dfrac{\frac{1}{x}-\frac{1}{x^{2n}}}{1+\frac{1}{x^{2n}}} = \dfrac{1}{x}$

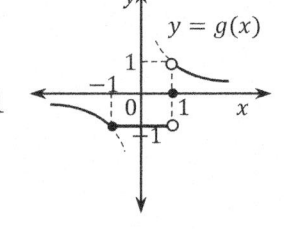

ii) When $x = 1$, $g(x) = \displaystyle\lim_{n\to\infty} \dfrac{x^{2n-1}-1}{x^{2n}+1} = \lim_{n\to\infty} \dfrac{1-1}{1+1} = 0$

iii) When $x = -1$, $g(x) = \displaystyle\lim_{n\to\infty} \dfrac{x^{2n-1}-1}{x^{2n}+1} = \lim_{n\to\infty} \dfrac{-1-1}{1+1} = -1$

iv) When $|x| < 1$, $g(x) = \displaystyle\lim_{n\to\infty} \dfrac{x^{2n-1}-1}{x^{2n}+1} = \lim_{n\to\infty} \dfrac{0-1}{0+1} = -1$

For $h(x)$, i) When $x > 0$, $h(x) = \dfrac{|x|}{x} = \dfrac{x}{x} = 1$

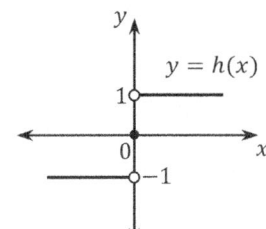

ii) When $x = 0$, $h(x) = 0$

iii) When $x < 0$, $h(x) = \dfrac{|x|}{x} = \dfrac{-x}{x} = -1$

Since $f(x)g(x)$ is continuous,

$\displaystyle\lim_{x \to 1^+} f(x)g(x) = f(1)$ and $\displaystyle\lim_{x \to 1^-} f(x)g(x) = -f(1)$.

∴ $\displaystyle\lim_{x \to 1} f(x)g(x) = f(1) \cdot g(1) = f(1) \cdot 0 = 0$ ∴ $f(1) = 0$ ······ ①

Since $f(x)h(x)$ is continuous,

$\displaystyle\lim_{x \to 0^+} f(x)h(x) = f(0)$ and $\displaystyle\lim_{x \to 0^-} f(x)h(x) = -f(0)$.

∴ $\displaystyle\lim_{x \to 0} f(x)h(x) = f(0) \cdot h(0) = f(0) \cdot 0 = 0$ ∴ $f(0) = 0$ ······ ②

By ① and ②, the quadratic function $f(x)$ with leading coefficient 1 is $f(x) = x(x - 1)$.

Therefore, $f(10) = 10 \cdot 9 = 90$

(3) For two functions $f(x) = -x^2 + 4x - 2$ and $(x) = 2\sin\dfrac{x}{2}$, let the maximum and

minimum values of $(f \circ g)(x)$ in the interval $[0, 2\pi]$ be M and n, respectively.

Find the value of $M + m$.

Let $g(x) = 2\sin\dfrac{x}{2} = t$, $0 \le x \le 2\pi$

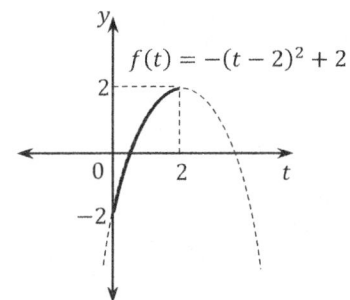

Since $0 \le 2\sin\dfrac{\pi}{2} \le 2$, $0 \le t \le 2$

∴ $(f \circ g)(x) = f(g(x)) = f(t) = -t^2 + 4t - 2$

$\qquad = -(t - 2)^2 + 2,\ 0 \le t \le 2$

Since $f(x)$ and $g(x)$ are continuous,

$(f \circ g)(x)$ is continuous in $[0, 2\pi]$ and has maximum and minimum values.

From the graph of $f(t)$, $M = 2$ and $m = -2$

Therefore, $M + m = 0$

(4) For $x > 0$, the function $f(x)$ such that $(4x - 1)f(x) = \ln 4x$ is continuous.

Find the value of $f\left(\dfrac{1}{4}\right)$.

Let $f(x) = \begin{cases} \dfrac{\ln 4x}{4x-1} & , x \ne \dfrac{1}{4} \\ a & , x = \dfrac{1}{4} \end{cases}$

Since $f(x)$ is continuous at $x > 0$, $f(x)$ must be continuous at $x = \dfrac{1}{4}$.

∴ $\displaystyle\lim_{x \to \frac{1}{4}} f(x) = f\left(\dfrac{1}{4}\right)$

Let $4x - 1 = t$. Then, $t \to 0$ as $x \to \dfrac{1}{4}$

$\therefore \displaystyle\lim_{x \to \frac{1}{4}} f(x) = \lim_{t \to 0} \dfrac{\ln(t+1)}{t} = 1$ Therefore, $f\left(\dfrac{1}{4}\right) = 1$

#24 For a continuous function $f(x)$ in $[a, b]$ with $f(a) = b$ and $f(b) = a$,

show that there is c $(a < c < b)$ such that $f(c) = c$.

Since $f(a) = b$ and $f(b) = a$, $f(a) > f(b)$ when $a < b$

By the intermediate-value theorem,

for the continuous function $f(x)$ on a closed interval $[a, b]$, there is k between $f(b)$ and $f(a)$

such that $f(c) = k, (a < c < b)$.

Since $f(b) < k < f(a)$ and $a < k < b$, $k = c$

$\therefore$ Therefore, there exists c such that $f(c) = c$ $(a < c < b)$.

#25 For a function $f(x) = |x| + \dfrac{|x|}{1+|x|} + \dfrac{|x|}{(1+|x|)^2} + \dfrac{|x|}{(1+|x|)^3} + \cdots\cdots$,

determine whether the following statements are true or false.

(1) $\displaystyle\lim_{x \to 0} f(\sin x) = f\left(\lim_{x \to 0} \sin x\right)$

i) When $x \neq 0$, $0 < \dfrac{1}{1+|x|} < 1$

$\therefore$ $f(x) = \dfrac{|x|}{1 - \frac{1}{1+|x|}} = \dfrac{|x|}{\frac{1+|x|-1}{1+|x|}} = 1 + |x|$

ii) When $x = 0$, $f(x) = 0$

By i) and ii), $f(x) = \begin{cases} 1 + |x| & , x \neq 0 \\ 0 & , x = 0 \end{cases}$

$\therefore$ $\displaystyle\lim_{x \to 0} f(\sin x) = \lim_{x \to 0} (1 + |\sin x|) = 1 + 0 = 1$

$f\left(\displaystyle\lim_{x \to 0} \sin x\right) = f(0) = 0$

$\therefore$ $\displaystyle\lim_{x \to 0} f(\sin x) \neq f\left(\lim_{x \to 0} \sin x\right)$ False

(2) $\displaystyle\lim_{h \to 0} \dfrac{f(h) - f(-h)}{h}$ **exists.**

Since $h \neq 0$, $f(h) - f(-h) = 1 + |h| - (1 + |-h|) = 0$

Since $\displaystyle\lim_{h \to 0} \dfrac{f(h) - f(-h)}{h} = 0$, $\displaystyle\lim_{h \to 0} \dfrac{f(h) - f(-h)}{h}$ exists.   True

(3) For any real number a, $\displaystyle\lim_{h\to0} f(a+h) = \lim_{h\to0} f(a-h)$

i) When $x = a \neq 0$,

Since $f(x)$ is continuous, $\displaystyle\lim_{h\to0} f(a+h) = \lim_{h\to0} f(a-h) = f(a)$

ii) When $x = a = 0$,

$\displaystyle\lim_{h\to0} f(a+h) = \lim_{h\to0} f(h) = \lim_{h\to0}(1+|h|) = 1$

$\displaystyle\lim_{h\to0} f(a-h) = \lim_{h\to0} f(-h) = \lim_{h\to0}(1+|-h|) = 1$

$\therefore \displaystyle\lim_{h\to0} f(a+h) = \lim_{h\to0} f(a-h) = 1$

By i) and ii), $\displaystyle\lim_{h\to0} f(a+h) = \lim_{h\to0} f(a-h)$ for any real number a.　　True

#26 For two functions $f(x)$ and $g(x)$,

determine whether the following statements are true or false.

(1) If $\displaystyle\lim_{x\to0} f(x)$ and $\displaystyle\lim_{x\to0} g(x)$ do not exist, $\displaystyle\lim_{x\to0}\{f(x)+g(x)\}$ does not exist.

Let $f(x) = \begin{cases} x+1 & , x \geq 0 \\ x-1 & , x < 0 \end{cases}$ and $g(x) = -f(x)$

Then $\displaystyle\lim_{x\to0^-} f(x) = -1$, $\displaystyle\lim_{x\to0^+} f(x) = 1$

Since $\displaystyle\lim_{x\to0^-} g(x) = 1$ and $\displaystyle\lim_{x\to0^+} g(x) = -1$, $\displaystyle\lim_{x\to0} f(x)$ and $\displaystyle\lim_{x\to0} g(x)$ do not exist.

But, $\displaystyle\lim_{x\to0}\{f(x)+g(x)\} = 0$

Therefore, $\displaystyle\lim_{x\to0}\{f(x)+g(x)\}$ exists.　　False

(2) If $y = f(x)$ is continuous at $x = 0$, then $y = |f(x)|$ is also continuous at $x = 0$.

Since $y = f(x)$ is continuous at $x = 0$, $\displaystyle\lim_{x\to0} f(x) = f(0) = a$ (a: constant)

i) When $a = 0$,

$\displaystyle\lim_{x\to0^+}|f(x)| = \lim_{x\to0^-}|f(x)| = f(0) = 0$ $\quad\therefore \displaystyle\lim_{x\to0}|f(x)| = |f(0)|$

ii) When $a > 0$,

$\displaystyle\lim_{x\to0}|f(x)| = \lim_{x\to0} f(x) = a = |f(0)|$

iii) When $a < 0$,

$\displaystyle\lim_{x\to0}|f(x)| = \lim_{x\to0}\{-f(x)\} = -a = |f(0)|$

Since $\displaystyle\lim_{x\to0}|f(x)| = |f(0)|$, $y = |f(x)|$ is continuous at $x = 0$ by i), ii) and iii)　　True

(3) If $y = |f(x)|$ is continuous at $x = 0$, then $y = f(x)$ is also continuous at $x = 0$.

Let $f(x) = \begin{cases} x+1 & , \ x \geq 0 \\ x-1 & , \ x < 0 \end{cases}$

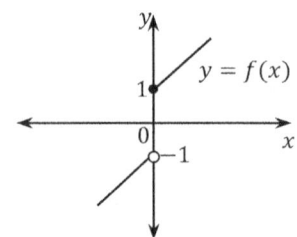
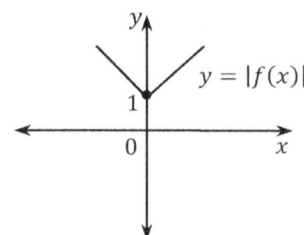

Since $\lim\limits_{x \to 0} |f(x)| = |f(0)| = 1$, $y = |f(x)|$ is continuous at $x = 0$.

But, $\lim\limits_{x \to 0^+} f(x) = 1$, $\lim\limits_{x \to 0^-} f(x) = -1$; $\lim\limits_{x \to 0^+} f(x) \neq \lim\limits_{x \to 0^-} f(x)$

Therefore, $y = f(x)$ is discontinuous at $x = 0$.   False

(4) When $f(x) = \begin{cases} 1 & , \ x \geq 0 \\ -1 & , \ x < 0 \end{cases}$ and $g(x) = |x|$, $(g \circ f)(x)$ is continuous at $x = 0$.

$\lim\limits_{x \to 0^-} (g \circ f)(x) = g(-1) = 1$

$\lim\limits_{x \to 0^+} (g \circ f)(x) = g(1) = 1$

$(g \circ f)(0) = g(f(0)) = g(1) = 1$

Since $\lim\limits_{x \to 0}(g \circ f)(x) = (g \circ f)(0) = 1$, $(g \circ f)(x)$ is continuous at $x = 0$.   True

(5) If $(g \circ f)(x)$ is continuous at $x = 0$, then $f(x)$ is continuous at $x = 0$.

Let $f(x) = \begin{cases} 1 & , \ x \geq 0 \\ -1 & , \ x < 0 \end{cases}$ and $g(x) = |x|$

Then by (4), $(g \circ f)(x)$ is continuous at $x = 0$.

But, $f(x)$ is discontinuous at $x = 0$.   False

(6) If $(f \circ f)(x)$ is continuous at $x = 0$, then $f(x)$ is continuous at $x = 0$.

Let $f(x) = \begin{cases} 1 & , \ x = 0 \\ -1 & , \ x \neq 0 \end{cases}$

When $x = 0$, $(f \circ f)(x) = f(1) = -1$

When $x \neq 0$, $(f \circ f)(x) = f(-1) = -1$

Since $(f \circ f)(x) = -1$, $(f \circ f)(x)$ is continuous at $x = 0$.

But, $f(x)$ is discontinuous at $x = 0$.   False

Chapter 3. The Derivative

#1 Find the difference quotient of each function f at the given interval.

(1) $f(x) = 2x + 3$ $[3, 6]$

$$\frac{\Delta f}{\Delta x} = \frac{(2 \cdot 6 + 3) - (2 \cdot 3 + 3)}{6 - 3} = \frac{15 - 9}{3} = 2$$

(2) $f(x) = x^2 - 2x + 3$ $[1, 3]$

$$\frac{\Delta f}{\Delta x} = \frac{(3^2 - 2 \cdot 3 + 3) - (1^2 - 2 \cdot 1 + 3)}{3 - 1} = \frac{6 - 2}{2} = 2$$

(3) $f(x) = \sqrt{x}$ $[4, 16]$

$$\frac{\Delta f}{\Delta x} = \frac{\sqrt{16} - \sqrt{4}}{16 - 4} = \frac{4 - 2}{12} = \frac{1}{6}$$

(4) $f(x) = x^3$ $[a, b]$

$$\frac{\Delta f}{\Delta x} = \frac{b^3 - a^3}{b - a} = \frac{(b - a)(b^2 + ab + a^2)}{b - a} = b^2 + ab + a^2$$

#2 When the graph of a function $y = f(x)$ is shown

as the Figure, let $g(x)$ be the inverse function of $f(x)$.

Find the difference quotient of $g(x)$ at the interval $[b, c]$.

$$\frac{\Delta g}{\Delta x} = \frac{g(c) - g(b)}{c - b} = \frac{f^{-1}(c) - f^{-1}(b)}{c - b}$$

Since $f(b) = c$ and $f(a) = b$, $f^{-1}(c) = b$ and $f^{-1}(b) = a$

$$\therefore \quad \frac{\Delta g}{\Delta x} = \frac{f^{-1}(c) - f^{-1}(b)}{c - b} = \frac{b - a}{c - b}$$

#3 Find the equation of the tangent to $y = x^2$ to the point $(x_0, f(x_0))$.

$$f(x_0) = x_0^2$$

$$f(x_0 + h) = (x_0 + h)^2 = x_0^2 + 2x_0 h + h^2$$

$$m = \lim_{h \to 0} \frac{f(x_0 + h) - f(x_0)}{h} = \lim_{h \to 0} \frac{x_0^2 + 2x_0 h + h^2 - x_0^2}{h} = \lim_{h \to 0}(2x_0 + h) = 2x_0 + 0 = 2x_0$$

$\therefore$ The equation of the tangent line is $y - f(x_0) = 2x_0(x - x_0)$.

#4 For a function $f(x) = ax^3 + bx + c$, the difference quotient of the function at the interval

$[0, 1]$ is 6 and the derivative of f at $x = 1$ is 2. Find the value of $a - b$.

$$\frac{\Delta f}{\Delta x} = \frac{f(1) - f(0)}{1 - 0} = (a + b + c) - c = a + b = 6 \quad \cdots\cdots \textcircled{1}$$

Since $f'(x) = 3ax^2 + b$ and $f'(1) = 2$, $3a + b = 2$ ...... ②

By ① and ②, $2a = -4$ $\quad\quad \therefore a = -2, \ b = 8$

Therefore, $a - b = -2 - 8 = -10$

#5 Find the derivative of $f(x) = \dfrac{1}{x-2}$ at $x = 1$ and $x = 3$.

Show that the derivative does not exist at $x = 2$, where the function is discontinuous.

$$f(x+h) - f(x) = \frac{1}{x+h-2} - \frac{1}{x-2} = \frac{x-2-(x+h-2)}{(x+h-2)(x-2)} = \frac{-h}{(x+h-2)(x-2)}$$

$$\therefore \ f'(x) = \lim_{h \to 0} \frac{f(x+h)-f(x)}{h} = \lim_{h \to 0} \frac{\frac{-h}{(x+h-2)(x-2)}}{h} = \lim_{h \to 0} \frac{-1}{(x+h-2)(x-2)}$$

$$= \frac{-1}{(x+0-2)(x-2)} = \frac{-1}{(x-2)^2}$$

$\therefore$ At $x = 2$, $f'(x)$ does not exist since the denominator is zero.

At $x = 1$, $f'(x) = \dfrac{-1}{(1-2)^2} = -1$

At $x = 3$, $f'(x) = \dfrac{-1}{(3-2)^2} = -1$

#6 Find the derivative of $f(x) = \dfrac{2x-3}{3x+4}$.

Examine the derivative at $x = -\dfrac{4}{3}$, where the function is discontinuous.

$$f(x+h) - f(x) = \frac{2(x+h)-3}{3(x+h)+4} - \frac{2x-3}{3x+4} = \frac{\{2(x+h)-3\}(3x+4)-(2x-3)\{3(x+h)+4\}}{\{3(x+h)+4\}(3x+4)}$$

$$= \frac{(6x+8-6x+9)h}{\{3(x+h)+4\}(3x+4)} = \frac{17h}{\{3(x+h)+4\}(3x+4)}$$

$$\therefore \ f'(x) = \lim_{h \to 0} \frac{f(x+h)-f(x)}{h} = \lim_{h \to 0} \frac{\frac{17h}{\{3(x+h)+4\}(3x+4)}}{h} = \lim_{h \to 0} \frac{17}{\{3(x+h)+4\}(3x+4)} = \frac{17}{\{3(x+0)+4\}(3x+4)}$$

$$= \frac{17}{(3x+4)^2}$$

$\therefore$ At $x = -\dfrac{4}{3}$, the derivative does not exist sine the denominator is zero.

#7 Examine the differentiability of the function at $x = 0$.

(1) $f(x) = |x|(x - 1)$

$$f'(0) = \lim_{h \to 0} \frac{f(0+h)-f(0)}{h} = \lim_{h \to 0} \frac{|h|(h-1)-0}{h} = \lim_{h \to 0} \frac{|h|(h-1)}{h}$$

i) If $h > 0$, then $f'(0) = \lim_{h \to 0} \dfrac{|h|(h-1)}{h} = \lim_{h \to 0} \dfrac{h(h-1)}{h} = \lim_{h \to 0}(h - 1) = (0 - 1) = -1$

ii) If $h < 0$, then $f'(0) = \lim\limits_{h \to 0} \dfrac{|h|(h-1)}{h} = \lim\limits_{h \to 0} \dfrac{(-h)(h-1)}{h} = \lim\limits_{h \to 0}(1-h) = (1-0) = 1$

$\therefore \lim\limits_{h \to 0^+} \dfrac{|h|(h-1)}{h} \neq \lim\limits_{h \to 0^-} \dfrac{|h|(h-1)}{h}$

Therefore, $f(x)$ is not differentiable at $x = 0$.

(2) $f(x) = \begin{cases} x^2 \cos\dfrac{1}{x}, & x \neq 0 \\ 0, & x = 0 \end{cases}$

$f'(0) = \lim\limits_{h \to 0} \dfrac{f(0+h)-f(0)}{h} = \lim\limits_{h \to 0} \dfrac{h^2 \cos\frac{1}{h}-0}{h} = \lim\limits_{h \to 0} h \cos\dfrac{1}{h} = 0$

$\therefore$ $f(x)$ is differentiable at $x = 0$.

(3) $f(x) = x - [x]$, ($[x]$; the greatest integer in x)

$f'(0) = \lim\limits_{h \to 0} \dfrac{f(0+h)-f(0)}{h} = \lim\limits_{h \to 0} \dfrac{h-[h]}{h} = \lim\limits_{h \to 0}\left(1 - \dfrac{[h]}{h}\right) = 1 - \lim\limits_{h \to 0} \dfrac{[h]}{h}$

Since $\lim\limits_{h \to 0^+} \dfrac{[h]}{h} = 0$ and $\lim\limits_{h \to 0^-} \dfrac{[h]}{h} = \infty$, $f(x)$ is not differentiable at $x = 0$.

#8 Examine each function is continuous at $x = 1$ but not differentiable.

(1) $f(x) = \sqrt{x-1}$

Since $\lim\limits_{x \to 1} f(x) = \lim\limits_{x \to 1} \sqrt{x-1} = 0$ and $f(1) = \sqrt{1-1} = 0$, $\lim\limits_{x \to 1} f(x) = f(1)$

$\therefore$ $f(x)$ is continuous at $x = 1$.

$\lim\limits_{h \to 0} \dfrac{f(1+h)-f(1)}{h} = \lim\limits_{h \to 0} \dfrac{\sqrt{(1+h)-1}-\sqrt{1-1}}{h} = \lim\limits_{h \to 0} \dfrac{\sqrt{h}-0}{h} = \lim\limits_{h \to 0} \dfrac{1}{\sqrt{h}} = \infty$

$\therefore f(x)$ is not differentiable. (Diverge)

(2) $f(x) = |x^2 - x|$

Since $\lim\limits_{x \to 1} f(x) = \lim\limits_{x \to 1} |x^2 - x| = 0$ and $f(1) = |1^2 - 1| = 0$, $\lim\limits_{x \to 1} f(x) = f(1)$

$\therefore$ $f(x)$ is continuous at $x = 1$.

$\lim\limits_{x \to 1^-} \dfrac{f(x)-f(1)}{x-1} = \lim\limits_{x \to 1^-} \dfrac{-(x^2-x)}{x-1} = \lim\limits_{x \to 1^-} \dfrac{-x(x-1)}{x-1} = \lim\limits_{x \to 1^-}(-x) = -1$

$\lim\limits_{x \to 1^+} \dfrac{f(x)-f(1)}{x-1} = \lim\limits_{x \to 1^+} \dfrac{x^2-x}{x-1} = \lim\limits_{x \to 1^+} \dfrac{x(x-1)}{x-1} = \lim\limits_{x \to 1^+} x = 1$

$\therefore \lim\limits_{x \to 1^-} \dfrac{f(x)-f(1)}{x-1} \neq \lim\limits_{x \to 1^+} \dfrac{f(x)-f(1)}{x-1}$

$\therefore f(x)$ is not differentiable at $x = 1$.

(3) $f(x) = \begin{cases} 2x^3, & x \geq 0 \\ 4x - 2, & x < 0 \end{cases}$

$\lim\limits_{x \to 1^-} f(x) = \lim\limits_{x \to 1^-}(4x - 2) = 4 \cdot 1 - 2 = 2$; $\lim\limits_{x \to 1^+} f(x) = \lim\limits_{x \to 1^+} 2x^3 = 2 \cdot 1^3 = 2$

Since $f(1) = 2 \cdot 1^3 = 2$, $\lim\limits_{x \to 1} f(x) = f(1)$ $\qquad \therefore$ $f(x)$ is continuous at $x = 1$.

$$\lim_{x\to 1^-}\frac{f(x)-f(1)}{x-1}=\lim_{x\to 1^-}\frac{(4x-2)-2}{x-1}=\lim_{x\to 1^-}\frac{4(x-1)}{x-1}=\lim_{x\to 1^-}4=4$$

$$\lim_{x\to 1^+}\frac{f(x)-f(1)}{x-1}=\lim_{x\to 1^+}\frac{2x^3-2}{x-1}=\lim_{x\to 1^+}\frac{2(x^3-1)}{x-1}=\lim_{x\to 1^+}\frac{2(x-1)(x^2+x+1)}{x-1}=\lim_{x\to 1^+}2(x^2+x+1)$$

$$=2(1+1+1)=6$$

$$\therefore\ \lim_{x\to 1^-}\frac{f(x)-f(1)}{x-1}\ne\lim_{x\to 1^+}\frac{f(x)-f(1)}{x-1}\qquad \therefore f(x)\text{ is not differentiable at }x=1.$$

#9 Find the limit.

(1) $\lim\limits_{x\to 0}\dfrac{\left[\frac{3}{2}+x\right]-\left[\frac{3}{2}\right]}{x}$ ([x]; the greatest integer in x)

Let $f(x)=[x]$

$$\lim_{x\to 0}\frac{\left[\frac{3}{2}+x\right]-\left[\frac{3}{2}\right]}{x}=\lim_{x\to 0}\frac{f\left(\frac{3}{2}+x\right)-f\left(\frac{3}{2}\right)}{x}=f'\left(\frac{3}{2}\right)$$

Since $1\le x<2\ \Rightarrow\ f(x)=[x]=1,\ f\left(\frac{3}{2}\right)=\left[\frac{3}{2}\right]=1$ (f is a constant function)

$$\therefore\ f'\left(\frac{3}{2}\right)=0\qquad \therefore\ \lim_{x\to 0}\frac{\left[\frac{3}{2}+x\right]-\left[\frac{3}{2}\right]}{x}=0$$

(2) When $f'(x)=10,\ \lim\limits_{n\to\infty}n\left\{f\left(\frac{2}{n}\right)-f(0)\right\}$

Let $\dfrac{2}{n}=x$ Then $n=\dfrac{2}{x}$

Since $x\to 0$ as $n\to\infty$,

$$\lim_{n\to\infty}n\left\{f\left(\frac{2}{n}\right)-f(0)\right\}=\lim_{x\to 0}\frac{2}{x}\{f(x)-f(0)\}=2\lim_{x\to 0}\frac{f(x)-f(0)}{x}=2f'(0)=2\cdot 10=20$$

(3) When $f(1)=1$ **and** $f'(1)=2,$ i) $\lim\limits_{x\to 1}\dfrac{x^3f(1)-f(x^2)}{x-1}$ ii) $\lim\limits_{x\to 1}\dfrac{xf(x)-1}{x^2-1}$

i) $$\lim_{x\to 1}\frac{x^3f(1)-f(x^2)}{x-1}=\lim_{x\to 1}\left\{\frac{(x^3-1)f(1)}{x-1}-\frac{f(x^2)-f(1)}{x-1}\right\}$$

$$=\lim_{x\to 1}\left\{(x^2+x+1)f(1)-\frac{f(x^2)-f(1)}{x^2-1}(x+1)\right\}$$

$$=3f(1)-f'(1)(1+1)=3\cdot 1-2\cdot 2=-1$$

ii) $$\lim_{x\to 1}\frac{xf(x)-1}{x^2-1}=\lim_{x\to 1}\left\{\frac{xf(x)-1}{x-1}\cdot\frac{1}{x+1}\right\}=\lim_{x\to 1}\left\{\frac{xf(x)-xf(1)+xf(1)-1}{x-1}\cdot\frac{1}{x+1}\right\}$$

$$=\lim_{x\to 1}\left[\left\{x\left(\frac{f(x)-f(1)}{x-1}\right)+\left(\frac{x\cdot 1-1}{x-1}\right)\right\}\cdot\frac{1}{x+1}\right]=\{1\cdot f'(1)+1\}\cdot\frac{1}{2}=\frac{3}{2}$$

(4) When $f'(0)=3,\ \lim\limits_{x\to 0}\dfrac{f(3x)-f(\sin x)}{x}$

$$\lim_{x\to 0}\frac{f(3x)-f(\sin x)}{x}=\lim_{x\to 0}\frac{f(3x)-f(0)-\{f(\sin x)-f(0)\}}{x}=\lim_{x\to 0}\left\{\frac{f(3x)-f(0)}{x}-\frac{f(\sin x)-f(0)}{x}\right\}$$

$$=\lim_{x\to 0}\left\{\frac{f(3x)-f(0)}{3x}\cdot 3-\frac{f(\sin x)-f(0)}{\sin x}\cdot\frac{\sin x}{x}\right\}=3f'(0)-f'(0)\cdot 1=3\cdot 3-3\cdot 1=6$$

(5) When $f(1) = \frac{1}{2}$ and $f'(1) = 4$, $\lim\limits_{x \to 1} \frac{f(x) - x^2 f(1)}{\sin(x-1)}$

$$\lim_{x \to 1} \frac{f(x) - x^2 f(1)}{\sin(x-1)} = \lim_{x \to 1} \frac{f(x) - f(1) + f(1) - x^2 f(1)}{\sin(x-1)} = \lim_{x \to 1} \left\{ \frac{f(x) - f(1) - f(1)\{x^2 - 1\}}{x-1} \cdot \frac{x-1}{\sin(x-1)} \right\}$$

$$= \lim_{x \to 1} \left[\left\{ \frac{f(x) - f(1)}{x-1} - f(1)(x+1) \right\} \cdot \frac{x-1}{\sin(x-1)} \right] = \left\{ f'(1) - \frac{1}{2} \cdot 2 \right\} \cdot 1 = (4-1) \cdot 1 = 3$$

(6) When $f(x) = \frac{1}{x}$, $\lim\limits_{x \to 1} \frac{f'(x) + 1}{x-1}$

$$f(x) = \frac{1}{x} \ \Rightarrow \ f'(x) = -\frac{1}{x^2}$$

$$\lim_{x \to 1} \frac{f'(x)+1}{x-1} = \lim_{x \to 1} \frac{-\frac{1}{x^2}+1}{x-1} = \lim_{x \to 1} \frac{\frac{-1+x^2}{x^2}}{x-1} = \lim_{x \to 1} \frac{x^2-1}{x^2(x-1)} = \lim_{x \to 1} \frac{(x+1)(x-1)}{x^2(x-1)} = \lim_{x \to 1} \frac{x+1}{x^2} = \frac{2}{1} = 2$$

(7) When $f(x) = x^4 - 2x^3 + x + 3$, $\lim\limits_{x \to 0} \frac{f(1+x) - f(1-x)}{x}$

$$\lim_{x \to 0} \frac{f(1+x) - f(1-x)}{x} = \lim_{x \to 0} \frac{f(1+x) - f(1) + f(1) - f(1-x)}{x} = \lim_{x \to 0} \left\{ \frac{f(1+x) - f(1)}{x} - \frac{f(1-x) - f(1)}{-x}(-1) \right\}$$

$$= f'(1) - f'(1)(-1) = 2f'(1)$$

$$f(x) = x^4 - 2x^3 + x + 3 \ \Rightarrow \ f'(x) = 4x^3 - 6x^2 + 1 \quad \therefore \ f'(1) = 4 \cdot 1^3 - 6 \cdot 1^2 + 1 = -1$$

$$\therefore \ \lim_{x \to 0} \frac{f(1+x) - f(1-x)}{x} = 2f'(1) = 2(-1) = -2$$

(8) For a function $f(x)$ such that i) $\lim\limits_{x \to 0} \frac{f(x)}{x} = \frac{1}{2}$ and ii) $f'(1) = 3$, $\lim\limits_{x \to 1} \frac{f(x) - f(1)}{f(x-1)}$

Let $x - 1 = h$ \qquad Then, $h \to 0$ as $x \to 1$

$$\lim_{x \to 1} \frac{f(x) - f(1)}{f(x-1)} = \lim_{h \to 0} \frac{f(1+h) - f(1)}{f(h)} = \lim_{h \to 0} \frac{f(1+h) - f(1)}{h} \cdot \frac{h}{f(h)} = f'(1) \lim_{h \to 0} \frac{h}{f(h)} = f'(1) \lim_{h \to 0} \frac{1}{\frac{f(h)}{h}}$$

$$= f'(1) \cdot \frac{1}{\frac{1}{2}} = 2f'(1) = 2 \cdot 3 = 6$$

#10 Find the derivative of each of the following functions.

(1) $f(x) = \sqrt{2x+1}$

$$f(x+h) - f(x) = \sqrt{2(x+h)+1} - \sqrt{2x+1} = \frac{\left(\sqrt{2(x+h)+1} - \sqrt{2x+1}\right)\left(\sqrt{2(x+h)+1} + \sqrt{2x+1}\right)}{\sqrt{2(x+h)+1} + \sqrt{2x+1}}$$

$$= \frac{2(x+h)+1 - (2x+1)}{\sqrt{2(x+h)+1} + \sqrt{2x+1}} = \frac{2h}{\sqrt{2(x+h)+1} + \sqrt{2x+1}}$$

$$f'(x) = \lim_{h \to 0} \frac{f(x+h) - f(x)}{h} = \lim_{h \to 0} \frac{2}{\sqrt{2(x+h)+1} + \sqrt{2x+1}} = \frac{2}{\sqrt{2(x+0)+1} + \sqrt{2x+1}} = \frac{2}{2\sqrt{2x+1}} = \frac{1}{\sqrt{2x+1}}$$

Another Approach:

$$f(x) = \sqrt{2x+1} = (2x+1)^{\frac{1}{2}}$$

$$f'(x) = \frac{1}{2}(2x+1)^{\frac{1}{2}-1}(2x+1)' = \frac{1}{2}(2x+1)^{-\frac{1}{2}} \cdot 2 = \frac{1}{\sqrt{2x+1}}$$

(2) $f(x) = \sqrt[3]{(x+1)(x^2+1)}$

$f^3(x) = (x+1)(x^2+1)$

$\dfrac{d}{dx} f^3(x) = \dfrac{d}{dx}\{(x+1)(x^2+1)\}$

$\qquad = (x+1)'(x^2+1) + (x+1)(x^2+1)' = 1(x^2+1) + (x+1)(2x) = 3x^2 + 2x + 1$

$\therefore\ 3f^2(x)f'(x) = 3x^2 + 2x + 1$

$\therefore\ f'(x) = \dfrac{3x^2+2x+1}{3f^2(x)} = \dfrac{3x^2+2x+1}{3\sqrt[3]{(x+1)^2(x^2+1)^2}}$

(3) $f(x) = \dfrac{2x^3+x-1}{x^2}$

$f(x) = \dfrac{2x^3+x-1}{x^2} = 2x + \dfrac{1}{x} - \dfrac{1}{x^2} = 2x + x^{-1} - x^{-2}$

$f'(x) = 2 + (-1)x^{-2} - (-2)x^{-3} = 2 - x^{-2} + 2x^{-3} = \dfrac{2x^3-x+2}{x^3}$

(4) $f(x) = \dfrac{1-x}{x^2+3}$

$f'(x) = \dfrac{(1-x)'(x^2+3) - (1-x)(x^2+3)'}{(x^2+3)^2} = \dfrac{(-1)(x^2+3) - (1-x)(2x)}{(x^2+3)^2} = \dfrac{x^2-2x-3}{(x^2+3)^2}$

(5) $f(x) = (x+1)^3(x^2-1)^2$

$f'(x) = 3(x+1)^2 \cdot (x^2-1)^2 + (x+1)^3 \cdot 2(x^2-1)^1 \cdot 2x$

$\qquad = 3(x+1)^2 \cdot (x^2-1)^2 + 4x(x+1)^3(x^2-1) = (x+1)^2(x^2-1)\{3(x^2-1) + 4x(x+1)\}$

$\qquad = (x+1)^3(x^2-1)\{3(x-1) + 4x\} = (x+1)^3(x^2-1)(7x-3)$

(6) $f(x) = \left(\dfrac{x}{x^2+1}\right)^3$

$f'(x) = 3\left(\dfrac{x}{x^2+1}\right)^2\left(\dfrac{x}{x^2+1}\right)' = 3\left(\dfrac{x}{x^2+1}\right)^2\left(\dfrac{1(x^2+1)-x(x^2+1)'}{(x^2+1)^2}\right) = 3\left(\dfrac{x}{x^2+1}\right)^2\left(\dfrac{x^2+1-x(2x)}{(x^2+1)^2}\right)$

$\qquad = 3\left(\dfrac{x}{x^2+1}\right)^2\left(\dfrac{-x^2+1}{(x^2+1)^2}\right) = \dfrac{3x^2}{(x^2+1)^2} \cdot \dfrac{-x^2+1}{(x^2+1)^2} = \dfrac{3x^2(-x^2+1)}{(x^2+1)^4}$

(7) $f(x) = \sqrt{\sqrt{x}+1}$

$f'(x) = \dfrac{(\sqrt{x}+1)'}{2\sqrt{\sqrt{x}+1}} = \dfrac{\frac{1}{2\sqrt{x}}}{2\sqrt{\sqrt{x}+1}} = \dfrac{1}{2\sqrt{x}\,(2\sqrt{\sqrt{x}+1})} = \dfrac{1}{4\sqrt{x}\,(\sqrt{\sqrt{x}+1})}$

(8) $f(x) = (x+3)\sqrt{x-4}$

$f'(x) = (x+3)'(\sqrt{x-4}) + (x+3)(\sqrt{x-4})' = \sqrt{x-4} + (x+3)\dfrac{1}{2\sqrt{x-4}}$

$\qquad = \dfrac{2(x-4)+(x+3)}{2\sqrt{x-4}} = \dfrac{3x-5}{2\sqrt{x-4}}$

(9) $f(x) = \frac{3-2x}{\sqrt{x^2+1}}$

$$f'(x) = \frac{(3-2x)'\left(\sqrt{x^2+1}\right)-(3-2x)\left(\sqrt{x^2+1}\right)'}{\left(\sqrt{x^2+1}\right)^2} = \frac{-2\left(\sqrt{x^2+1}\right)-(3-2x)\frac{2x}{2\sqrt{x^2+1}}}{\left(\sqrt{x^2+1}\right)^2} = \frac{\frac{-4\left(x^2+1\right)-2x(3-2x)}{2\sqrt{x^2+1}}}{\left(\sqrt{x^2+1}\right)^2}$$

$$= \frac{-4x^2-4+4x^2-6x}{2\sqrt{x^2+1}(\sqrt{x^2+1})^2} = \frac{-4-6x}{2(x^2+1)\sqrt{x^2+1}} = \frac{-3x-2}{(x^2+1)\sqrt{x^2+1}}$$

#11 Find the values of the constants a and b.

(1) For a function $f(x) = x^2 + ax + b$, $xf'(x) - 4f(x) + 2x^2 - 8 = 0$ is always true.

Find the values of the constants a and b.

Since $f'(x) = 2x + a$, $xf'(x) = 2x^2 + ax$

$\therefore xf'(x) - 4f(x) + 2x^2 - 8 = 2x^2 + ax - 4(x^2 + ax + b) + 2x^2 - 8 = -3ax - 4b - 8 = 0$

Since the equation is always true, $-3a = 0$ and $-4b - 8 = 0$

$\therefore a = 0$, $b = -2$

(2) For a function $f(x) = x^4 + ax^2 + bx$, $\lim\limits_{x \to 2} \frac{f(x)-f(2)}{x-2} = 10$ and $\lim\limits_{x \to 1} \frac{f(x)-f(1)}{x^2-1} = -2$.

Find the values of the constants a and b.

$\lim\limits_{x \to 2} \frac{f(x)-f(2)}{x-2} = f'(2) = 10$

$\lim\limits_{x \to 1} \frac{f(x)-f(1)}{x^2-1} = \lim\limits_{x \to 1} \frac{f(x)-f(1)}{x-1} \cdot \frac{1}{x+1} = f'(1) \cdot \frac{1}{2} = -2$; $f'(1) = -4$

$f(x) = x^4 + ax^2 + bx \Rightarrow f'(x) = 4x^3 + 2ax + b$

$\therefore f'(2) = 4 \cdot 8 + 2a \cdot 2 + b = 32 + 4a + b = 10$; $4a + b = -22$ ······①

$f'(1) = 4 \cdot 1 + 2a \cdot 1 + b = 4 + 2a + b = -4$; $2a + b = -8$ ······②

By ① and ②, $2a = -14$; $a = -7$, $b = 6$

(3) When a polynomial $x^5 - 5x + a$ is divided by $(x-b)^2$, there is no remainder. ($a > 0$,

$b > 0$) Find the values of the constants a and b.

Note: If a polynomial $f(x)$ is divided by $(x-k)^2$ with no remainder,

then $f(k) = 0$ and $f'(k) = 0$

Let $Q(x)$ be the quotient. Then, $x^5 - 5x + a = Q(x)(x-b)^2$ ······①

Substituting $x = b$ into ①, $b^5 - 5b + a = 0$ ······②

From ①, $\frac{d}{dx}\{x^5 - 5x + a\} = \frac{d}{dx}\{Q(x)(x-b)^2\}$

$\therefore 5x^4 - 5 = Q'(x)(x-b)^2 + Q(x) \cdot 2(x-b)$ ······③

Substituting $x = b$ into ③, $5b^4 - 5 = 0$; $b^4 = 1$

Since $b > 0$, $b = 1$ $\qquad$ By ②, $1 - 5 + a = 0$; $a = 4$ $\qquad$ $\therefore a = 4$, $b = 1$

(4) For a graph of the function $f(x) = ax^2 + b$, the slope of the tangent line at $(1, 2)$ is -3. Find the values of the constants a and b.

Since the graph of the function passes through the point $(1, 2)$, $2 = a + b$; $b = 2 - a$

Since the slope of the tangent line at $x = 1$ is -3, $f'(1) = -3$

$$f'(x) = \lim_{x \to 1} \frac{f(x) - f(1)}{x - 1} = \lim_{x \to 1} \frac{ax^2 + b - (a + b)}{x - 1} = \lim_{x \to 1} \frac{ax^2 - a}{x - 1} = \lim_{x \to 1} \frac{a(x + 1)(x - 1)}{x - 1}$$

$$= \lim_{x \to 1} a(x + 1) = a(1 + 1) = 2a$$

$\therefore$ $2a = -3$; $a = -\dfrac{3}{2}$, $b = 2 - a = 2 + \dfrac{3}{2} = \dfrac{7}{2}$

$\therefore$ $a = -\dfrac{3}{2}$, $b = \dfrac{7}{2}$

(5) For a polynomial $f(x)$ such that $\lim\limits_{h \to 0} \dfrac{f(4+h) + a}{h} = b$, θ is the angle between the tangent line at $(4, -3)$ on the graph of $y = f(x)$ and x-axis. When $\tan \theta = \dfrac{1}{3}$, find the values of a and b.

Since $\lim\limits_{h \to 0} \dfrac{f(4+h) + a}{h} = b$ (i.e., the limit exists) and $\lim\limits_{h \to 0} h = 0$, $\lim\limits_{h \to 0} f(4 + h) + a = 0$

$\therefore$ $f(4) + a = 0$; $a = -f(4)$

Since $(4, -3)$ lies on the graph of $f(x)$, $f(4) = -3$ $\quad \therefore$ $a = 3$

Since $\tan \theta$ is the slope of the tangent line at $(4, -3)$, $\tan \theta = f'(4)$ $\quad \therefore$ $f'(4) = \dfrac{1}{3}$

$\therefore$ $f'(4) = \lim\limits_{h \to 0} \dfrac{f(4+h) - f(4)}{h} = \lim\limits_{h \to 0} \dfrac{f(4+h) + a}{h} = \lim\limits_{h \to 0} b = b = \dfrac{1}{3}$

$\therefore$ $a = 3$, $b = \dfrac{1}{3}$

(6) When a function $f(x) = \begin{cases} x^3 + ax^2 + bx, & x \geq 1 \\ 2x^2 + 1, & x < 1 \end{cases}$ is differentiable at any real number x, find the values of the constants a and b.

Since f is differentiable at any real number x, f is continuous at $x = 1$.

$\therefore$ $\lim\limits_{x \to 1} f(x) = f(1)$ $\quad$ That is, $\lim\limits_{x \to 1^+} f(x) = \lim\limits_{x \to 1^-} f(x) = f(1)$

Since $\lim\limits_{x \to 1^+} f(x) = \lim\limits_{x \to 1^+} (x^3 + ax^2 + bx) = 1 + a + b \cdots\cdots$① and

$\qquad \lim\limits_{x \to 1^-} f(x) = \lim\limits_{x \to 1^-} (2x^2 + 1) = 2 + 1 = 3 \cdots\cdots$②, $\quad$ by ① and ②, $a + b = 2 \cdots\cdots$③

Since f is differentiable at $x = 1$, $\lim\limits_{h \to 0^+} \dfrac{f(1+h) - f(1)}{h} = \lim\limits_{h \to 0^-} \dfrac{f(1+h) - f(1)}{h}$

$$\lim_{h \to 0^+} \frac{f(1+h) - f(1)}{h} = \lim_{h \to 0^+} \frac{\{(1+h)^3 + a(1+h)^2 + b(1+h)\} - (1 + a + b)}{h}$$

$$= \lim_{h \to 0^+} \frac{1 + 3h + 3h^2 + h^3 + a + 2ah + ah^2 + b + bh - 1 - a - b}{h} = \lim_{h \to 0^+} \frac{h^3 + h^2(3 + a) + h(3 + 2a + b)}{h}$$

$$= \lim_{h \to 0^+} (h^2 + h(3 + a) + 3 + 2a + b)$$

$$= 3 + 2a + b \quad \cdots\cdots ④$$

$$\lim_{h \to 0} \frac{f(1+h)-f(1)}{h} = \lim_{h \to 0} \frac{\{2(1+h)^2+1\}-3}{h} = \lim_{h \to 0} \frac{2h^2+4h}{h} = \lim_{h \to 0^-} (2h + 4) = 4 \quad \cdots\cdots ⑤$$

∴ By ④ and ⑤, $3 + 2a + b = 4$; $2a + b = 1$ $\cdots\cdots ⑥$

Therefore, by ③ and ⑥, $a = -1$, $b = 3$

#12 Compute $\frac{dy}{dx}$.

(1) $y^2 = 3x$

$\frac{d}{dx}(y^2) = \frac{d}{dx}(3x)$; $(2y)\frac{dy}{dx} = 3$ $\quad$ ∴ $\frac{dy}{dx} = \frac{3}{2y}$ $(y \neq 0)$

(2) $y^3 = x^2$

$\frac{d}{dx}(y^3) = \frac{d}{dx}(x^2)$; $(3y^2)\frac{dy}{dx} = 2x$ $\quad$ ∴ $\frac{dy}{dx} = \frac{2x}{3y^2}$ $(y \neq 0)$

(3) $\sqrt{x} + \sqrt{y} = 2$

$\frac{d}{dx}(\sqrt{x} + \sqrt{y}) = \frac{d}{dx}(2)$; $\frac{1}{2\sqrt{x}} + \frac{1}{2\sqrt{y}}\frac{dy}{dx} = 0$ $\quad$ ∴ $\frac{dy}{dx} = -\frac{\sqrt{y}}{\sqrt{x}}$ $(x \neq 0)$

(4) $x^3 + y^3 - 3xy = 0$

$\frac{d}{dx}(x^3 + y^3 - 3xy) = \frac{d}{dx}(0)$; $\left\{3x^2 + 3y^2\frac{dy}{dx} - 3\left(y + x\frac{dy}{dx}\right)\right\} = 0$

∴ $\frac{dy}{dx}(3y^2 - 3x) = -3x^2 + 3y$ $\quad$ ∴ $\frac{dy}{dx} = \frac{-3x^2+3y}{3y^2-3x} = \frac{-x^2+y}{y^2-x}$ $(y^2 \neq x)$

(5) $x = y\sqrt{1 + y}$

$\frac{d}{dy}(x) = \frac{d}{dy}(y\sqrt{1 + y})$

∴ $\frac{dx}{dy} = 1 \cdot \sqrt{1 + y} + y \cdot \frac{1}{2\sqrt{y+1}} = \frac{2(y+1)+y}{2\sqrt{y+1}} = \frac{3y+2}{2\sqrt{y+1}}$

∴ $\frac{dy}{dx} = \frac{1}{\frac{dx}{dy}} = \frac{2\sqrt{y+1}}{3y+2}$

(6) $x = 2t - 2, \ y = t^2$

$\frac{d}{dt}(x) = \frac{d}{dt}(2t - 2)$; $\frac{dx}{dt} = 2$

$\frac{d}{dt}(y) = \frac{d}{dt}(t^2)$; $\frac{dy}{dt} = 2t$

∴ $\frac{dy}{dx} = \frac{dy}{dt} \cdot \frac{dt}{dx} = 2t \cdot \frac{1}{2} = t = \frac{x+2}{2}$

#13 Find the value.

(1) When the graph of a function $y = f(x)$ is shown

as the Figure, the equation of the tangent line at $P(a, 5)$

on the graph is $y = 3x + b$. Find the value of the limit:

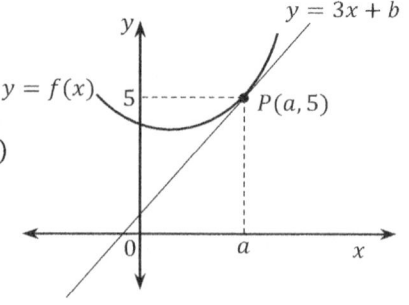

$$\lim_{n \to \infty} n \left\{ f\left(a + \frac{3}{n}\right) - f(a) \right\} \cdot f\left(a + \frac{2}{n}\right)$$

Since $P(a, 5)$ lies on the graph of the function $y = f(x)$, $f(a) = 5$.

Since the slope of the tangent line at $x = a$ is 3, $f'(a) = 3$.

Let $\dfrac{3}{n} = h$ Then, $h \to 0$ as $n \to \infty$

$$\lim_{n \to \infty} n \left\{ f\left(a + \frac{3}{n}\right) - f(a) \right\} \cdot f\left(a + \frac{2}{n}\right) = \lim_{h \to 0} \frac{3}{h} \{f(a + h) - f(a)\} \cdot f\left(a + \frac{2}{3}h\right)$$

$$= 3 \lim_{h \to 0} \left\{ \frac{f(a+h) - f(a)}{h} \right\} \cdot \lim_{h \to 0} f\left(a + \frac{2}{3}h\right) = 3f'(a) \cdot f(a + 0)$$

$$= 3 \cdot 3 \cdot f(a) = 3 \cdot 3 \cdot 5 = 45$$

(2) When a polynomial $x^5 - 2x + 4$ is divided by $(x - 1)^2$, find the remainder.

Let $Q(x)$ and $R(x)$ be the qotient and remainder, respectively.

Since the polynomial is divided by degree of 2, $R(x)$ will be a linear form; i.e., $R(x) = ax + b$.

Thus, $x^5 - 2x + 4 = Q(x)(x - 1)^2 + ax + b$ $\cdots\cdots$ ①

Substituting $x = 1$ into ①, $1 - 2 + 4 = a + b$ $\therefore$ $a + b = 3$

From ①, $\dfrac{d}{dx}(x^5 - 2x + 4) = \dfrac{d}{dx}\{Q(x)(x - 1)^2 + ax + b\}$

$\therefore$ $5x^4 - 2 = Q'(x)(x - 1)^2 + Q(x) \cdot 2(x - 1) + a$ $\cdots\cdots$ ②

Substituting $x = 1$ into ②, $5 - 2 = a$ $\therefore$ $a = 3$

$\therefore$ $a = 3$, $b = 0$

Therefore, the remainder is $3x + 0 = 3x$.

(3) When a function $f(-x) = f(x)$ is differentiable at $x = 0$, find the value of $f'(0)$.

$$f'(0) = \lim_{h \to 0} \frac{f(0+h) - f(0)}{h} = \lim_{h \to 0} \frac{f(0-h) - f(0)}{h} = \lim_{h \to 0} \frac{f(0-h) - f(0)}{-h} \cdot (-1) = -f'(0)$$

$\therefore$ $2f'(0) = 0$ $\therefore$ $f'(0) = 0$

(4) For any real number x, a polynomial function $f(x)$ satisfies $f(-x) = f(x)$ and

$$\lim_{h \to 0} \frac{f(3 - 2h) - f(3 + 3h)}{h} = 10. \text{ Find the value of } f'(-3).$$

$$\lim_{h \to 0} \frac{f(3 - 2h) - f(3 + 3h)}{h} = \lim_{h \to 0} \frac{f(3 - 2h) - f(3) + f(3) - f(3 + 3h)}{h} = \lim_{h \to 0} \frac{f(3 - 2h) - f(3)}{h} - \frac{f(3 + 3h) - f(3)}{h}$$

$$= \lim_{h \to 0} \frac{f(3 - 2h) - f(3)}{-2h} \cdot (-2) - \lim_{h \to 0} \frac{f(3 + 3h) - f(3)}{3h} \cdot 3$$

$$= -2f'(3) - 3f'(3) = -5f'(3) = 10 \qquad \therefore f'(3) = -2$$

$$\therefore \; f'(-3) = \lim_{h\to 0}\frac{f(-3+h)-f(-3)}{h} = \lim_{h\to 0}\frac{f(3-h)-f(3)}{h} \quad (\because \; f(-x)=f(x))$$

$$= \lim_{h\to 0}\frac{f(3-h)-f(3)}{-h}\cdot(-1) = -f'(3) = -(-2) = 2$$

(5) For a function $f(x)$ with degree 4 and leading coefficient 1, $f(x)$ satisfies:

(i) $f(x) = f(-x)$ for any real number x and

(ii) $\lim\limits_{h\to 1}\dfrac{f(2h)+1}{h-1} = 10.$

Find the value of $f'(-2)$.

Since $\lim\limits_{h\to 1}\dfrac{f(2h)+1}{h-1} = 10$, the limit exists and $\lim\limits_{h\to 1}(h-1) = 0$

Thus, $\lim\limits_{h\to 1}\{f(2h)+1\} = 0$

$\therefore \; f(2\cdot 1) + 1 = 0 ; \; f(2) = -1 \; ; \; 1 = -f(2)$

$$\lim_{h\to 1}\frac{f(2h)+1}{h-1} = \lim_{h\to 1}\frac{f(2h)-f(2)}{2h-2}\cdot 2 = 2f'(2) = 10 \quad \therefore \; f'(2) = 5$$

$$f'(-2) = \lim_{h\to 1}\frac{f(-2+h)-f(-2)}{h} = \lim_{h\to 1}\frac{f(2-h)-f(2)}{h} = \lim_{h\to 1}\frac{f(2-h)-f(2)}{-h}\cdot(-1) = -f'(2) = -5$$

(6) For a function defined by $f(x) = \begin{cases} 2\sin x + x^2\cos\frac{1}{x^2}, & x\neq 0 \\ 0, & x = 0 \end{cases}$, find the value of $f'(0)$.

$$f'(0) = \lim_{h\to 0}\frac{f(0+h)-f(0)}{h} = \lim_{h\to 0}\frac{f(h)-f(0)}{h} = \lim_{h\to 0}\frac{2\sin h + h^2\cos\frac{1}{h^2}-0}{h} = \lim_{h\to 0}\left(\frac{2\sin h}{h} + \frac{h^2\cos\frac{1}{h^2}}{h}\right)$$

$$= 2\lim_{h\to 0}\left(\frac{\sin h}{h}\right) + \lim_{h\to 0}\left(h\cos\frac{1}{h^2}\right) = 2\cdot 1 + 0 = 2$$

(7) When $f(x) = (x+\sqrt{1+x^2})^{10}$, find the value of $f'(1)f'(-1)$.

$$f'(x) = 10(x+\sqrt{1+x^2})^9\left(x+\sqrt{1+x^2}\right)' = 10(x+\sqrt{1+x^2})^9\left(1+\frac{(1+x^2)'}{2\sqrt{1+x^2}}\right)$$

$$= 10(x+\sqrt{1+x^2})^9\left(1+\frac{2x}{2\sqrt{1+x^2}}\right) = 10(x+\sqrt{1+x^2})^9\left(\frac{\sqrt{1+x^2}+x}{\sqrt{1+x^2}}\right)$$

$$= \frac{10\left(x+\sqrt{1+x^2}\right)^{10}}{\sqrt{1+x^2}}$$

(8) When a function $f(x)$ is a polynomial for any x, $f(x) = g'(x)$ and $\{f(x)+g(x)\}' = x^3 + 3x^2 + 4x + 5$. Find the value of $f(-1)$.

Note that: $\{f(x)+g(x)\}' = f'(x) + g'(x)$

Since $f(x) = g'(x)$, $\{f(x)+g(x)\}' = f'(x) + f(x) = x^3 + 3x^2 + 4x + 5$

Since the degree of $f(x)$ is greater than the degree of $f'(x)$, $f(x)$ is a polynomial with degree 3 and leading coefficient 1.

Let $f(x) = x^3 + ax^2 + bx + c$. Then, $f'(x) = 3x^2 + 2ax + b$.

$\therefore \; f'(x) + f(x) = x^3 + (3+a)x^2 + (2a+b)x + b + c = x^3 + 3x^2 + 4x + 5$

$\therefore$ $3 + a = 3$, $2a + b = 4$, $b + c = 5$

$\therefore$ $a = 0$, $b = 4$, $c = 1$ $\qquad \therefore f(x) = x^3 + 4x + 1$

Therefore, $f(-1) = -1 - 4 + 1 = -4$

(9) For a composite function $F(x) = f\big(g(x)\big)$ at which $f'(x) = \frac{1}{x^2+1}$, $g'(x) = \frac{1}{x^4+1}$, and $g(0) = 2$, find the value of $F'(0)$.

Since $F'(x) = f'\big(g(x)\big)g'(x)$,

$$F'(0) = f'\big(g(0)\big)g'(0) = f'(2) \cdot \frac{1}{0^4+1} = f'(2) = \frac{1}{2^2+1} = \frac{1}{5}$$

(10) Find the value of positive integer n such that $\lim\limits_{x \to 1} \dfrac{x^{2n}+x^n-2}{x-1} = 15$.

$$\lim_{x \to 1} \frac{x^{2n}+x^n-2}{x-1} = \lim_{x \to 1} \frac{(x^n-1)(x^n+2)}{x-1} = \lim_{x \to 1} \frac{(x-1)(x^{n-1}+x^{n-2}+\cdots\cdots+x+1)(x^n+2)}{x-1}$$

$$= \lim_{x \to 1}(x^{n-1} + x^{n-2} + \cdots\cdots + x + 1)(x^n + 2) = n(1 + 2) = 3n = 15$$

$\therefore$ $n = 5$

Another approach:

Let $f(x) = x^{2n} + x^n$

Then, $f(1) = 1 + 1 = 2$

$\therefore$ $\lim\limits_{x \to 1} \dfrac{x^{2n}+x^n-2}{x-1} = \lim\limits_{x \to 1} \dfrac{f(x)-f(1)}{x-1} = f'(1) = 15$

Since $f'(x) = 2nx^{2n-1} + nx^{n-1}$, $f'(1) = 2n + n = 3n = 15$ $\qquad \therefore n = 5$

(11) For a differentiable function $f(x)$ such that $f(x + y) = f(x) + f(y) + 2xy$, the tangent line at $x = 0$ is the x-axis. Find the value of $f'(1)$.

Substituting $x = 0$ and $y = 0$, $f(0 + 0) = f(0) + f(0) + 2 \cdot 0 = 2f(0)$

$\therefore$ $f(0) = 2f(0)$; $f(0) = 0$

Since the tangent line at $x = 0$ is the x-axis, the slope of the line is 0; i.e., $f'(0) = 0$

$$f'(1) = \lim_{h \to 0} \frac{f(1+h)-f(1)}{h} = \lim_{h \to 0} \frac{f(1)+f(h)+2h-f(1)}{h} = \lim_{h \to 0} \frac{f(h)+2h}{h}$$

$$= \lim_{h \to 0} \left\{\frac{f(h)}{h} + 2\right\} = \lim_{h \to 0} \frac{f(0+h)-f(0)}{h} + 2 = f'(0) + 2 = 0 + 2 = 2$$

(12) A function $f(x)$ is differentiable and its inverse $f^{-1}(x)$ exists. When $f\big(3f^{-1}(x) - 4x\big) = x$ for any real number x, find the value of $f^{-1}(5)$.

$f\big(3f^{-1}(x) - 4x\big) = x$ $\Rightarrow$ $3f^{-1}(x) - 4x = f^{-1}(x)$; $2f^{-1}(x) = 4x$; $f^{-1}(x) = 2x$

$\therefore$ $f(2x) = x$

Let $2x = t$ then $x = \dfrac{t}{2}$

Since $f(2x) = x$, $f(t) = \dfrac{t}{2}$ $\qquad \therefore f(x) = \dfrac{x}{2}$

$$f'(x) = \lim_{h \to 0} \frac{f(x+h)-f(x)}{h} = \lim_{h \to 0} \frac{\frac{x+h}{2}-\frac{x}{2}}{h} = \lim_{h \to 0} \frac{\frac{h}{2}}{h} = \lim_{h \to 0} \frac{h}{2h} = \lim_{h \to 0} \frac{1}{2} = \frac{1}{2}$$

$$\therefore \ f^{-1}(5) = \frac{1}{2}$$

(13) A differentiable function $f(x)$ satisfies $f(x + y) = f(x)f(y)$ for any real numbers x and y. When $f'(0) = 4$ and $f(a) = 2$, find the value of $f'(3a)$. (a; constant)

$f(x + y) = f(x)f(y) \ \Rightarrow \ f(0 + 0) = f(0)f(0) \ ; \ f(0) = \{f(0)\}^2$

$\therefore \ f(0)\{f(0) - 1\} = 0 \ ; \ f(0) = 0$ or $f(0) = 1$

If $f(0) = 0$, then

$$f'(0) = \lim_{h \to 0} \frac{f(0+h)-f(0)}{h} = \lim_{h \to 0} \frac{f(0)f(h)-f(0)}{h} = f(0) \cdot \lim_{h \to 0} \frac{f(h)-1}{h} = 0 \cdot \lim_{h \to 0} \frac{f(h)-1}{h} = 0$$

But $f'(0) = 4$ is given. $\therefore \ f(0) \ne 0$.

$\therefore \ f(0) = 1$

$$f'(3a) = \lim_{h \to 0} \frac{f(3a+h)-f(3a)}{h} = \lim_{h \to 0} \frac{f(3a)f(h)-f(3a)}{h} = f(3a) \lim_{h \to 0} \frac{f(h)-1}{h}$$

Since $f(3a) = f(a + a + a) = f(a)f(a)f(a) = \{f(a)\}^3$ and $f(0) = 1$,

$$f'(3a) = f(3a) \lim_{h \to 0} \frac{f(h)-1}{h} = \{f(a)\}^3 \lim_{h \to 0} \frac{f(h)-f(0)}{h} = \{f(a)\}^3 f'(0) = 2^3 \cdot 4 = 32$$

(14) For two polynomials $f(x)$ and $g(x)$ such that

    (i) $f(0) = 1, \ f'(0) = -2, \ g(0) = 3$, and

    (ii) $\lim\limits_{x \to 0} \dfrac{f(x)g(x)-3}{x} = 0$,

    find the value of $g'(0)$.

Let $F(x) = f(x)g(x)$

Since $f(0) = 1$ and $g(0) = 3$, $F(0) = 1 \cdot 3 = 3$

$$\lim_{x \to 0} \frac{f(x)g(x)-3}{x} = \lim_{x \to 0} \frac{F(x)-F(0)}{x-0} = F'(0) = 0 \quad \text{by (ii)}$$

Since $F'(x) = f'(x)g(x) + f(x)g'(x)$,

$F'(0) = f'(0)g(0) + f(0)g'(0) = -2 \cdot 3 + 1 \cdot g'(0) = -6 + g'(0) = 0$

$\therefore \ g'(0) = 6$

(15) For any real number x, a polynomial $f(x)$ satisfies $(x^n - 2)f'(x) = f(x)$ and $f(4) = 5$, find the value of $f(10)$.

Let $f(x) = a_m x^m + a_{m-1} x^{m-1} + a_{m-2} x^{m-2} + \cdots\cdots + a_1 x + a_0 \quad (a_m \ne 0)$

Then, $f(x)$ has degree m.

Since $f'(x) = a_m \cdot m x^{m-1} + a_{m-1} \cdot (m - 1)x^{m-2} + a_{m-2} \cdot (m - 2)x^{m-3} + \cdots\cdots + a_1,$

$f'(x)$ has degree of $(m - 1)$.

Since $(x^n - 2)f'(x) = f(x)$,

$$(x^n - 2)\{a_m \cdot mx^{m-1} + a_{m-1} \cdot (m-1)x^{m-2} + a_{m-2} \cdot (m-2)x^{m-3} + \cdots\cdots + a_1\}$$

$$= a_m x^m + a_{m-1}x^{m-1} + a_{m-2}x^{m-2} + \cdots\cdots + a_1 x + a_0$$

Since the highest degrees of both sides are the same, $n + (m-1) = m \qquad \therefore n = 1$

Since the leading coefficients of both sides are the same, $ma_m = a_m$

$\therefore a_m(m-1) = 0 \qquad \therefore m = 1 \; (\because a_m \neq 0)$

$\therefore f(x)$ is the form $f(x) = ax + b \; (a \neq 0, \; a, b;\text{constants})$

$\therefore f'(x) = a$

$\therefore f(x) = (x^n - 2)f'(x) = (x-2)\cdot a \; ; \; f(4) = (4-2)\cdot a = 2a = 5 \qquad \therefore a = \dfrac{5}{2}$

$\therefore f(x) = \dfrac{5}{2}(x-2) \qquad \therefore f(10) = \dfrac{5}{2}(10-2) = 20$

(16) For a polynomial $f(x)$ such that

    (i) $f(1) = 5$, and

    (ii)   for any real number x, $(f \circ f)(x) = f(x)f'(x) + 3$,

    find the value of $f(10)$.

Let $f(x) = a_n x^n + a_{n-1}x^{n-1} + a_{n-2}x^{n-2} + \cdots\cdots + a_1 x + a_0 \; (a_n \neq 0)$ (degree of n)

Then, $f'(x) = a_n \cdot nx^{n-1} + a_{n-1}\cdot(n-1)x^{n-2} + a_{n-2}\cdot(n-2)x^{n-3} + \cdots\cdots + a_1$

$$\text{(degree of } n-1)$$

Note that $(f \circ f)(x) = f(x)f'(x) + 3 \;\cdots\cdots ①$

Since the highest degrees of both sides of ① are the same, $n^2 = n + (n-1)$

$\therefore n^2 - 2n + 1 = 0 \; ; \; (n-1)^2 = 0 \; ; \; n = 1$

$\therefore f(x)$ is the form $f(x) = ax + b \; (a \neq 0, \; a, b;\text{constants})$

By ①, $a(ax + b) + b = (ax + b)\cdot a + 3$

$\therefore a^2 x + ab + b = a^2 x + ab + 3 \qquad \therefore b = 3$

Since $f(1) = 5$ by (i), $f(1) = a\cdot 1 + b = a + 3 = 5 \qquad \therefore a = 2$

$\therefore f(x) = 2x + 3 \qquad\qquad \therefore f(10) = 23$

(17) For positive integers a and b,

    a function $f(x) = \lim\limits_{n \to \infty} \dfrac{ax^{n+b}+2x-1}{x^n + 1} \; (x > 0)$ is differentiable at $x = 1$.

    Find the value of $a - b$.

i) When $0 < x < 1$,

$$f(x) = \lim_{n \to \infty} \frac{ax^{n+b}+2x-1}{x^n+1} = 2x - 1$$

ii) When $x = 1$,

$$f(x) = \lim_{n \to \infty} \frac{ax^{n+b}+2x-1}{x^n+1} = \frac{a\cdot 1 + 2\cdot 1 - 1}{1+1} = \frac{a+1}{2}$$

iii) When $x > 1$,

$$f(x) = \lim_{n\to\infty} \frac{ax^{n+b}+2x-1}{x^n+1} = \lim_{n\to\infty} \frac{ax^b+\frac{2}{x^{n-1}}-\frac{1}{x^n}}{1+\frac{1}{x^n}} = ax^b$$

Since $f(x)$ is differentiable at $x = 1$, $f(x)$ is continuous at $x = 1$.

$\therefore \quad \lim_{x\to1^-} f(x) = \lim_{x\to1^+} f(x) = f(1) \quad ; \quad 2\cdot 1 - 1 = a\cdot 1^b = \frac{a+1}{2}$

$\therefore \quad 1 = a = \frac{a+1}{2} \quad ; \quad a = 1 \quad \cdots\cdots ①$

Since $f(x)$ is differentiable at $x = 1$, $\lim_{x\to1^-} \frac{f(x)-f(1)}{x-1} = \lim_{x\to1^+} \frac{f(x)-f(1)}{x-1}$

When $0 < x < 1$, $f'(x) = 2$

When $x > 1$, $f'(x) = abx^{b-1}$

$\therefore \quad 2 = ab\cdot 1^{b-1} \quad ; \quad 2 = ab \quad \cdots\cdots ②$

By ① and ②, $\quad a = 1,\ b = 2 \qquad \therefore a - b = -1$

(18) For any positive integer n, a differentiable function $f(x)$ satisfies $f(nx) = n^3 f(x)$

and $f'(1) = 20$. Find the value of $f'\left(\frac{1}{2}\right)$.

$$f(x) = f\left(n\cdot\frac{x}{n}\right) = n^3 f\left(\frac{x}{n}\right) \quad \therefore f\left(\frac{x}{n}\right) = \frac{1}{n^3}f(x)$$

$$f'\left(\frac{1}{2}\right) = \lim_{h\to0} \frac{f\left(\frac{1}{2}+h\right)-f\left(\frac{1}{2}\right)}{h} = \lim_{h\to0} \frac{f\left(\frac{1+2h}{2}\right)-f\left(\frac{1}{2}\right)}{h} = \lim_{h\to0} \frac{\frac{1}{2^3}f(1+2h)-\frac{1}{2^3}f(1)}{h} = \lim_{h\to0} \frac{f(1+2h)-f(1)}{2^3 h}$$

$$= \frac{1}{2^3}\lim_{h\to0} \frac{f(1+2h)-f(1)}{2h}\cdot 2 = \frac{1}{2^2}\lim_{h\to0} \frac{f(1+2h)-f(1)}{2h} = \frac{1}{4}f'(1) = \frac{1}{4}\cdot 20 = 5$$

#14 For a polynomial function $f(x)$ and positive integers m and n,

$$\lim_{x\to\infty} \frac{f(x)}{x^m} = 1, \quad \lim_{x\to\infty} \frac{f'(x)}{x^{m-1}} = a, \quad \lim_{x\to0} \frac{f(x)}{x^n} = b, \quad \text{and} \quad \lim_{x\to0} \frac{f'(x)}{x^{n-1}} = 10 \ (a, b; \text{real numbers})$$

State whether the following statements are true or false.

(1) $m \geq n$ **(2) $ab \geq 10$** **(3) If $f(x)$ has degree of 3, then $am = bn$.**

$\lim_{x\to\infty} \frac{f(x)}{x^m} = 1 \Rightarrow f(x)$ has degree of m and leading coefficient 1.

$\lim_{x\to0} \frac{f(x)}{x^n} = b \Rightarrow$ The lowest degree of $f(x)$ is n and the term's coefficient is b.

$\therefore \quad f(x) = 1\cdot x^m + \cdots\cdots + b\cdot x^n$

$\therefore \quad f'(x) = m\cdot x^{m-1} + \cdots\cdots + b\cdot nx^{n-1}$

$\lim_{x\to\infty} \frac{f'(x)}{x^{m-1}} = \lim_{x\to\infty} \frac{m\cdot x^{m-1}+\cdots\cdots+b\cdot nx^{n-1}}{x^{m-1}} = m = a$

$\lim_{x\to0} \frac{f'(x)}{x^{n-1}} = \lim_{x\to0} \frac{m\cdot x^{m-1}+\cdots\cdots+b\cdot nx^{n-1}}{x^{n-1}} = bn = 10 \quad ; \quad b = \frac{10}{n}$

(1) Since the highest degree of $f(x)$ is m and the lowest degree of it is n, $m \geq n$ (True)

(2) Since $ab = m \cdot \dfrac{10}{n} \geq n \cdot \dfrac{10}{n} = 10$, $ab \geq 10$ (True)

(3) If $f(x)$ has degree of 3, $m = 3$ $\therefore a = 3$ $\therefore am = 9$, $bn = \dfrac{10}{n} \cdot n = 10$

$\therefore am \neq bn$ (False)

#15 For a function $f(x) = \begin{cases} 1 - x & , x < 0 \\ x^2 - 1 & , 0 \leq x < 1 \\ \frac{2}{3}(x^3 - 1), & x \geq 1 \end{cases}$,

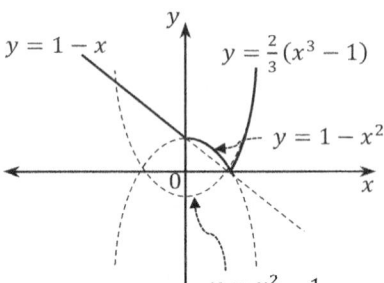

$y = 1 - x$ $y = \frac{2}{3}(x^3 - 1)$

$y = 1 - x^2$

$y = x^2 - 1$

state whether the following statements are true or false.

(1) $f(x)$ is differentiable at $x = 1$.

$$\lim_{x \to 1^-} \frac{f(x) - f(1)}{x - 1} = \lim_{x \to 1^-} \frac{(x^2 - 1) - 0}{x - 1} = \lim_{x \to 1^-} \frac{(x-1)(x+1)}{x - 1} = \lim_{x \to 1^-} (x + 1) = 1 + 1 = 2$$

$$\lim_{x \to 1^+} \frac{f(x) - f(1)}{x - 1} = \lim_{x \to 1^+} \frac{\frac{2}{3}(x^3 - 1) - 0}{x - 1} = \frac{2}{3}\lim_{x \to 1^+} \frac{(x-1)(x^2 + x + 1)}{x - 1} = \frac{2}{3}\lim_{x \to 1^+} (x^2 + x + 1) = \frac{2}{3}(1 + 1 + 1) = 2$$

$\therefore f'(1) = \lim_{x \to 1} \dfrac{f(x) - f(1)}{x - 1} = 2$

$\therefore f(x)$ is differentiable at $x = 1$.   (True)

(2) $|f(x)|$ is differentiable at $x = 0$.

$$\lim_{x \to 0^-} \frac{|f(x)| - |f(0)|}{x - 0} = \lim_{x \to 0^-} \frac{(1-x) - |-1|}{x} = \lim_{x \to 0^-} \frac{1 - x - 1}{x} = \lim_{x \to 0^-} \frac{-x}{x} = -1$$

$$\lim_{x \to 0^+} \frac{|f(x)| - |f(0)|}{x - 0} = \lim_{x \to 0^+} \frac{(1-x^2) - |-1|}{x} = \lim_{x \to 0^+} \frac{1 - x^2 - 1}{x} = \lim_{x \to 0^+} \frac{-x^2}{x} = \lim_{x \to 0^+} (-x) = 0$$

$\therefore \lim_{x \to 0^-} \dfrac{|f(x)| - |f(0)|}{x - 0} \neq \lim_{x \to 0^+} \dfrac{|f(x)| - |f(0)|}{x - 0}$

$\therefore |f(x)|'$ does not exist. ; $|f(x)|$ is not differentiable at $x = 0$.   (False)

(3) 2 is the minimum value of the positive integer k at which $x^k f(x)$ is differentiable at $x = 0$.

$$\lim_{h \to 0^-} \frac{h^k f(h)}{h} = \lim_{h \to 0^-} \frac{h^k(1 - h)}{h} = \lim_{h \to 0^-} h^{k-1}(1 - h)$$

$$\lim_{h \to 0^+} \frac{h^k f(h)}{h} = \lim_{h \to 0^+} \frac{h^k(h^2 - 1)}{h} = \lim_{h \to 0^+} h^{k-1}(h^2 - 1)$$

If $x^k f(x)$ is differentiable at $x = 0$, then $\lim_{h \to 0^-} \dfrac{h^k f(h)}{h} = \lim_{h \to 0^+} \dfrac{h^k f(h)}{h}$

$\therefore \lim_{h \to 0^-} h^{k-1}(1 - h) = \lim_{h \to 0^+} h^{k-1}(h^2 - 1)$

$\therefore k \geq 2$

$\therefore$ The minimum value of k is 2.   (True)

#16 Find the derivative of the function.

(1) $y = \sin\sqrt{1-x^2}$

$y' = (\cos\sqrt{1-x^2}) \cdot (\sqrt{1-x^2})' = (\cos\sqrt{1-x^2}) \cdot \dfrac{(1-x^2)'}{2\sqrt{1-x^2}} = (\cos\sqrt{1-x^2}) \cdot \dfrac{-2x}{2\sqrt{1-x^2}}$

$= \dfrac{-x(\cos\sqrt{1-x^2})}{\sqrt{1-x^2}}$

(2) $y = \cos(\sin x)$

$y' = \{-\sin(\sin x)\}(\sin x)' = \{-\sin(\sin x)\}\cos x = -(\cos x)\sin(\sin x)$

(3) $y = \cos x^o$

Since $\pi = 180^o$, $1^o = \dfrac{\pi}{180}$; $x^o = \dfrac{\pi}{180}x$

$\therefore\ y = \cos\dfrac{\pi}{180}x$

$\therefore\ y' = -\sin\dfrac{\pi}{180}x \cdot \left(\dfrac{\pi}{180}x\right)' = -\dfrac{\pi}{180}\sin\dfrac{\pi}{180}x = -\dfrac{\pi}{180}\sin x^o$

(4) $y = (2x^2 + 1)\sin 2x$

$y' = (2x^2 + 1)'\sin 2x + (2x^2 + 1)(\sin 2x)' = 4x \cdot \sin 2x + (2x^2 + 1)(\cos 2x)\cdot 2$

$= 4x\sin 2x + (4x^2 + 2)\cos 2x$

(5) $y = \sin^3 x\cos 3x$

$y' = (\sin^3 x)'\cos 3x + \sin^3 x(\cos 3x)' = 3\sin^2 x(\cos x)\cos 3x + \sin^3 x(-\sin 3x)\cdot 3$

$= 3\sin^2 x(\cos x\cos 3x - \sin x\sin 3x) = 3\sin^2 x\cos 4x$

(6) $y = (\sec x + \tan x)^3$

$y' = 3(\sec x + \tan x)^2(\sec x + \tan x)' = 3(\sec x + \tan x)^2(\sec x\tan x + \sec^2 x)$

$= 3(\sec x + \tan x)^2\sec x(\tan x + \sec x) = 3(\sec x + \tan x)^3\sec x$

(7) $y = \sin^2(2\pi x - a)$

$y' = 2\sin(2\pi x - a) \cdot (\sin(2\pi x - a))' = 2\sin(2\pi x - a) \cdot \cos(2\pi x - a)\cdot 2\pi$

$= \sin 2(2\pi x - a)\cdot 2\pi = 2\pi\sin 2(2\pi x - a)$

(8) $y = \sec^3(2x + 5)$

$y' = 3\sec^2(2x + 5)\,(\sec(2x + 5))' = 3\sec^2(2x + 5)\,\{\sec(2x + 5)\tan(2x + 5)\}(2x + 5)'$

$= 3\sec^2(2x + 5)\,\{\sec(2x + 5)\tan(2x + 5)\}\cdot 2 = 6\sec^3(2x + 5)\tan(2x + 5)$

(9) $y = 3e^x$

$y' = 3(e^x)' = 3e^x$

(10) $y = 2\cdot 3^x$

$y' = 2\cdot(3^x)' = 2\cdot 3^x\log 3 = 2(\log 3)\cdot 3^x$

(11) $y = xe^x$

$y' = (x)'e^x + x(e^x)' = e^x + xe^x = e^x(1 + x)$

(12) $y = e^x \sin x$

$y' = (e^x)' \sin x + e^x (\sin x)' = e^x \sin x + e^x \cos x = e^x (\sin x + \cos x)$

(13) $y = \frac{e^x}{e^x+1}$

$y' = \left(\frac{e^x}{e^x+1}\right)' = \frac{(e^x)'(e^x+1) - e^x(e^x+1)'}{(e^x+1)^2} = \frac{e^x(e^x+1) - e^x \cdot e^x}{(e^x+1)^2} = \frac{e^x}{(e^x+1)^2}$

(14) $y = \frac{x}{3^x}$

$y' = \frac{(x)'3^x - x(3^x)'}{(3^x)^2} = \frac{3^x - x(3^x \log 3)}{(3^x)^2} = \frac{1 - x \log 3}{3^x}$

(15) $y = (x^2 + e^x)^3$

$y' = 3(x^2 + e^x)^2(x^2 + e^x)' = 3(x^2 + e^x)^2(2x + e^x)$

(16) $y = \log x + x$

$y' = \frac{1}{x} + 1$

(17) $y = x \log x - x$

$y' = (x)' \log x + x(\log x)' - 1 = \log x + x \cdot \frac{1}{x} - 1 = \log x$

(18) $y = 3 \log_2 x - x^3$

$y' = 3 \frac{1}{x} \frac{1}{\log 2} - 3x^2 = \frac{3}{x \log 2} - 3x^2$

(19) $y = e^x \log x$

$y' = (e^x)' \log x + e^x (\log x)' = e^x \log x + e^x \cdot \frac{1}{x} = e^x \left(\log x + \frac{1}{x}\right)$

(20) $y = \frac{\log x}{e^x}$

$y' = \frac{(\log x)' e^x - \log x (e^x)'}{(e^x)^2} = \frac{\frac{1}{x}e^x - e^x \log x}{(e^x)^2} = \frac{1}{e^x}\left(\frac{1}{x} - \log x\right)$

(21) $y = (\log_2 x)^3$

$y' = 3(\log_2 x)^2(\log_2 x)' = 3(\log_2 x)^2\left(\frac{1}{x} \cdot \frac{1}{\log 2}\right) = \frac{3(\log_2 x)^2}{x \log 2}$

(22) $y = \log(\tan x + \sec x)$

$y' = \frac{1}{\tan x + \sec x}(\tan x + \sec x)' = \frac{1}{\tan x + \sec x}(\sec^2 x + \sec x \tan x)$

$\quad = \frac{\sec x}{\tan x + \sec x}(\sec x + \tan x) = \sec x$

(23) $y = \log(x + \sqrt{x^2 + 1})$

$y' = \frac{\left(x + \sqrt{x^2+1}\right)'}{x + \sqrt{x^2+1}} = \frac{1 + \frac{(x^2+1)'}{2\sqrt{x^2+1}}}{x + \sqrt{x^2+1}} = \frac{1 + \frac{2x}{2\sqrt{x^2+1}}}{x + \sqrt{x^2+1}} = \frac{\frac{\sqrt{x^2+1} + x}{\sqrt{x^2+1}}}{x + \sqrt{x^2+1}} = \frac{1}{\sqrt{x^2+1}}$

(24) $y = \log\frac{x-1}{x+1}$

$y = \log|x - 1| - \log|x + 1|$

$y' = \frac{(x-1)'}{x-1} - \frac{(x+1)'}{x+1} = \frac{1}{x-1} - \frac{1}{x+1} = \frac{x+1-(x-1)}{x^2-1} = \frac{2}{x^2-1}$

(25) $y = a^{\sin x}$

$y' = a^{\sin x} \log a \, (\sin x)' = (\log a) a^{\sin x} \cdot \cos x$

(26) $y = \sin 2^x$

$y' = \cos 2^x \cdot (2^x)' = \cos 2^x \cdot (2^x \cdot \log 2 \cdot x') = (\log 2) 2^x \cos 2^x$

(27) $y = e^{x^2} \sin x$

$y' = \left(e^{x^2}\right)' \sin x + e^{x^2}(\sin x)' = e^{x^2}(x^2)' \sin x + e^{x^2} \cos x = e^{x^2}(2x \sin x + \cos x)$

(28) $y = \log_{10}(x^2 + 1)$

$y' = \frac{1}{x^2+1} \cdot \frac{1}{\log 10}\left(x^2 + 1\right)' = \frac{2x}{(x^2+1)\log 10}$

(29) $y = x \log(x^2 + 1)$

$y' = x' \log(x^2 + 1) + x\{\log(x^2 + 1)\}' = \log(x^2 + 1) + x\left(\frac{2x}{x^2+1}\right) = \log(x^2 + 1) + \frac{2x^2}{x^2+1}$

(30) $y = e^x \log(\sin x)$

$y' = (e^x)' \log(\sin x) + e^x\{\log(\sin x)\}' = e^x \cdot \log(\sin x) + e^x \cdot \frac{(\sin x)'}{\sin x}$

$\quad = e^x \cdot \log(\sin x) + e^x \cdot \frac{\cos x}{\sin x} = e^x\{\log(\sin x) + \cot x\}$

(31) $y = e^{x^x}, \ x > 0$

Let $f(x) = x^x$

Then, $\log f(x) = \log x^x = x \log x$

$\therefore \ \frac{d}{dx}\{\log f(x)\} = \frac{d}{dx}\{x \log x\}$

$\therefore \ \frac{f'(x)}{f(x)} = x' \log x + x(\log x)' = \log x + x \cdot \frac{1}{x} = \log x + 1$

$\therefore \ f'(x) = f(x)(\log x + 1) = x^x(\log x + 1)$

$\therefore \ y' = e^{x^x} \cdot (x^x)' = e^{x^x} \cdot f'(x) = e^{x^x} \cdot x^x \cdot (\log x + 1)$

#17 Compute $\frac{dy}{dx}$.

(1) $x = \sin y$

$\frac{d}{dx}(x) = \frac{d}{dx}(\sin y) \ ; \ \ 1 = \cos y \cdot \frac{dy}{dx}$

$\therefore \ \frac{dy}{dx} = \frac{1}{\cos y}$

Note that: $\sin^2 y + \cos^2 y = 1$; $\cos^2 y = 1 - \sin^2 y$ $\therefore \cos y = \pm\sqrt{1 - \sin^2 y} = \pm\sqrt{1 - x^2}$

$\therefore \dfrac{dy}{dx} = \dfrac{1}{\cos y} = \dfrac{1}{\pm\sqrt{1-x^2}}$

(2) $x = \cos y$ $(0 < y < \pi)$

$\dfrac{d}{dx}(x) = \dfrac{d}{dx}(\cos y)$; $1 = -\sin y \cdot \dfrac{dy}{dx}$ $\therefore \dfrac{dy}{dx} = -\dfrac{1}{\sin y} = -\dfrac{1}{\sqrt{1-x^2}}$

(3) $\sin x + \sin y = 1$

$\dfrac{d}{dx}(\sin x + \sin y) = \dfrac{d}{dx}(1)$; $\cos x + \cos y \cdot \dfrac{dy}{dx} = 0$ $\therefore \dfrac{dy}{dx} = -\dfrac{\cos x}{\cos y}$

(4) $\sin x + \cos y = 1$

$\dfrac{d}{dx}(\sin x + \cos y) = \dfrac{d}{dx}(1)$; $\cos x - \sin y \cdot \dfrac{dy}{dx} = 0$ $\therefore \dfrac{dy}{dx} = \dfrac{\cos x}{\sin y}$

(5) $\cos(x + y) + \cos(x - y) = 1$

$\cos(x + y) + \cos(x - y) = \cos x \cos y - \sin x \sin y + \cos x \cos y + \sin x \sin y = 2\cos x \cos y = 1$

$\therefore \dfrac{d}{dx}(2\cos x \cos y) = \dfrac{d}{dx}(1)$

$\therefore 2\left\{(-\sin x)\cos y + \cos x (-\sin y)\dfrac{dy}{dx}\right\} = 0$

$\therefore \dfrac{dy}{dx} = -\dfrac{\sin x \cos y}{\cos x \sin y}$

(6) $x = 2\cos^3\theta$, $y = 2\sin^3\theta$

$\dfrac{dx}{d\theta} = 2\cdot 3\cdot \cos^2\theta\,(-\sin\theta)$; $\dfrac{dy}{d\theta} = 2\cdot 3\cdot \sin^2\theta\,(\cos\theta)$

$\dfrac{dy}{dx} = \dfrac{dy}{d\theta}\cdot\dfrac{d\theta}{dx} = 6\sin^2\theta\,(\cos\theta)\cdot\dfrac{1}{6\cos^2\theta(-\sin\theta)} = -\dfrac{\sin\theta}{\cos\theta} = -\tan\theta$

(7) $x = 3\cos t$, $y = 2\sin t$

$\dfrac{dx}{dt} = -3\sin t$; $\dfrac{dy}{dt} = 2\cos t$

$\dfrac{dy}{dx} = \dfrac{dy}{dt}\cdot\dfrac{dt}{dx} = 2\cos t \cdot\dfrac{1}{-3\sin t} = -\dfrac{2}{3}\cdot\dfrac{\cos t}{\sin t} = -\dfrac{2}{3}\cot t$

(8) $y^x = x^y$ $(x > 0,\ y > 0)$

$\log y^x = \log x^y$

$x\log y = y\log x$

$\dfrac{d}{dx}(x\log y) = \dfrac{d}{dx}(y\log x)$

$1\cdot\log y + x\cdot\dfrac{1}{y}\cdot\dfrac{dy}{dx} = \dfrac{dy}{dx}\cdot\log x + y\cdot\dfrac{1}{x}$

$\therefore \dfrac{dy}{dx}\left(\dfrac{x}{y} - \log x\right) = \dfrac{y}{x} - \log y$

$\therefore \dfrac{dy}{dx} = \dfrac{\frac{y - x\log y}{x}}{\frac{x - y\log x}{y}} = \dfrac{y(y - x\log y)}{x(x - y\log x)} = \dfrac{y^2 - xy\log y}{x^2 - xy\log x}$

(9) $x = \dfrac{1-t^2}{1+t^2}, \ y = \dfrac{2t}{1+t^2}$

$\dfrac{dx}{dt} = \dfrac{-2t(1+t^2)-(1-t^2)2t}{(1+t^2)^2} = \dfrac{-4t}{(1+t^2)^2}$; $\dfrac{dy}{dt} = \dfrac{2(1+t^2)-2t\cdot 2t}{(1+t^2)^2} = \dfrac{-2t^2+2}{(1+t^2)^2}$

$\dfrac{dy}{dx} = \dfrac{dy}{dt}\cdot\dfrac{dt}{dx} = \dfrac{-2t^2+2}{(1+t^2)^2}\cdot\dfrac{(1+t^2)^2}{-4t} = \dfrac{-2t^2+2}{-4t} = \dfrac{t^2-1}{2t}$

#18 Calculate the second derivatives of each function.

(1) $y = \sqrt{x^2 + 1}$

$y' = \dfrac{2x}{2\sqrt{x^2+1}} = \dfrac{x}{\sqrt{x^2+1}}$

$y'' = \dfrac{1\cdot\sqrt{x^2+1}-x\left(\sqrt{x^2+1}\right)'}{\left(\sqrt{x^2+1}\right)^2} = \dfrac{\sqrt{x^2+1}-x\left(\frac{2x}{2\sqrt{x^2+1}}\right)}{x^2+1} = \dfrac{\sqrt{x^2+1}-\frac{x^2}{\sqrt{x^2+1}}}{x^2+1} = \dfrac{\frac{x^2+1-x^2}{\sqrt{x^2+1}}}{x^2+1} = \dfrac{1}{(x^2+1)\sqrt{x^2+1}}$

(2) $y = \cos^3 x$

$y' = 3\cos^2 x\,(\cos x)' = 3\cos^2 x\,(-\sin x)$

$y'' = (3\cos^2 x)'(-\sin x) + 3\cos^2 x\,(-\sin x)' = 6\cos x\,(-\sin x)(-\sin x) + 3\cos^2 x\,(-\cos x)$

$= 6\cos x \sin^2 x - 3\cos^3 x = 3(2\cos x \sin^2 x - \cos^3 x)$

(3) $y = e^{x^2}$

$y' = e^{x^2}(x^2)' = e^{x^2}(2x)$

$y'' = \left(e^{x^2}\right)'(2x) + e^{x^2}(2x)' = e^{x^2}\cdot 2x\cdot 2x + e^{x^2}\cdot 2 = 4x^2 e^{x^2} + 2e^{x^2} = 2e^{x^2}(2x^2+1)$

(4) $y = \dfrac{x}{\log x}$

$y' = \dfrac{x'\log x - x(\log x)'}{(\log x)^2} = \dfrac{\log x - x\left(\frac{1}{x}\right)}{(\log x)^2} = \dfrac{\log x - 1}{(\log x)^2}$

$y'' = \dfrac{(\log x-1)'(\log x)^2 - (\log x-1)\{(\log x)^2\}'}{(\log x)^4} = \dfrac{\frac{1}{x}(\log x)^2 - (\log x-1)\left(2\log x\cdot\left(\frac{1}{x}\right)\right)}{(\log x)^4}$

$= \dfrac{\frac{1}{x}\log x(\log x-2\log x+2)}{(\log x)^4} = \dfrac{-\log x+2}{x(\log x)^3}$

#19 For a function $f(x) = e^{ax}\sin x$,

(1) Find the values of $f'(\pi)$ and $f''(\pi)$.

(2) Find the value of the constant a such that $f''(x) - 2f'(x) + 2f(x) = 0$.

$f(x) = e^{ax}\sin x$

$f'(x) = (e^{ax})'\sin x + e^{ax}(\sin x)' = e^{ax}\cdot a\cdot\sin x + e^{ax}\cdot\cos x = e^{ax}(a\sin x + \cos x)$

$f''(x) = (e^{ax})'(a\sin x + \cos x) + e^{ax}(a\sin x + \cos x)'$

$= e^{ax}\cdot a\cdot(a\sin x + \cos x) + e^{ax}(a\cos x - \sin x) = ae^{ax}(a\sin x + \cos x) + e^{ax}(a\cos x - \sin x)$

(1) $f'(\pi) = e^{a\pi}(a \sin \pi + \cos \pi) = e^{a\pi}(0 - 1) = -e^{a\pi}$

$\quad f''(\pi) = ae^{a\pi}(a \sin \pi + \cos \pi) + e^{a\pi}(a \cos \pi - \sin \pi) = -ae^{a\pi} - ae^{a\pi} = -2ae^{a\pi}$

(2) $f''(x) - 2f'(x) + 2f(x)$

$\quad = ae^{ax}(a \sin x + \cos x) + e^{ax}(a \cos x - \sin x) - 2\{e^{ax}(a \sin x + \cos x)\} + 2e^{ax} \sin x = 0$

Since $e^{ax} > 0$, we divide both sides of the equation by e^{ax} and simplify:

$(a - 1)^2 \sin x + 2(a - 1) \cos x = 0$

Since the quality holds for any real number x, $a - 1 = 0$ $\quad \therefore a = 1$

#20 Find the value.

(1) For a differentiable function $f(x)$, let $g(x) = \dfrac{x}{1-f(x)}$ and $g'(0) = \dfrac{1}{3}$.

Find the value of $f(0)$. $(f(x) \neq 1)$

let $g(x) = \dfrac{x}{1-f(x)}$ $\Rightarrow$ $g'(x) = \dfrac{1\cdot(1-f(x)) - x(1-f(x))'}{(1-f(x))^2} = \dfrac{1-f(x)+xf'(x)}{(1-f(x))^2}$

$g'(0) = \dfrac{1-f(0)+0\cdot f'(0)}{(1-f(0))^2} = \dfrac{1-f(0)}{(1-f(0))^2} = \dfrac{1}{1-f(0)} = \dfrac{1}{3}$

$\therefore \ 1 - f(0) = 3$ $\qquad \therefore f(0) = -2$

(2) For a function $f(x) = x^2 e^{-x} \sin x$, find the value of $\displaystyle\lim_{x \to 0} \dfrac{f'(x)}{x^2}$.

$f'(x) = (x^2)' e^{-x} \sin x + x^2 (e^{-x})' \sin x + x^2 e^{-x} (\sin x)'$

$\quad = 2xe^{-x} \sin x + x^2(-e^{-x}) \sin x + x^2 e^{-x}(\cos x) = e^{-x}(2x \sin x - x^2 \sin x + x^2 \cos x)$

$\displaystyle\lim_{x \to 0} \dfrac{f'(x)}{x^2} = \lim_{x \to 0} \dfrac{e^{-x}(2x \sin x - x^2 \sin x + x^2 \cos x)}{x^2} = \lim_{x \to 0} e^{-x} \left(\dfrac{2 \sin x}{x} - \sin x + \cos x \right) = 1(2 - 0 + 1) = 3$

(3) For a function $f(x) = \displaystyle\lim_{h \to 0} \dfrac{e^{x+h}-e^x}{\sqrt{x+h}-\sqrt{x}}$, find the value of $f'(1)$.

$f(x) = \displaystyle\lim_{h \to 0} \dfrac{e^{x+h}-e^x}{\sqrt{x+h}-\sqrt{x}} = \lim_{h \to 0} \dfrac{(e^{x+h}-e^x)(\sqrt{x+h}+\sqrt{x})}{(\sqrt{x+h}-\sqrt{x})(\sqrt{x+h}+\sqrt{x})} = \lim_{h \to 0} \dfrac{(e^{x+h}-e^x)(\sqrt{x+h}+\sqrt{x})}{x+h-x}$

$\quad = \displaystyle\lim_{h \to 0} \dfrac{e^x(e^h-1)(\sqrt{x+h}+\sqrt{x})}{h} = \lim_{h \to 0} e^x \cdot \dfrac{e^h-1}{h}(\sqrt{x+h}+\sqrt{x}) = e^x \cdot 1 \cdot (\sqrt{x}+\sqrt{x}) = 2e^x\sqrt{x}$

$f'(x) = 2\left\{(e^x)'\sqrt{x} + e^x(\sqrt{x})'\right\} = 2\left(e^x\sqrt{x} + e^x \dfrac{1}{2\sqrt{x}}\right) = 2e^x\left(\sqrt{x} + \dfrac{1}{2\sqrt{x}}\right)$

$f'(1) = 2e^1\left(\sqrt{1} + \dfrac{1}{2\sqrt{1}}\right) = 2e \cdot \dfrac{3}{2} = 3e$

(4) For the graph of the curve $x^4 + y^3 + axy + b = 0$, the value of $\dfrac{dy}{dx}$ at $(1, 0)$ on the graph is -2. Find the value of $a + b$. $(a, b; \text{constants})$

$\dfrac{d}{dx}(x^4 + y^3 + axy + b) = \dfrac{d}{dx}(0)$; $4x^3 + 3y^2 \dfrac{dy}{dx} + a\left(1 \cdot y + x\dfrac{dy}{dx}\right) + 0 = 0$

$\dfrac{dy}{dx}(3y^2 + ax) = -4x^3 - ay$ $\qquad \therefore \ \dfrac{dy}{dx} = \dfrac{-4x^3-ay}{3y^2+ax}$ $\quad \cdots\cdots$ ①

Substituting $(x = 1, y = 0)$ into ①, $\dfrac{-4-0}{0+a} = \dfrac{-4}{a} = -2$ ∴ $a = 2$

Since the point $(1, 0)$ lies on the graph, $1^4 + 0 + a \cdot 1 \cdot 0 + b = 0$; $1 + b = 0$ ∴ $b = -1$

Therefore, $a + b = 2 + (-1) = 1$

(5) For the function $f(x) = \dfrac{e^x \cos x}{1+\sin x}$, find the value of $f'\left(\dfrac{\pi}{2}\right)$.

$f(x) = \dfrac{e^x \cos x}{1+\sin x} \Rightarrow \log|f(x)| = \log\left|\dfrac{e^x \cos x}{1+\sin x}\right| = \log|e^x \cos x| - \log|1 + \sin x|$

$$= \log|e^x| + \log|\cos x| - \log|1 + \sin x|$$

∴ $\dfrac{d}{dx}\log|f(x)| = \dfrac{d}{dx}\{\log|e^x| + \log|\cos x| - \log|1 + \sin x|\}$

∴ $\dfrac{f'(x)}{f(x)} = \dfrac{e^x}{e^x} + \dfrac{-\sin x}{\cos x} - \dfrac{\cos x}{1+\sin x} = 1 - \dfrac{\sin x}{\cos x} - \dfrac{\cos x}{1+\sin x}$

∴ $f'(x) = f(x)\left(1 - \dfrac{\sin x}{\cos x} - \dfrac{\cos x}{1+\sin x}\right) = \dfrac{e^x \cos x}{1+\sin x} \cdot \dfrac{\cos x - 1}{\cos x} = \dfrac{e^x(\cos x - 1)}{1+\sin x}$

∴ $f'\left(\dfrac{\pi}{2}\right) = \dfrac{e^{\frac{\pi}{2}}(\cos\frac{\pi}{2}-1)}{1+\sin\frac{\pi}{2}} = \dfrac{e^{\frac{\pi}{2}}(0-1)}{1+1} = -\dfrac{1}{2}e^{\frac{\pi}{2}}$

(6) For the function $f(x) = x^{\sqrt{x}}$ $(x > 0)$, find the range of x such that $f'(x) > 0$.

$f(x) = x^{\sqrt{x}} \Rightarrow \log f(x) = \log x^{\sqrt{x}} = \sqrt{x}\log x$

∴ $\dfrac{d}{dx}(\log f(x)) = \dfrac{d}{dx}(\sqrt{x}\log x)$

∴ $\dfrac{f'(x)}{f(x)} = \left(\sqrt{x}\right)'\log x + \sqrt{x}(\log x)' = \dfrac{\log x}{2\sqrt{x}} + \sqrt{x}\cdot\dfrac{1}{x} = \dfrac{\log x}{2\sqrt{x}} + \dfrac{1}{\sqrt{x}} = \dfrac{\log x + 2}{2\sqrt{x}}$

Since $f'(x) > 0$, $x^{\sqrt{x}}\left(\dfrac{\log x + 2}{2\sqrt{x}}\right) > 0$

Since $\sqrt{x} > 0$, $x^{\sqrt{x}} > 0$

∴ $\dfrac{\log x + 2}{2\sqrt{x}} > 0$; $\log x + 2 > 0$; $\log x > -2$; $\log x > \log e^{-2}$

∴ $x > e^{-2}$

(7) For the function $f(x) = \log(\log x)$, find the value of $\displaystyle\lim_{h\to 0}\dfrac{f(e+h)-f(e-h)}{h}$.

$\displaystyle\lim_{h\to 0}\dfrac{f(e+h)-f(e-h)}{h} = \lim_{h\to 0}\dfrac{f(e+h)-f(e)+f(e)-f(e-h)}{h} = \lim_{h\to 0}\left\{\dfrac{f(e+h)-f(e)}{h} + \dfrac{f(e)-f(e-h)}{h}\right\}$

$$= \lim_{h\to 0}\left\{\dfrac{f(e+h)-f(e)}{h} - \dfrac{f(e-h)-f(e)}{h}\right\} = \lim_{h\to 0}\left\{\dfrac{f(e+h)-f(e)}{h} + \dfrac{f(e-h)-f(e)}{-h}\right\}$$

$$= f'(e) + f'(e) = 2f'(e)$$

$f(x) = \log(\log x) \Rightarrow f'(x) = \dfrac{(\log x)'}{\log x} = \dfrac{\frac{1}{x}}{\log x} = \dfrac{1}{x\log x}$

∴ $\displaystyle\lim_{h\to 0}\dfrac{f(e+h)-f(e-h)}{h} = 2f'(e) = \dfrac{2}{e\log e} = \dfrac{2}{e}$

(8) Find the value of the limit: $\lim\limits_{x \to a} \dfrac{x^2 e^a - a^2 e^x}{x-a}$.

Let $f(x) = e^x$

Then, $f'(x) = e^x$; $f'(a) = e^a$

$$\lim_{x \to a} \frac{x^2 e^a - a^2 e^x}{x-a} = \lim_{x \to a} \frac{x^2 e^a - a^2 e^a + a^2 e^a - a^2 e^x}{x-a} = \lim_{x \to a} \frac{e^a(x^2-a^2) + a^2(e^a - e^x)}{x-a}$$

$$= \lim_{x \to a} e^a(x+a) - \lim_{x \to a} \frac{a^2(e^x - e^a)}{x-a} = 2ae^a - a^2 \lim_{x \to a} \frac{f(x)-f(a)}{x-a} = 2ae^a - a^2 f'(a)$$

$$= 2ae^a - a^2 e^a = ae^a(2-a)$$

(9) For $A = \lim\limits_{x \to 0} \dfrac{2}{x} \log\left(\dfrac{e^x + e^{2x} + e^{3x} + \cdots\cdots + e^{nx}}{n}\right)$, **express** A **as** n.

Let $f(x) = \log(e^x + e^{2x} + e^{3x} + \cdots\cdots + e^{nx})$

Then, $f(0) = \log n$

$$\therefore A = 2 \lim_{x \to 0} \frac{1}{x}\{\log(e^x + e^{2x} + e^{3x} + \cdots\cdots + e^{nx}) - \log n\} = 2 \lim_{x \to 0} \frac{f(x)-f(0)}{x} = 2f'(0)$$

Since $f'(x) = \dfrac{e^x + 2e^{2x} + 3e^{3x} + \cdots\cdots + ne^{nx}}{e^x + e^{2x} + e^{3x} + \cdots\cdots + e^{nx}}$, $f'(0) = \dfrac{1+2+3+\cdots\cdots+n}{1+1+1+\cdots\cdots+1} = \dfrac{\frac{n(n+1)}{2}}{n} = \dfrac{n(n+1)}{2n} = \dfrac{n+1}{2}$

$$\therefore A = 2f'(0) = 2 \cdot \frac{n+1}{2} = n+1$$

(10) For the function $f(x) = \sin^2 \dfrac{x}{2}$, **find the value of** $\lim\limits_{x \to \pi} \dfrac{f'(x)}{x-\pi}$.

$f(x) = \sin^2 \dfrac{x}{2} \Rightarrow f'(x) = 2\sin\dfrac{x}{2}\left(\sin\dfrac{x}{2}\right)' = 2\sin\dfrac{x}{2}\left(\cos\dfrac{x}{2}\right)\cdot\dfrac{1}{2} = \sin\dfrac{x}{2}\cos\dfrac{x}{2}$

Note that: $\sin 2\theta = 2\sin\theta\cos\theta$; $\sin\theta = 2\sin\dfrac{\theta}{2}\cos\dfrac{\theta}{2}$; $\dfrac{1}{2}\sin\theta = \sin\dfrac{\theta}{2}\cos\dfrac{\theta}{2}$

$\therefore f'(x) = \sin\dfrac{x}{2}\cos\dfrac{x}{2} = \dfrac{1}{2}\sin x$ $\qquad \therefore f'(\pi) = \dfrac{1}{2}\sin\pi = \dfrac{1}{2}\cdot 0 = 0$

$f''(x) = \left(\dfrac{1}{2}\sin x\right)' = \dfrac{1}{2}\cos x$ $\qquad \therefore f''(\pi) = \dfrac{1}{2}\cos\pi = -\dfrac{1}{2}$

Since $\lim\limits_{x \to \pi}\dfrac{f'(x)}{x-\pi} = \lim\limits_{x \to \pi}\dfrac{f'(x)-0}{x-\pi} = \lim\limits_{x \to \pi}\dfrac{f'(x)-f'(\pi)}{x-\pi} = f''(\pi) = -\dfrac{1}{2}$

(11) When a differentiable function $f(x)$ **satisfies** $f(2x^2 + 3x) - f(5x) = (x^2 - 4x)^3 + 2x$

for any real number x, **find the value of** $f'(5)$.

$\dfrac{d}{dx}\{f(2x^2 + 3x) - f(5x)\} = \dfrac{d}{dx}\{(x^2 - 4x)^3 + 2x\}$

$\therefore (4x+3)f'(2x^2 + 3x) - 5f'(5x) = 3(x^2 - 4x)^2(2x-4) + 2$

Since the equality satisfies for any real number x, we can substitute $x = 1$ into the expression.

Thus, we have: $(4+3)f'(5) - 5f'(5) = 3(-3)^2(-2) + 2$

$\therefore 2f'(5) = -52$

$\therefore f'(5) = -26$

(12) For a function $f(x) = x^{2\ln x}$, find the limit: $\displaystyle\lim_{x\to 0}\frac{f(e+2x)-f(e-2x)}{x}$

$$\lim_{x\to 0}\frac{f(e+2x)-f(e-2x)}{x} = \lim_{x\to 0}\frac{f(e+2x)-f(e)+f(e)-f(e-2x)}{x} = \lim_{x\to 0}\left\{\frac{f(e+2x)-f(e)}{x} - \frac{f(e-2x)-f(e)}{x}\right\}$$

$$= \lim_{x\to 0}\left\{\frac{f(e+2x)-f(e)}{2x}\cdot 2 + \frac{f(e-2x)-f(e)}{-2x}\cdot 2\right\}$$

$$= 2f'(e) + 2f'(e) = 4f'(e)$$

$$f(x) = x^{2\ln x} \Rightarrow \ln f(x) = \ln x^{2\ln x} = \ln x^{\ln x^2} = \ln x^2\,(\ln x) = (2\ln x)(\ln x) = 2(\ln x)^2$$

$$\therefore \frac{d}{dx}\{\ln f(x)\} = \frac{d}{dx}\{2(\ln x)^2\}$$

$$\therefore \frac{f'(x)}{f(x)} = 4\ln x\cdot\frac{1}{x} \;;\; f'(x) = 4f(x)\frac{\ln x}{x} \;;\; f'(e) = 4f(e)\frac{\ln e}{e} = 4f(e)\frac{1}{e}$$

Since $f(e) = e^{2\ln e} = e^2$, $f'(e) = 4\cdot e^2\cdot\frac{1}{e} = 4e$

$$\therefore \lim_{x\to 0}\frac{f(e+2x)-f(e-2x)}{x} = 4f'(e) = 4\cdot 4e = 16e$$

(13) For a differentiable function $f(x)$, let $g(x)$ be the inverse function of $f(x)$.

When $\displaystyle\lim_{h\to 0}\frac{g(2+2h)-g(2-h)}{h} = 15$ and $f(5) = 2$, find the value of $f'(5)$.

$$\lim_{h\to 0}\frac{g(2+2h)-g(2-h)}{h} = \lim_{h\to 0}\frac{g(2+2h)-g(2)+g(2)-g(2-h)}{h} = \lim_{h\to 0}\left\{\frac{g(2+2h)-g(2)}{h} - \frac{g(2-h)-g(2)}{h}\right\}$$

$$= \lim_{h\to 0}\left\{\frac{g(2+2h)-g(2)}{2h}\cdot 2 + \frac{g(2-h)-g(2)}{-h}\right\} = 2g'(2) + g'(2) = 3g'(2) = 15$$

$$\therefore g'(2) = 5$$

$$\therefore f'(5) = \frac{1}{g'(f(5))} = \frac{1}{g'(2)} = \frac{1}{5}$$

(14) For a differentiable function $f(x)$, let $g(x)$ be the inverse function of $f(x)$.

When $\displaystyle\lim_{x\to 1}\frac{f(x)-3}{x-1} = 2$, find the value of $g'(3)$.

Since $\displaystyle\lim_{x\to 1}\frac{f(x)-3}{x-1}$ exists and $\displaystyle\lim_{x\to 1}(x-1) = 0$, $\displaystyle\lim_{x\to 1}\{f(x)-3\} = 0$

$$\therefore f(1)-3 = 0 \;;\; f(1) = 3 \qquad \therefore 1 = f^{-1}(3) = g(3)$$

$$\therefore \lim_{x\to 1}\frac{f(x)-3}{x-1} = \lim_{x\to 1}\frac{f(x)-f(1)}{x-1} = f'(1) = 2$$

$$\therefore g'(3) = \frac{1}{f'(g(3))} = \frac{1}{f'(1)} = \frac{1}{2}$$

(15) For a function $f(x) = 4\sin\frac{x}{2}$ $\left(0\le x\le\frac{\pi}{2}\right)$, let $g(x)$ be the inverse of $f(x)$.

Find the value of $g'(2)$.

Let $g(2) = a$ Then, $2 = g^{-1}(a) = f(a)$

$$\therefore f(a) = 4\sin\frac{a}{2} = 2 \;;\; \sin\frac{a}{2} = \frac{1}{2}$$

Since $0\le x\le\frac{\pi}{2}$, $\frac{a}{2} = \frac{\pi}{6}$ $\qquad\therefore a = \frac{\pi}{3} \;;\; g(2) = \frac{\pi}{3}$

$$f'(x) = 4\cos\frac{x}{2} \cdot \frac{1}{2} = 2\cos\frac{x}{2}$$

$$g'(2) = \frac{1}{f'(g(2))} = \frac{1}{f'\left(\frac{\pi}{3}\right)} = \frac{1}{2\cos\frac{\pi}{6}} = \frac{1}{2 \cdot \frac{\sqrt{3}}{2}} = \frac{1}{\sqrt{3}} = \frac{\sqrt{3}}{3}$$

(16) For a function $f(x) = \ln(\ln x)$ $(x > 1)$, let $g(x)$ be the inverse function of $f(x)$.

Find the value of $\dfrac{g'(0)}{g(0)}$.

Let $g(0) = a$.

Then $0 = g^{-1}(a) = f(a)$

Since $f(a) = \ln(\ln a) = 0$, $\ln a = 1$; $a = e$

$$f'(x) = \frac{(\ln x)'}{\ln x} = \frac{\frac{1}{x}}{\ln x} = \frac{1}{x\ln x} \ ; \ f'(a) = \frac{1}{a\ln a}$$

Since $f'(a) = \dfrac{1}{g'(f(a))} = \dfrac{1}{g'(0)} = \dfrac{1}{a\ln a}$, $g'(0) = a\ln a$

$$\therefore \ \frac{g'(0)}{g(0)} = \frac{a\ln a}{a} = \ln a = \ln e = 1$$

(17) For a function $f(x) = \ln(e^x + 1)$, let $g(x)$ be the inverse function of $f(x)$.

Find the value of $\dfrac{1}{f'(a)} - \dfrac{1}{g'(a)}$. $(a > 0)$

$$f(x) = \ln(e^x + 1) \ \Rightarrow \ f'(x) = \frac{(e^x+1)'}{e^x+1} = \frac{e^x}{e^x+1} \qquad \therefore \ \frac{1}{f'(a)} = \frac{e^a+1}{e^a}$$

Let $g(a) = b$

Then, $a = g^{-1}(b) = f(b)$

$\therefore f(b) = \ln(e^b + 1) = a$; $e^b + 1 = e^a$

Since $g'(a) = \dfrac{1}{f'(g(a))} = \dfrac{1}{f'(b)} = \dfrac{e^b+1}{e^b}$, $\dfrac{1}{f'(a)} - \dfrac{1}{g'(a)} = \dfrac{e^a+1}{e^a} - \dfrac{e^a-1}{e^a} = \dfrac{2}{e^a}$

(18) For the graph of a curve $x^3 + y^3 + 6xy = 1$,

find the slope of the tangent line at $(1, -2)$ on the graph.

$$\frac{d}{dx}(x^3 + y^3 + 6xy) = \frac{d}{dx}(1) \ ; \ 3x^2 + 3y^2 \cdot \frac{dy}{dx} + 6\left(1 \cdot y + x \cdot \frac{dy}{dx}\right) = 0$$

$$\therefore \ \frac{dy}{dx}(y^2 + 2x) = -x^2 - 2y$$

$$\therefore \ \frac{dy}{dx} = -\frac{x^2+2y}{y^2+2x}, \quad y^2 + 2x \neq 0$$

Therefore, the slope of the tangent line at $(1, -2)$ is $-\dfrac{1^2+2(-2)}{(-2)^2+2\cdot1} = \dfrac{3}{6} = \dfrac{1}{2}$

(19) Find the slope of the curve $x = y^2 - 3y$ at the points where it crosses the y-axis.

The points of crossing the y-axis are $(0, 0)$ and $(0, 3)$.

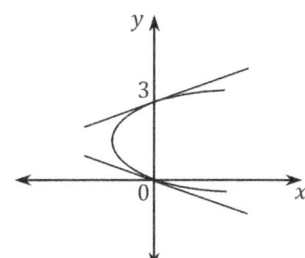

Since $\dfrac{dx}{dy} = 2y - 3$, $\dfrac{dy}{dx} = \dfrac{1}{\frac{dx}{dy}} = \dfrac{1}{2y-3}$

$\therefore$ At $(0, 0)$, the slope is $-\dfrac{1}{3}$; At $(0, 3)$, the slope is $\dfrac{1}{3}$.

(20) For any real number x, a function satisfies

  i) $f(1) = 3$, $f'(1) = 2$ and ii) $\displaystyle\lim_{x \to 1}\dfrac{f'(f(x))-1}{x-1} = 4$. Find the value of $f''(3)$.

Since $\displaystyle\lim_{x \to 1}\dfrac{f'(f(x))-1}{x-1}$ exists and $\displaystyle\lim_{x \to 1}(x - 1) = 0$, $\displaystyle\lim_{x \to 1}\{f'(f(x)) - 1\} = 0$

$\therefore$ $f'(f(1)) = 1$

$\displaystyle\lim_{x \to 1}\dfrac{f'(f(x))-1}{x-1} = \lim_{x \to 1}\dfrac{f'(f(x))-f'(f(1))}{x-1} = \lim_{x \to 1}\left\{\dfrac{f'(f(x))-f'(f(1))}{f(x)-f(1)} \cdot \dfrac{f(x)-f(1)}{x-1}\right\}$

$\qquad\qquad = f''(f(1)) \cdot f'(1) = f''(3) \cdot 2 = 2f''(3) = 4$

$\therefore$ $f''(3) = \dfrac{4}{2} = 2$

(21) For a function $f(x) = \dfrac{ax+1}{x^2+bx}$ defined on any non-zero real number x, the sum and

  product of two roots of $f'(x) = 0$ are -1 and $\dfrac{1}{2}$, respectively. When the equation

  $f'(x) + c = 0$ has a double root, find the value of c. ($ab(\ne 0)$; real number)

$f(x) = \dfrac{ax+1}{x^2+bx}$

$\Rightarrow$ $f'(x) = \dfrac{(ax+1)'(x^2+bx)-(ax+1)(x^2+bx)'}{(x^2+bx)^2} = \dfrac{a(x^2+bx)-(ax+1)(2x+b)}{(x^2+bx)^2}$

$\qquad = \dfrac{ax^2+abx-(2ax^2+2x+abx+b)}{(x^2+bx)^2} = \dfrac{-ax^2-2x-b}{(x^2+bx)^2}$

$f'(x) = 0 \iff -ax^2 - 2x - b = 0 \iff ax^2 + 2x + b = 0 \cdots\cdots①$

Let α and β be the two roots of the equation ①.

Then, $\alpha + \beta = -\dfrac{2}{a} = -1$; $a = 2$

$\qquad \alpha\beta = \dfrac{b}{a} = \dfrac{1}{2}$; $b = 1$

$f'(x) + c = 0 \iff ax^2 + 2x + b + c = 0$

$\therefore$ $2x^2 + 2x + 1 + c = 0 \cdots\cdots②$

Let D be the discriminant of the equation ②.

To have a double root, $\dfrac{D}{4} = 0$

$\therefore$ $1^2 - 2(1 + c) = 0$; $-1 - 2c = 0$ $\qquad\therefore$ $c = -\dfrac{1}{2}$

(22) The graph of a function $f(x) = \ln x$ is shown as the Figure.

For a point P on the graph and a point Q on the x-axis,

the line segment $\overline{PQ}$ and the x-axis are perpendicular.

For a point $A(1, 0)$, let $\overline{PQ} = a$ and $\overline{PA} = f(a)$.

Find the value of $f'(\ln 3)$.

Since $\overline{PQ} = a$, the y-coordinate of the point P is a.

Since P lies on the graph of $f(x) = \ln x$, $a = \ln x$ $\quad \therefore \ x = e^a$

$\therefore P = P(e^a, a)$

$\therefore$ The distance between the two points $P(e^a, a)$ and $A(1, 0)$ is $\overline{PA} = \sqrt{(1 - e^a)^2 + (0 - a)^2}$.

Or, by the Pythagorean theorem, $f(a) = \overline{PA} = \sqrt{(e^a - 1)^2 + a^2}$

$\therefore f'(a) = \dfrac{\left((e^a - 1)^2 + a^2\right)'}{2\sqrt{(e^a - 1)^2 + a^2}} = \dfrac{2(e^a - 1)e^a + 2a}{2\sqrt{(e^a - 1)^2 + a^2}} = \dfrac{(e^a - 1)e^a + a}{\sqrt{(e^a - 1)^2 + a^2}}$

$\therefore f'(\ln 3) = \dfrac{(e^{\ln 3} - 1)e^{\ln 3} + \ln 3}{\sqrt{(e^{\ln 3} - 1)^2 + (\ln 3)^2}} = \dfrac{(3 - 1) \cdot 3 + \ln 3}{\sqrt{(3 - 1)^2 + (\ln 3)^2}} = \dfrac{6 + \ln 3}{\sqrt{4 + (\ln 3)^2}}$

#21 Find the value of $a + b$.

(1) When a function $f(x) = \begin{cases} a \cos \dfrac{\pi}{2} x + 1, & x > 1 \\ e^{x-1} + b & , x \le 1 \end{cases}$ is differentiable at $x = 1$,

find the value of $a + b$.

Since $f(x)$ is differentiable at $x = 1$, $f(x)$ is continuous at $x = 1$.

$\therefore \displaystyle\lim_{x \to 1^-} f(x) = \lim_{x \to 1^+} f(x) = f(1)$

That is, $\displaystyle\lim_{x \to 1^-}(e^{x-1} + b) = \lim_{x \to 1^+}(a \cos \frac{\pi}{2} x + 1) = e^0 + b$

$\therefore \ 1 + b = 1 \ ; \ b = 0$

$f'(x) = \begin{cases} -\dfrac{\pi}{2} a \sin \dfrac{\pi}{2} x, & x > 1 \\ e^{x-1} & , x < 1 \end{cases}$

Since $f(x)$ is differentiable at $x = 1$, $f'(1)$ exists.

$\therefore \ \displaystyle\lim_{x \to 1^-} e^{x-1} = \lim_{x \to 1^+}\left(-\frac{\pi}{2} a \sin \frac{\pi}{2} x\right) \ ; \ 1 = -\frac{\pi}{2}a \ ; \ a = -\frac{2}{\pi}$

Therefore, $a + b = -\dfrac{2}{\pi} + 0 = -\dfrac{2}{\pi}$

(2) When a function $f(x) = \begin{cases} \log_2(x-1) & , \ x \geq 2 \\ a \tan \pi(x-1) + b & , \ x < 2 \end{cases}$ **is differentiable at** $x = 2$,

find the value of $a + b$.

Since $f(x)$ is differentiable at $x = 2$, $f(x)$ is continuous at $x = 2$.

$\therefore \ \lim\limits_{x \to 2^-} f(x) = \lim\limits_{x \to 2^+} f(x) = f(2)$

That is, $\lim\limits_{x \to 2^-} \{a \tan \pi(x-1) + b\} = \lim\limits_{x \to 2^+} (\log_2(x-1)) = f(2)$

$\therefore \ a \tan \pi + b = 0 \ ; \ b = 0$

Since $f(x)$ is differentiable at $x = 2$, $f'(2)$ exists.

$f'(x) = \begin{cases} \dfrac{1}{x-1} \cdot \dfrac{1}{\log 2} & , \ x > 2 \\ a\pi \sec^2(\pi(x-1)) & , \ x < 2 \end{cases}$

$\therefore \ \lim\limits_{x \to 2^-} a\pi \sec^2(\pi(x-1)) = \lim\limits_{x \to 2^+} \dfrac{1}{x-1} \cdot \dfrac{1}{\log 2}$

$\therefore \ a\pi \sec^2 \pi = \dfrac{1}{\log 2} \ ; \ a = \dfrac{1}{\pi \log 2}$

Therefore, $a + b = \dfrac{1}{\pi \log 2} + 0 = \dfrac{1}{\pi \log 2}$

(3) When a function $f(x) = \begin{cases} a \ln(3x-2) + 2, & x > 1 \\ e^{x^2-1} + b & , \ x \leq 1 \end{cases}$ **is differentiable at** $x = 1$,

find the value of $a + b$.

Since $f(x)$ is differentiable at $x = 1$, $f(x)$ is continuous at $x = 1$.

$\therefore \ \lim\limits_{x \to 1^-} f(x) = \lim\limits_{x \to 1^+} f(x) = f(1)$

That is, $\lim\limits_{x \to 1^-} (e^{x^2-1} + b) = \lim\limits_{x \to 1^+} \{a \ln(3x-2) + 2\} = e^0 + b$

$\therefore \ 1 + b = 2 \ ; \ b = 1$

Since $f(x)$ is differentiable at $x = 1$, $f'(1)$ exists.

$f'(x) = \begin{cases} \dfrac{3a}{3x-2} & , \ x > 1 \\ 2x e^{x^2-1} & , \ x < 1 \end{cases}$

$\therefore \ \lim\limits_{x \to 1^-} 2x e^{x^2-1} = \lim\limits_{x \to 1^+} \dfrac{3a}{3x-2}$

$\therefore \ 2 = 3a \ ; \ a = \dfrac{2}{3}$

Therefore, $a + b = \dfrac{2}{3} + 1 = \dfrac{5}{3}$

(4) When the limit is $\lim\limits_{x\to a}\dfrac{b\log x}{x^2-a^2}=1$, find the value of $a+b$.

Since the limit exists and $\lim\limits_{x\to a}(x^2-a^2)=0$, $\quad \lim\limits_{x\to a}(b\log x)=0 \qquad \therefore\ b\log a=0$

Since $\lim\limits_{x\to a}\dfrac{b\log x}{x^2-a^2}=1$, $b\neq 0$

$\therefore\ \log a=0\ ;\ a=1$

$$\lim_{x\to a}\frac{b\log x}{x^2-a^2}=\lim_{x\to 1}\left(\frac{\log x}{x-1}\cdot\frac{b}{x+1}\right)=\lim_{x\to 1}\left(\frac{\log x-\log 1}{x-1}\cdot\frac{b}{x+1}\right)=\lim_{x\to 1}\left(\frac{\log x-\log 1}{x-1}\right)\lim_{x\to 1}\left(\frac{b}{x+1}\right)$$

Let $f(x)=\log x$

Then $f(1)=\log 1$ and $f'(x)=\dfrac{1}{x}$

$\therefore\ \lim\limits_{x\to a}\dfrac{b\log x}{x^2-a^2}=\lim\limits_{x\to 1}\dfrac{f(x)-f(1)}{x-1}\cdot\lim\limits_{x\to 1}\dfrac{b}{x+1}=f'(1)\cdot\dfrac{b}{2}=1\cdot\dfrac{b}{2}=1\ ;\ b=2$

Therefore, $a+b=1+2=3$

(5) When a function $f(x)=\dfrac{ax+b}{x^2+1}$ satisfies

i) $\lim\limits_{x\to 3}\dfrac{f(x)-f(3)}{x-3}=-\dfrac{11}{50}$ and ii) $\lim\limits_{x\to 1}\dfrac{f(x)-f(1)}{x^2-1}=0$, find the value of $a+b$.

$\lim\limits_{x\to 3}\dfrac{f(x)-f(3)}{x-3}=f'(3)=-\dfrac{11}{50}$

$\lim\limits_{x\to 1}\dfrac{f(x)-f(1)}{x^2-1}=\lim\limits_{x\to 1}\left\{\dfrac{f(x)-f(1)}{x-1}\cdot\dfrac{1}{x+1}\right\}=f'(1)\cdot\dfrac{1}{2}=0\ ;\ f'(1)=0$

$\therefore\ f(x)=\dfrac{ax+b}{x^2+1}\ \Rightarrow\ f'(x)=\dfrac{(ax+b)'(x^2+1)-(ax+b)(x^2+1)'}{(x^2+1)^2}=\dfrac{a(x^2+1)-2x(ax+b)}{(x^2+1)^2}=\dfrac{-ax^2-2bx+a}{(x^2+1)^2}$

$f'(3)=\dfrac{-9a-6b+a}{100}=-\dfrac{11}{50}\ ;\ -9a-6b+a=-22\ ;\ -8a-6b=-22$

$\therefore\ 4a+3b=11$

$f'(1)=\dfrac{-a-2b+a}{4}=-\dfrac{2b}{4}=-\dfrac{b}{2}=0 \qquad \therefore\ b=0,\ a=\dfrac{11}{4}$

Therefore, $a+b=\dfrac{11}{4}+0=\dfrac{11}{4}$

Chapter 4. Applications of the Derivative

#1 Find the values of constants a, b, and c.

(1) For the graph of a function $y = x^3 + ax + b$, the equation of the tangent line at $(1, 3)$ on the graph is $y = 2x + c$.

Let $f(x) = x^3 + ax + b$. Then, $f'(x) = 3x^2 + a$

Since the slope of the tangent line at $(1, 3)$ is $f'(1) = 2$, $3 \cdot 1^2 + a = 2$ $\therefore a = -1$

Since $(1, 3)$ lies on the graph, $f(1) = 3$

$\therefore 1^3 + a \cdot 1 + b = 1^3 + (-1) \cdot 1 + b = 3$; $b = 3$

Since the point $(1, 3)$ lies on the tangent line $y = 2x + c$, $3 = 2 \cdot 1 + c$ $\therefore c = 1$

Therefore, $a = -1$, $b = 3$, $c = 1$

(2) For the graph of a function $y = ax^3 + bx^2 + cx$ and two points $(1, 0)$ and $(2, 4)$ on the graph, the tangent lines at the points are parallel to each other.

Let $f(x) = ax^3 + bx^2 + cx$. Then, $f'(x) = 3ax^2 + 2bx + c$

Since the tangent lines at $(1, 0)$ and $(2, 4)$ are parallel, the slopes of the lines are the same.

$\therefore f'(1) = f'(2)$

$\therefore 3a + 2b + c = 12a + 4b + c$; $9a + 2b = 0$ $\cdots\cdots$ ①

Since $f(1) = 0$ and $f(2) = 4$, $a + b + c = 0$ and $8a + 4b + 2c = 4$

$$\begin{array}{r} 4a + 2b + c = 2 \\ - \underline{\quad a + b + c = 0 \quad} \\ 3a + b = 2 \end{array} \cdots\cdots ②$$

By ① and ②, $3a = -4$ $\therefore a = -\dfrac{4}{3}$

$b = 2 - 3a = 2 - 3 \cdot \left(-\dfrac{4}{3}\right) = 6$

$c = -a - b = -\left(-\dfrac{4}{3}\right) - 6 = -\dfrac{14}{3}$

Therefore, $a = -\dfrac{4}{3}$, $b = 6$, $c = -\dfrac{14}{3}$

(3) For the graph of $y = x^3 + ax^2 + bx + c$ passing through $(1, 2)$, the graph and the line $y = 3x - 2$ intersect only at a point $(2, 4)$.

Since $(1, 2)$ lies on the graph, $2 = 1 + a + b + c$ $\therefore a + b + c = 1$ $\cdots\cdots$ ①

Let $f(x) = x^3 + ax^2 + bx + c$ Then, $f'(x) = 3x^2 + 2ax + b$

Since $f'(2) = 3$, $3 \cdot 2^2 + 2a \cdot 2 + b = 12 + 4a + b = 3$ $\therefore 4a + b = -9$ $\cdots\cdots$ ②

Since $(2, 4)$ lies on the graph, $4 = 2^3 + a \cdot 2^2 + 2b + c$ $\therefore 4a + 2b + c = -4$ $\cdots\cdots$ ③

③ − ① ; $3a + b = -5$ ······ ④

② − ④ ; $a = -4$

By ④, $b = -3a - 5 = -3(-4) - 5 = 7$

By ①, $c = 1 - a - b = 1 - (-4) - 7 = -2$

Therefore, $a = -4$, $b = 7$, $c = -2$

(4) The two graphs of $y = a + \sin x$ and $y = \cos^2 x + bx$ intersect only at one point on a line $y = x + c$. $\left(-\dfrac{\pi}{2} < x < \dfrac{\pi}{2}\right)$

Let $f(x) = a + \sin x$ and $g(x) = \cos^2 x + bx$

Then, $f'(x) = \cos x$ and $g'(x) = 2\cos x \,(-\sin x) + b$

Let $x = t$ be the point of tangency of $f(x)$ and $g(x)$.

Then, the coordinates of the point is $(t,\ t + c)$.

Since $f(t) = g(t)$ and $f'(t) = g'(t) = 1$,

$a + \sin t = \cos^2 t + bt$ ······ ①   and $\cos t = 2\cos t\,(-\sin t) + b = 1$ ······ ②

By ① and ②, $t = 0$ $\left(\because -\dfrac{\pi}{2} < x < \dfrac{\pi}{2}\right)$ and $a = b = 1$

∴ $f(x) = 1 + \sin x$, $g(x) = \cos^2 x + x$

Since the point of tangency lies on the graph of $f(x)$, the point is $(t,\ t + c) = (0, 1)$

Therefore, $a = b = c = 1$

#2 Find the equation of the tangent line to the graph at the point indicated.

(1) $y = x^2$ at $(2, 4)$

$y = x^2 \Rightarrow y' = 2x$

$y'_{x=2} = 2 \cdot 2 = 4$

∴ The equation of the tangent line at $(2, 4)$ is $y - 4 = 4(x - 2)$; $y = 4x - 4$

Alternative Approach:

Let m be the slope of the tangent line passing through $(2, 4)$.

Then, the equation of the tangent line is $y - 4 = m(x - 2)$; $y = mx - 2m + 4$ ······ ①

Substituting ① into $y = x^2$, $x^2 = mx - 2m + 4$; $x^2 - mx + 2m - 4 = 0$ ······ ②

Let D be the discriminant of the equation ②. Then, $D = 0$

∴ $D = m^2 - 4(2m - 4) = m^2 - 8m + 16 = (m - 4)^2 = 0$ ∴ $m = 4$

Therefore, $y = 4x - 2 \cdot 4 + 4 = 4x - 4$

(2) $y = x^3 + x^2 - 3x + 4$ at $(1, 5)$

$y' = 3x^2 + 2x - 3$

$y'_{x=1} = 3 \cdot 1^2 + 2 \cdot 1 - 3 = 2$

∴ The equation of the tangent line at $(1,5)$ is $y - 5 = 2(x - 1)$; $y = 2x + 3$

(3) $y = \frac{\ln x}{x}$ at $(0, 0)$

$y' = \frac{\frac{1}{x} \cdot x - (\ln x) \cdot 1}{x^2} = \frac{1 - \ln x}{x^2}$

Then, the point of tangency at $x = t$ is $\left(t, \frac{\ln t}{t}\right)$.

The slope of the tangent line at $x = t$ is $y'_{x=t} = \frac{1 - \ln t}{t^2}$

∴ The equation of the tangent line at $\left(t, \frac{\ln t}{t}\right)$ is $y - \frac{\ln t}{t} = \frac{1 - \ln t}{t^2}(x - t)$

That is, $y = \frac{1 - \ln t}{t^2} x - \frac{1 - \ln t}{t^2} t + \frac{\ln t}{t} = \frac{1 - \ln t}{t^2} x - \frac{1}{t} + \frac{2 \ln t}{t} \cdots \cdots$ ①

Since $(0, 0)$ lies on the graph ①, $0 = -\frac{1}{t} + \frac{2 \ln t}{t}$; $\ln t = \frac{1}{2}$; $t = e^{\frac{1}{2}}$

Therefore, the equation of the tangent line at $(0, 0)$ is

$y = \frac{1 - \ln e^{\frac{1}{2}}}{e} x - \frac{1}{e^{\frac{1}{2}}} + \frac{2 \ln e^{\frac{1}{2}}}{e^{\frac{1}{2}}} = \frac{1 - \frac{1}{2}}{e} x - \frac{1}{e^{\frac{1}{2}}} + \frac{2 \cdot \frac{1}{2}}{e^{\frac{1}{2}}} = \frac{1}{2e} x - \frac{1}{e^{\frac{1}{2}}} + \frac{1}{e^{\frac{1}{2}}} = \frac{1}{2e} x$

(4) $x^2 + y^2 = r^2$ at (x_1, y_1)

$\frac{d}{dx}(x^2 + y^2) = \frac{d}{dx}(r^2)$; $2x + 2y\frac{dy}{dx} = 0$ ∴ $\frac{dy}{dx} = -\frac{x}{y}$ $(y \neq 0)$

∴ The equation of the tangent line at (x_1, y_1) is $y - y_1 = -\frac{x_1}{y_1}(x - x_1)$

That is, $y_1 y = -x_1 x + x_1^2 + y_1^2$; $x_1 x + y_1 y = x_1^2 + y_1^2 = r^2$

Therefore, the equation of the tangent line at (x_1, y_1) is $x_1 x + y_1 y = r^2$

(5) $y^2 = 4px$ at (x_1, y_1)

$\frac{d}{dx}(y^2) = \frac{d}{dx}(4px)$; $2y\frac{dy}{dx} = 4p$ ∴ $\frac{dy}{dx} = \frac{2p}{y}$ $(y \neq 0)$

∴ The equation of the tangent line at (x_1, y_1) is $y - y_1 = \frac{2p}{y_1}(x - x_1)$

That is, $y_1 y - y_1^2 = 2p(x - x_1)$; $y_1 y - 4px_1 = 2p(x - x_1)$; $y_1 y = 2p(x + x_1)$

Therefore, the equation of the tangent line at (x_1, y_1) is $y_1 y = 2p(x + x_1)$

(6) $\dfrac{x^2}{a^2} + \dfrac{y^2}{b^2} = 1$ at (x_1, y_1) $(a, b \, ; \text{constants})$

$\dfrac{d}{dx}\left(\dfrac{x^2}{a^2} + \dfrac{y^2}{b^2}\right) = \dfrac{d}{dx}(1)$; $\dfrac{2x}{a^2} + \dfrac{2y}{b^2} \cdot \dfrac{dy}{dx} = 0$ $\qquad \therefore \dfrac{dy}{dx} = -\dfrac{\frac{2x}{a^2}}{\frac{2y}{b^2}} = -\dfrac{b^2 x}{a^2 y}$ $(y \neq 0)$

$\therefore$ The equation of the tangent line at (x_1, y_1) is $y - y_1 = -\dfrac{b^2 x_1}{a^2 y_1}(x - x_1)$

That is, $a^2 y_1 y - a^2 y_1{}^2 = -b^2 x_1 (x - x_1)$

Dividing by $a^2 b^2$, $\dfrac{y_1 y}{b^2} - \dfrac{y_1{}^2}{b^2} = -\dfrac{x_1}{a^2}(x - x_1)$; $\dfrac{y_1 y}{b^2} - \dfrac{y_1{}^2}{b^2} = -\dfrac{x_1 x}{a^2} + \dfrac{x_1{}^2}{a^2}$

Therefore, the equation of the tangent line at (x_1, y_1) is

$\dfrac{x_1 x}{a^2} + \dfrac{y_1 y}{b^2} = \dfrac{x_1{}^2}{a^2} + \dfrac{y_1{}^2}{b^2}$ $\qquad \therefore \dfrac{x_1 x}{a^2} + \dfrac{y_1 y}{b^2} = 1$

(7) $x \cos y + y \cos x = -2\pi$ at (π, π)

$\dfrac{d}{dx}(x \cos y + y \cos x) = \dfrac{d}{dx}(-2\pi)$; $\cos y + x(-\sin y)\dfrac{dy}{dx} + \dfrac{dy}{dx}\cos x + y(-\sin x) = 0$ $\therefore$

$\dfrac{dy}{dx}(\cos x - x \sin y) = y \sin x - \cos y$

$\therefore \dfrac{dy}{dx} = \dfrac{y \sin x - \cos y}{\cos x - x \sin y}$

The slope of the tangent line at (π, π) is $\dfrac{\pi \sin \pi - \cos \pi}{\cos \pi - \pi \sin \pi} = \dfrac{-(-1)}{-1} = -1$

$\therefore$ The equation of the tangent line at (π, π) is $y - \pi = -1(x - \pi)$; $y = -x + 2\pi$

(8) $x^2 + 5ye^x + y^3 = 4$ at $(0, 1)$

$\dfrac{d}{dx}(x^2 + 5ye^x + y^3) = \dfrac{d}{dx}(4)$

$2x + 5\dfrac{dy}{dx}e^x + 5ye^x + 3y^2 \dfrac{dy}{dx} = 0$; $\dfrac{dy}{dx}(5e^x + 3y^2) = -2x - 5ye^x$ $\qquad \therefore \dfrac{dy}{dx} = \dfrac{-2x - 5ye^x}{5e^x + 3y^2}$

The slope of the tangent line at $(0, 1)$ is $\dfrac{0 - 5 \cdot 1 \cdot e^0}{5e^0 + 3 \cdot 1^2} = \dfrac{-5}{5 + 3} = -\dfrac{5}{8}$

$\therefore$ The equation of the tangent line at $(0, 1)$ is $y - 1 = -\dfrac{5}{8}(x - 0)$; $y = -\dfrac{5}{8}x + 1$

(9) $y = x^{2x}$ $(x > 0)$ at $(1, 1)$

$\log y = \log x^{2x} = 2x \log x$

$\dfrac{d}{dx}(\log y) = \dfrac{d}{dx}(2x \log x)$

$\dfrac{1}{y} \cdot \dfrac{dy}{dx} = 2\left(\log x + x \cdot \dfrac{1}{x}\right)$ $\qquad \therefore \dfrac{dy}{dx} = 2x^{2x}(\log x + 1)$

The slope of the tangent line at $(1, 1)$ is $2(\log 1 + 1) = 2$

$\therefore$ The equation of the tangent line at $(1, 1)$ is $y - 1 = 2(x - 1)$; $y = 2x - 1$

(10) $x = \cos^3 \theta$, $y = \sin^3 \theta$ at $\theta = \frac{\pi}{6}$

$\frac{dx}{d\theta} = 3\cos^2 \theta \, (-\sin\theta)$, $\quad \frac{dy}{d\theta} = 3\sin^2 \theta \, (\cos\theta)$

$\frac{dy}{dx} = \frac{dy}{d\theta} \cdot \frac{d\theta}{dx} = 3\sin^2 \theta \, (\cos\theta) \cdot \frac{1}{3\cos^2 \theta(-\sin\theta)} = -\frac{\sin\theta}{\cos\theta} = -\tan\theta$

The slope of the tangent line at $\theta = \frac{\pi}{6}$ is $-\tan\frac{\pi}{6} = -\frac{1}{\sqrt{3}}$

At $\theta = \frac{\pi}{6}$, $x = \cos^3 \frac{\pi}{6} = \left(\frac{\sqrt{3}}{2}\right)^3 = \frac{3\sqrt{3}}{8}$, $y = \sin^3 \frac{\pi}{6} = \left(\frac{1}{2}\right)^3 = \frac{1}{8}$

∴ The equation of the tangent line at $\left(\frac{3\sqrt{3}}{8}, \frac{1}{8}\right)$ is $y - \frac{1}{8} = -\frac{1}{\sqrt{3}}\left(x - \frac{3\sqrt{3}}{8}\right)$

That is, $y = -\frac{1}{\sqrt{3}}x + \frac{3}{8} + \frac{1}{8} = -\frac{\sqrt{3}}{3}x + \frac{1}{2}$

(11) $y = \left|x^2 - x[x]\right|$ at $x = 2.5$ ($[x]$ = **The greatest integer in** x)

$2 \leq x < 3 \implies [x] = 2$

$x^2 - x[x] = x^2 - 2x = x(x-2) \geq 0$

∴ $y = \left|x^2 - x[x]\right| = x^2 - 2x$ $\qquad$ ∴ $y' = 2x - 2$

When $x = 2.5$, $y = \left(\frac{5}{2}\right)^2 - 2\left(\frac{5}{2}\right) = \frac{25}{4} - 5 = \frac{5}{4}$ and $y' = 2\left(\frac{5}{2}\right) - 2 = 5 - 2 = 3$

∴ The equation of the tangent line at $\left(2.5, \frac{5}{4}\right)$ is $y - \frac{5}{4} = 3\left(x - \frac{5}{2}\right)$

That is, $y = 3x - \frac{15}{2} + \frac{5}{4} = 3x - \frac{25}{4}$

#3 Find the equation of the normal line to the curve at the point indicated.

(1) $y = x^3$ at $(-1, -1)$

$y' = 3x^2$; $\quad y'_{x=-1} = 3$

∴ The equation of the tangent line at $(-1, -1)$ is $y - (-1) = 3(x - (-1))$; $y = 3x + 2$

Since the normal line and tangent line are perpendicular, the slope of the normal line is $-\frac{1}{3}$.

∴ The equation of the normal line at $(-1, -1)$ is $y - (-1) = -\frac{1}{3}(x - (-1))$; $y = -\frac{1}{3}x - \frac{4}{3}$

(2) $y = x^3 - 3x^2 + 4$ at $(1, 2)$

$y' = 3x^2 - 6x$; $\quad y'_{x=1} = 3 - 6 = -3$

∴ The equation of the tangent line at $(1, 2)$ is $y - 2 = -3(x - 1)$; $y = -3x + 5$

∴ The slope of the normal line is $\frac{1}{3}$.

∴ The equation of the normal line at $(1, 2)$ is $y - 2 = \frac{1}{3}(x - 1)$; $y = \frac{1}{3}x + \frac{5}{3}$

(3) $y = e^{x-1}$ at $(3, e)$

$y' = e^{x-1}$; $y'_{x=3} = e^{3-1} = e^2$

∴ The equation of the tangent line at $(3, e)$ is $y - e = e^2(x - 3)$; $y = e^2 x - 3e^2 + e$

∴ The slope of the normal line is $-\dfrac{1}{e^2}$

∴ The equation of the normal line at $(3, e)$ is $y - e = -\dfrac{1}{e^2}(x - 3)$; $y = -\dfrac{1}{e^2}x + \dfrac{3}{e^2} + e$

(4) $y = \log(1 + x)$ at $(1, 0)$

$y' = \dfrac{1}{1+x}$; $y'_{x=1} = \dfrac{1}{2}$

∴ The equation of the tangent line at $(1, 0)$ is $y - 0 = \dfrac{1}{2}(x - 1)$; $y = \dfrac{1}{2}x - \dfrac{1}{2}$

∴ The slope of the normal line is -2

∴ The equation of the normal line at $(1, 0)$ is $y - 0 = -2(x - 1)$; $y = -2x + 2$

(5) $y = \dfrac{x}{1-x}$ at $(3, 1)$

$y' = \dfrac{1(1-x)-x(-1)}{(1-x)^2} = \dfrac{1}{(1-x)^2}$; $y'_{x=3} = \dfrac{1}{(1-3)^2} = \dfrac{1}{4}$

∴ The equation of the tangent line at $(3,1)$ is $y - 1 = \dfrac{1}{4}(x - 3)$; $y = \dfrac{1}{4}x + \dfrac{1}{4}$

∴ The slope of the normal line is -4

∴ The equation of the normal line at $(3, 1)$ is $y - 1 = -4(x - 3)$; $y = -4x + 13$

(6) $y = \tan^2 x$ at $\left(\dfrac{\pi}{4}, 1\right)$

$y' = 2 \tan x \sec^2 x$; $y'_{x=\frac{\pi}{4}} = 2 \tan\dfrac{\pi}{4} \sec^2 \dfrac{\pi}{4} = 2 \cdot 1 \cdot 2 = 4$

∴ The equation of the tangent line at $\left(\dfrac{\pi}{4}, 1\right)$ is $y - 1 = 4\left(x - \dfrac{\pi}{4}\right)$; $y = 4x - \pi + 1$

∴ The slope of the normal line is $-\dfrac{1}{4}$

∴ The equation of the normal line at $\left(\dfrac{\pi}{4}, 1\right)$ is $y - 1 = -\dfrac{1}{4}\left(x - \dfrac{\pi}{4}\right)$; $y = -\dfrac{1}{4}x + \dfrac{\pi}{16} + 1$

#4 Find the value.

(1) When the graphs of $y = ax^2$ and $y = \log x$ intersect only at one point, find the value of the constant a.

$y = ax^2 \Rightarrow y' = 2ax$

$y = \log x \Rightarrow y' = \dfrac{1}{x}$

Let (α, β) be the point of tangency. Then, $2a\alpha = \dfrac{1}{\alpha}$ ⋯⋯ ①

Since (α, β) lies on the two graphs, $\beta = a\alpha^2$ and $\beta = \log \alpha$; i.e., $a\alpha^2 = \log \alpha$ ······ ②

By ①, $2a\alpha^2 = 1$; $a\alpha^2 = \dfrac{1}{2}$

By ②, $\log \alpha = a\alpha^2 = \dfrac{1}{2}$; $\alpha = e^{\frac{1}{2}}$

Therefore, $a = \dfrac{1}{2\alpha^2} = \dfrac{1}{2e}$

(2) When the two graphs of $y = a - 2\cos^2 x$ and $y = 2\sin x$ intersect only at one point,
   find the value of the constant a. $(0 < x < 2\pi)$

Let $f(x) = a - 2\cos^2 x$ Then, $f'(x) = -4\cos x\,(-\sin x) = 4\cos x \sin x$

Let $g(x) = 2\sin x$ Then, $g'(x) = 2\cos x$

Let the two graphs intersect at $x = t$. Then, $f(t) = g(t)$ and $f'(t) = g'(t)$

$\therefore$ $a - 2\cos^2 t = 2\sin t$; $a = 2\cos^2 t + 2\sin t$ ······ ①

    $4\cos t \sin t = 2\cos t$; $2\cos t\,(2\sin t - 1) = 0$; $\cos t = 0$ or $\sin t = \dfrac{1}{2}$

i) When $\cos t = 0$,

$t = \dfrac{\pi}{2}$ or $t = \dfrac{3\pi}{2}$ ($\because 0 < x < 2\pi$)

If $t = \dfrac{\pi}{2}$, then $a = 2\cos^2 \dfrac{\pi}{2} + 2\sin \dfrac{\pi}{2} = 0 + 2 = 2$

If $t = \dfrac{3\pi}{2}$, then $a = 2\cos^2 \dfrac{3\pi}{2} + 2\sin \dfrac{3\pi}{2} = 0 - 2 = -2$

ii) When $\sin t = \dfrac{1}{2}$,

$t = \dfrac{\pi}{6}$ or $t = \dfrac{5\pi}{6}$ ($\because 0 < x < 2\pi$)

If $t = \dfrac{\pi}{6}$, then $a = 2\cos^2 \dfrac{\pi}{6} + 2\sin \dfrac{\pi}{6} = \dfrac{3}{2} + 1 = \dfrac{5}{2}$

If $t = \dfrac{5\pi}{6}$, then $a = 2\cos^2 \dfrac{5\pi}{6} + 2\sin \dfrac{5\pi}{6} = \dfrac{3}{2} + 1 = \dfrac{5}{2}$

Therefore, by i) and ii), $a = 2$, $a = -2$, or $a = \dfrac{5}{2}$

(3) Find the value of the positive constant a so that the tangent lines of the two graphs
   $f(x) = 3x^2$ and $g(x) = a - 4x^2$ are perpendicular to each other.

Suppose the two graphs intersect at $x = t$.

Then, the intersection point on the graph of $f(x) = 3x^2$ is $(t,\ 3t^2)$ and the intersection point
on the graph of $g(x) = a - 4x^2$ is $(t, a - 4t^2)$.

Since they are the same point, $3t^2 = a - 4t^2$; $a = 7t^2$

Since $f'(x) = 6x$, the slope of the tangent line at $x = t$ on the graph of $f(x)$ is $6t$.

Since $g'(x) = -8x$, the slope of the tangent line at $x = t$ on the graph of $g(x)$ is $-8t$.

Since the tangent lines of the two graphs are perpendicular to each other,

$6t \times (-8t) = -1$; $-48t^2 = -1$ $\therefore$ $t^2 = \frac{1}{48}$ Therefore, $a = 7t^2 = 7 \times \frac{1}{48} = \frac{7}{48}$

(4) When the tangent lines of the two graphs $y = \log(2x + 3)$ and $y = a - \log x$ are
perpendicular to each other, find the value of the constant a.

$y = \log(2x + 3) \Rightarrow y' = \frac{2}{2x+3}$

$y = a - \log x \Rightarrow y' = -\frac{1}{x}$

Let t be the x-coordinate of the intersection point.

Then, $\frac{2}{2t+3} \times \left(-\frac{1}{t}\right) = -1$; $(2t + 3)t = 2$; $2t^2 + 3t - 2 = 0$; $(2t - 1)(t + 2) = 0$

$\therefore$ $t = \frac{1}{2}$ or $t = -2$

Since $t > 0$, $t = \frac{1}{2}$

Since the two graphs pass through the intersection point, $\log(2t + 3) = a - \log t$

$\therefore$ $\log(2 \cdot \frac{1}{2} + 3) = a - \log \frac{1}{2}$; $\log 4 = a - \log \frac{1}{2}$

$\therefore$ $a = \log 4 + \log \frac{1}{2} = \log \left(4 \cdot \frac{1}{2}\right) = \log 2$

(5) Find the equation of the tangent line at $\left(\frac{1}{2}, \frac{1}{2}\right)$ to the curve $x^3 + y^3 - 3xy = 0$.

$\frac{d}{dx}(x^3 + y^3 - 3xy) = \frac{d}{dx}(0)$

$\therefore$ $3x^2 + 3y^2 \frac{dy}{dx} - 3\left(y + x\frac{dy}{dx}\right) = 0$; $\frac{dy}{dx}(3y^2 - 3x) = 3y - 3x^2$

$\therefore$ $\frac{dy}{dx} = \frac{3y - 3x^2}{3y^2 - 3x} = \frac{y - x^2}{y^2 - x}$

$\therefore$ The slope of the tangent line at $\left(\frac{1}{2}, \frac{1}{2}\right)$ is $\frac{\frac{1}{2} - \left(\frac{1}{2}\right)^2}{\left(\frac{1}{2}\right)^2 - \frac{1}{2}} = -1$

Therefore, the equation of the tangent line at $\left(\frac{1}{2}, \frac{1}{2}\right)$ is $y - \frac{1}{2} = -1\left(x - \frac{1}{2}\right)$

That is, $y = -x + 1$

(6) When the tangent line at $(1, e)$ on the graph of $y = e^x$ intersects only at one point on
the graph of $y = 3\sqrt{x - a}$, find the value of the real number a.

$y = e^x \Rightarrow y' = e^x$ $\therefore$ $y'_{x=1} = e^1 = e$

$\therefore$ The equation of the tangent line at $(1, e)$ is $y - e = e(x - 1)$; $y = ex$ $\cdots\cdots$ ①

Substituting ① into $y = 3\sqrt{x - a}$, $ex = 3\sqrt{x - a}$

Squaring both sides, $e^2x^2 = 9(x - a)$; $e^2x^2 - 9x + 9a = 0$ $\cdots\cdots$ ②

Let D be the discriminant of the equation ②.

Then, $D = 0$ ∴ $81 - 36ae^2 = 0$; $a = \frac{81}{36e^2} = \frac{9}{4e^2}$

(7) For the tangent line at $(0,0)$ on the graph of $y = \frac{2}{3}\sin 2x + 3x^2$, let θ be the angle between the tangent line and x-axis. Find the slope of the line which bisects the angle. $(0 < \theta < \frac{\pi}{2})$

$y = \frac{2}{3}\sin 2x + 3x^2 \Rightarrow y' = 2 \cdot \frac{2}{3}\cos 2x + 6x = \frac{4}{3}\cos 2x + 6x$ ∴ $y'_{x=0} = \frac{4}{3}$

∴ $\tan \theta = \frac{4}{3}$

Let $\theta = 2\alpha$ Then, $\tan 2\alpha = \frac{4}{3}$; $\frac{2\tan\alpha}{1-\tan^2\alpha} = \frac{4}{3}$

∴ $4\tan^2\alpha + 6\tan\alpha - 4 = 0$; $2\tan^2\alpha + 3\tan\alpha - 2 = 0$; $(2\tan\alpha - 1)(\tan\alpha + 2) = 0$

∴ $\tan\alpha = \frac{1}{2}$ or $\tan\alpha = -2$

Since $0 < \alpha < \frac{\pi}{4}$, $\tan\alpha = \frac{1}{2}$

Therefore, the slope of the bisecting line is $\frac{1}{2}$.

(8) Let $P(0, y_1)$ and $Q(3, y_2)$ be points on the graph of $y = x^3 - 4x$. For the line l passing through the two points P and Q, find the equation of the tangent line, which is parallel to the line l, of the graph.

$y = x^3 - 4x \Rightarrow y' = 3x^2 - 4$

The slope of the line passing through $P(0, y_1)$ and $Q(3, y_2)$ is $\frac{y_2-y_1}{3-0} = \frac{15-0}{3-0} = 5$

∴ $3x^2 - 4 = 5$; $3x^2 = 9$; $x^2 = 3$; $x = \pm\sqrt{3}$

∴ Points of tangency are $(\sqrt{3}, -\sqrt{3})$ and $(-\sqrt{3}, \sqrt{3})$

∴ The equation of the tangent line at $(\sqrt{3}, -\sqrt{3})$ is $y + \sqrt{3} = 5(x - \sqrt{3})$; $y = 5x - 6\sqrt{3}$

The equation of the tangent line at $(-\sqrt{3}, \sqrt{3})$ is $y - \sqrt{3} = 5(x + \sqrt{3})$; $y = 5x + 6\sqrt{3}$

(9) When the graph of $y = x^3 - 4x + 3$ has two tangent lines which are parallel to the line $y = 2x + 3$, find the distance between the two tangent lines.

$y = x^3 - 4x + 3 \Rightarrow y' = 3x^2 - 4$

Since the tangent lines are parallel to the line $y = 2x + 3$, the slope of the tangent line is 2.

∴ $3x^2 - 4 = 2$; $3x^2 = 6$; $x^2 = 2$; $x = \pm\sqrt{2}$

∴ The points of tangency are $(\sqrt{2}, -2\sqrt{2} + 3)$ and $(-\sqrt{2}, 2\sqrt{2} + 3)$

∴ The equation of the tangent line at $(\sqrt{2}, -2\sqrt{2} + 3)$ is $y - (-2\sqrt{2} + 3) = 2(x - \sqrt{2})$

That is, $y = 2x - 4\sqrt{2} + 3$

The equation of the tangent line at $(-\sqrt{2},\ 2\sqrt{2} + 3)$ is $y - (2\sqrt{2} + 3) = 2(x + \sqrt{2})$

That is, $y = 2x + 4\sqrt{2} + 3$

Let d be the distance between the two tangent lines.

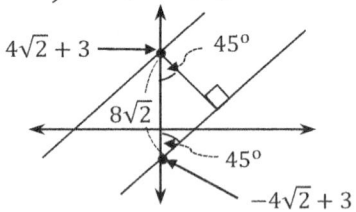

Then, $\sin 45° = \dfrac{d}{8\sqrt{2}}$ $\quad \therefore\ d = 8\sqrt{2}\sin 45° = 8\sqrt{2} \cdot \dfrac{1}{\sqrt{2}} = 8$

(10) **When the graph of $y = x^4$ and a circle with center on y-axis intersect only at one**

point $(1, 1)$, find the value of the radius of the circle.

Let $(0, a)$ be the center of the circle and r be the radius of it.

Then, the equation of the circle is $x^2 + (y - a)^2 = r^2$

Since $(1, 1)$ lies on the circle, $1^2 + (1 - a)^2 = r^2$

$y = x^4 \ \Rightarrow\ y' = 4x^3 \quad \therefore\ y'_{x=1} = 4$

$\therefore$ The equation of the tangent line at $(1, 1)$ is $y - 1 = 4(x - 1)$; $y = 4x - 3$

Since the line passing through $(0, a)$ and $(1, 1)$ is the normal line at $(1, 1)$,

the slope of the normal line is $\dfrac{1-a}{1-0} = -\dfrac{1}{4}$ $\quad \therefore\ 1 - a = -\dfrac{1}{4}$; $a = \dfrac{5}{4}$

$\therefore\ r^2 = 1^2 + (1 - a)^2 = 1 + \left(1 - \dfrac{5}{4}\right)^2 = \dfrac{17}{16}$

Since $r > 0$, $r = \dfrac{\sqrt{17}}{4}$

(11) **When the graph of $y = x^2 + 3$ has two tangent lines at a point $P(1, 0)$, and Q and R**

are the points of tangency, find the area of the triangle $\triangle PQR$.

$y = x^2 + 3 \ \Rightarrow\ y' = 2x$

$\therefore$ The equation of the tangent line at $(t,\ t^2 + 3)$ on the graph is $y - (t^2 + 3) = 2t(x - t)$

Since the line passes through the point $P(1, 0)$, $0 - (t^2 + 3) = 2t(1 - t)$

$\therefore\ t^2 - 2t - 3 = 0$; $(t - 3)(t + 1) = 0$; $t = 3$ or $t = -1$

$\therefore\ Q = Q(-1, 4),\ R = R(3, 12)$

Therefore, the area of the triangle $\triangle PQR$ is

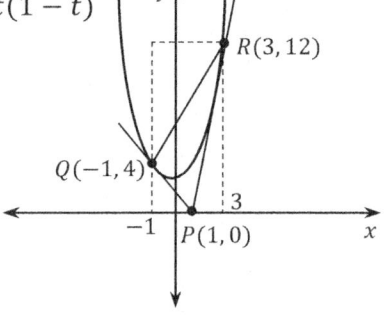

$(4 \times 12) - \left\{\left(\dfrac{1}{2} \cdot 2 \cdot 4\right) + \left(\dfrac{1}{2} \cdot 2 \cdot 12\right) + \left(\dfrac{1}{2} \cdot 8 \cdot 4\right)\right\}$

$= 48 - (4 + 12 + 16) = 16$

(12) **When the tangent line at $P(a, e^{2a})$ on the graph of $y = e^{2x}$ crosses the x-axis at Q,**

let R be the point on x-axis at which the line segment $\overline{PR}$ and x-axis are

perpendicular.

Find the value of a so that the area of the triangle $\triangle PQR$ is 64.

$y = e^{2x} \Rightarrow y' = 2e^{2x}$

The equation of the tangent line at $P(a, e^{2a})$ is $y - e^{2a} = 2e^{2a}(x - a) \cdots\cdots$ ①

Since ① crosses the x-axis $(y = 0)$, $0 - e^{2a} = 2e^{2a}(x - a)$

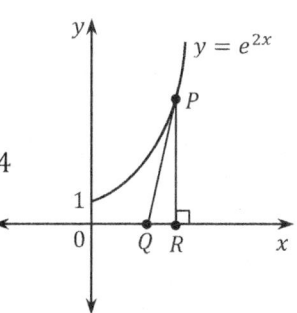

$\therefore \quad x - a = -\dfrac{1}{2} \ ; \quad x = a - \dfrac{1}{2}$

$\therefore \quad Q = Q\left(a - \dfrac{1}{2},\ 0\right),\ R = R(a, 0)$

$\therefore$ The area of the triangle $\triangle PQR$ is $\dfrac{1}{2} \cdot \left(a - \left(a - \dfrac{1}{2}\right)\right) \cdot e^{2a} = 64$

$\therefore \quad \dfrac{1}{4} e^{2a} = 64 \ ; \quad e^{2a} = 64 \cdot 4 = 4^4$

$\therefore \quad 2a = \log 4^4 = 4 \log 4$

Therefore, $a = 2 \log 4$

#5 For a function $f(x) = e^{2x-3}$, let $g(x)$ be the inverse of $f(x)$.

Find the equation of the tangent line at $x = 1$ on the graph of $y = g(x)$.

$y = e^{2x-3} \Rightarrow \ln y = \ln e^{2x-3} = 2x - 3$

$\therefore \quad x = \dfrac{3 + \ln y}{2}$

Changing x and y, $\quad y = \dfrac{3 + \ln x}{2} \qquad \therefore g(x) = \dfrac{3 + \ln x}{2} \qquad \therefore g'(x) = \dfrac{1}{2x}$

$\therefore$ The slope of the tangent line at $x = 1$ is $\dfrac{1}{2 \cdot 1} = \dfrac{1}{2}$

The point of tangency is $(1, g(1)) = \left(1, \dfrac{3}{2}\right)$

$\therefore$ The equation of the tangent line at $x = 1$ is $y - \dfrac{3}{2} = \dfrac{1}{2}(x - 1) \ ; \ y = \dfrac{1}{2}x + 1$

Alternative Approach:

Let $g(1) = t$

Then, $1 = g^{-1}(t) = f(t)$

$f(x) = e^{2x-3} \Rightarrow f(t) = e^{2t-3} = 1 \qquad \therefore 2t - 3 = 0 \ ; \ t = \dfrac{3}{2}$

$f'(x) = 2e^{2x-3} \qquad \therefore g'(1) = \dfrac{1}{f'(g(1))} = \dfrac{1}{f'(t)} = \dfrac{1}{f'\left(\frac{3}{2}\right)} = \dfrac{1}{2e^{2\left(\frac{3}{2}\right)-3}} = \dfrac{1}{2e^0} = \dfrac{1}{2}$

$\therefore$ The equation of the tangent line at $(1, g(1)) = (1, t) = \left(1, \dfrac{3}{2}\right)$ is $y - \dfrac{3}{2} = \dfrac{1}{2}(x - 1)$

That is, $y = \dfrac{1}{2}x + 1$

#6 Find the number c guaranteed by the Mean Value Theorem for the following function on the given interval.

(1) $f(x) = x^2$, $[0, 2]$

$f(x) = x^2 \Rightarrow f'(x) = 2x$

$\frac{f(2) - f(0)}{2 - 0} = 2c \quad (0 < c < 2)$ $\qquad \therefore \frac{2^2 - 0}{2 - 0} = 2c \qquad \therefore c = 1$

(2) $f(x) = 2\sqrt{x - 1}$, $[1, 3]$

$f(x) = 2\sqrt{x - 1} \Rightarrow f'(x) = 2 \cdot \frac{1}{2\sqrt{x-1}} = \frac{1}{\sqrt{x-1}}$

$\frac{f(3) - f(1)}{3 - 1} = \frac{1}{\sqrt{c-1}} \quad (1 < c < 3)$

$\therefore \frac{2\sqrt{2} - 0}{3 - 1} = \frac{1}{\sqrt{c-1}} \; ; \; \sqrt{2} = \frac{1}{\sqrt{c-1}} \; ; \; 2(c - 1) = 1 \qquad \therefore c = \frac{3}{2}$

(3) $f(x) = \log x$, $[1, e]$

$f(x) = \log x \Rightarrow f'(x) = \frac{1}{x}$

$\frac{f(e) - f(1)}{e - 1} = \frac{1}{c} \quad (1 < c < e)$

$\therefore \frac{\log e - \log 1}{e - 1} = \frac{1}{c} \; ; \; \frac{1 - 0}{e - 1} = \frac{1}{c} \qquad \therefore c = e - 1$

#7 Find the number c guaranteed by the Rolle's Theorem for the continuous function $f(x) = e^x + e^{-x}$ on a closed interval $[-2, 2]$.

The function $f(x) = e^x + e^{-x}$ is continuous on the closed interval $[-2, 2]$ and differentiable on an open interval $(-2, 2)$.

Since $f(-2) = f(2) = e^2 + e^{-2}$, there is at least one c in $(-2, 2)$ such that $f'(c) = 0$.

Since $f'(x) = e^x - e^{-x}$, $f'(c) = e^c - e^{-c} = 0$

Multiplying e^c, $e^{2c} - e^0 = 0 \; ; \; e^{2c} = e^0 \qquad \therefore c = 0$

#8 Find the limit using the Mean Value Theorem.

(1) $\displaystyle\lim_{x \to 2} \frac{3^x - 3^2}{x - 2}$

Let $f(x) = 3^x$

Then, f is continuous on $[2, x]$ and differentiable on $(2, x)$.

By the Mean Value Theorem, there is at least one c $(2 < c < x)$ such that $\frac{f(x) - f(2)}{x - 2} = f'(c)$

Since $f'(x) = 3^x \log 3$, $\dfrac{f(x)-f(2)}{x-2} = 3^c \log 3$

Since $c \to 2$ as $x \to 2$, $\displaystyle\lim_{x\to 2} \dfrac{3^x-3^2}{x-2} = \lim_{c\to 2} 3^c \log 3 = 3^2 \log 3 = 9 \log 3$

Alternative Approach:

Let $x - 2 = h$ Then, $x = h + 2$

Since $h \to 0$ as $x \to 2$, $\displaystyle\lim_{x\to 2}\dfrac{3^x-3^2}{x-2} = \lim_{h\to 0}\dfrac{3^{h+2}-3^2}{h} = \lim_{h\to 0}\dfrac{9(3^h-1)}{h} = 9\log 3$

$$\left(\because \lim_{h\to 0}\dfrac{a^h-1}{h} = \log a,\ a > 1\right)$$

Alternative Approach:

Let $f(x) = 3^x$ Then, $f'(x) = 3^x \log 3$

$\displaystyle\lim_{x\to 2}\dfrac{3^x-3^2}{x-2} = \lim_{x\to 2}\dfrac{f(x)-f(2)}{x-2} = f'(2) = 3^2 \log 3 = 9 \log 3$

Alternative Approach:

$\displaystyle\lim_{x\to 2}\dfrac{3^x-3^2}{x-2}$ $\left(\text{Form of } \dfrac{0}{0}\right)$

By L'Hôpital's Rule, $\displaystyle\lim_{x\to 2}\dfrac{3^x-3^2}{x-2} = \lim_{x\to 2}\dfrac{(3^x-3^2)'}{(x-2)'} = \lim_{x\to 2}\dfrac{3^x \log 3}{1} = 3^2 \log 3 = 9 \log 3$

(2) $\displaystyle\lim_{x\to 0^+}\dfrac{\sin x - \sin(\sin x)}{x - \sin x}$

Let $f(x) = \sin x$

Then, f is continuous on $[\sin x, x]$, $x > 0$, and differentiable on $(\sin x, x)$.

By Mean Value Theorem, there is at least one c $(\sin x < c < x)$ such that $\dfrac{f(x)-f(\sin x)}{x-\sin x} = f'(c)$

$\therefore\ \dfrac{\sin x - \sin(\sin x)}{x - \sin x} = \cos c$

Since $c \to 0^+$ as $x \to 0^+$, $\displaystyle\lim_{x\to 0^+}\dfrac{\sin x - \sin(\sin x)}{x-\sin x} = \lim_{x\to 0^+}\cos c = 1$

#9 For $x > 0$, prove the inequality.

(1) $\dfrac{1}{x+1} < \log(x+1) - \log x < \dfrac{1}{x}$

Let $f(x) = \log x$

Then, f is differentiable and continuous when $x > 0$.

$\therefore$ f is continuous on a closed interval $[x, x+1]$ and differentiable on an open interval
$(x, x+1)$, $x > 0$.

By Mean Value Theorem, there is at least one c $(x < c < x+1)$ such that $\dfrac{\log(x+1)-\log x}{x+1-x} = f'(c)$.

Since $f'(x) = \frac{1}{x}$, $\frac{\log(x+1) - \log x}{1} = \frac{1}{c}$

Since $x < c < x+1$ and $x > 0$, $\frac{1}{x+1} < \frac{1}{c} < \frac{1}{x}$

Therefore, $\frac{1}{x+1} < \log(x+1) - \log x < \frac{1}{x}$

(2) $0 < \frac{1}{x} \log \frac{e^x - 1}{x} < 1$

Let $f(x) = e^x$

Then, f is differentiable and continuous for all x.

∴ f is continuous on a closed interval $[0, x]$ and differentiable on an open interval $(0, x)$.

By Mean Value Theorem, there is at least one c $(0 < c < x)$ such that $\frac{f(x) - f(0)}{x - 0} = f'(c)$.

∴ $\frac{e^x - e^0}{x - 0} = e^c$; $\frac{e^x - 1}{x} = e^c$

∴ $\log\left(\frac{e^x - 1}{x}\right) = \log e^c = c$

Since $0 < c < x$, $0 < \log\left(\frac{e^x - 1}{x}\right) < x$

Since $x > 0$, $0 < \frac{1}{x} \log\left(\frac{e^x - 1}{x}\right) < 1$

#10 Use L'Hôpital's Rule to complete the following limits.

(1) $\lim\limits_{x \to a} \dfrac{a \sin x - x \sin a}{x - a}$

$\lim\limits_{x \to a} \dfrac{a \sin x - x \sin a}{x - a}$ $\left(\text{Form of } \dfrac{0}{0}\right) = \lim\limits_{x \to a} \dfrac{(a \sin x - x \sin a)'}{(x - a)'} = \lim\limits_{x \to a} \dfrac{a \cos x - \sin a}{1} = a \cos a - \sin a$

(2) $\lim\limits_{x \to 0^+} \left(\dfrac{1}{x} - \dfrac{1}{\sin x}\right)$

$\lim\limits_{x \to 0^+} \dfrac{1}{x} = \infty, \quad \lim\limits_{x \to 0^+} \dfrac{1}{\sin x} = \infty$

$\lim\limits_{x \to 0^+} \left(\dfrac{1}{x} - \dfrac{1}{\sin x}\right)$ (Form of $\infty - \infty$) $= \lim\limits_{x \to 0^+} \dfrac{\sin x - x}{x \sin x}$ $\left(\text{Form of } \dfrac{0}{0}\right) = \lim\limits_{x \to 0^+} \dfrac{(\sin x - x)'}{(x \sin x)'}$

$\qquad = \lim\limits_{x \to 0^+} \dfrac{\cos x - 1}{\sin x + x \cos x}$ $\left(\text{Form of } \dfrac{0}{0}\right) = \lim\limits_{x \to 0^+} \dfrac{(\cos x - 1)'}{(\sin x + x \cos x)'} = \lim\limits_{x \to 0^+} \dfrac{-\sin x}{\cos x + (\cos x - x \sin x)}$

$\qquad = \lim\limits_{x \to 0^+} \dfrac{-\sin x}{2 \cos x - x \sin x} = \dfrac{0}{2 - 0} = 0$

(3) $\lim\limits_{x \to 0} \dfrac{1 - \cos x - \frac{1}{2} x^2}{x^4}$

$\lim\limits_{x \to 0} \dfrac{1 - \cos x - \frac{1}{2} x^2}{x^4}$ $\left(\text{Form of } \dfrac{0}{0}\right) = \lim\limits_{x \to 0} \dfrac{\left(1 - \cos x - \frac{1}{2} x^2\right)'}{(x^4)'} = \lim\limits_{x \to 0} \dfrac{\sin x - x}{4x^3}$ $\left(\text{Form of } \dfrac{0}{0}\right)$

$$= \lim_{x\to 0}\frac{\cos x-1}{12x^2}\left(\text{Form of }\frac{0}{0}\right)=\lim_{x\to 0}\frac{(\cos x-1)'}{(12x^2)'}=\lim_{x\to 0}\frac{-\sin x}{24x}\left(\text{Form of }\frac{0}{0}\right)$$

$$=\lim_{x\to 0}\frac{-\cos x}{24}=\frac{-\cos 0}{24}=-\frac{1}{24}$$

(4) $\lim_{x\to 0}\dfrac{x-\log(1+x)}{x^2}$

$$\lim_{x\to 0}\frac{x-\log(1+x)}{x^2}\left(\text{Form of }\frac{0}{0}\right)=\lim_{x\to 0}\frac{\{x-\log(1+x)\}'}{(x^2)'}=\lim_{x\to 0}\frac{1-\frac{1}{1+x}}{2x}=\lim_{x\to 0}\frac{\frac{x}{1+x}}{2x}$$

$$=\lim_{x\to 0}\frac{x}{2x(1+x)}\left(\text{Form of }\frac{0}{0}\right)=\lim_{x\to 0}\frac{x'}{\{2x(1+x)\}'}=\lim_{x\to 0}\frac{1}{2(1+x)+2x}=\lim_{x\to 0}\frac{1}{4x+2}=\frac{1}{2}$$

(5) $\lim_{x\to\infty}\dfrac{e^{-x}}{x^{-2}}$

$$\lim_{x\to\infty}\frac{e^{-x}}{x^{-2}}=\lim_{x\to\infty}\frac{x^2}{e^x}=\left(\text{Form of }\frac{\infty}{\infty}\right)=\lim_{x\to\infty}\frac{(x^2)'}{(e^x)'}=\lim_{x\to\infty}\frac{2x}{e^x}\left(\text{Form of }\frac{\infty}{\infty}\right)=\lim_{x\to\infty}\frac{(2x)'}{(e^x)'}$$

$$=\lim_{x\to\infty}\frac{2}{e^x}=0$$

(6) $\lim_{x\to\infty}\dfrac{e^{3x}}{x^3}$

$$\lim_{x\to\infty}\frac{e^{3x}}{x^3}\left(\text{Form of }\frac{\infty}{\infty}\right)=\lim_{x\to\infty}\frac{(e^{3x})'}{(x^3)'}=\lim_{x\to\infty}\frac{3e^{3x}}{3x^2}=\lim_{x\to\infty}\frac{e^{3x}}{x^2}\left(\text{Form of }\frac{\infty}{\infty}\right)$$

$$=\lim_{x\to\infty}\frac{3e^{3x}}{2x}\left(\text{Form of }\frac{\infty}{\infty}\right)=\lim_{x\to\infty}\frac{9e^{3x}}{2}=\infty$$

(7) $\lim_{x\to 0}\dfrac{a^x-b^x}{x}\ (a>0,\ b>0)$

$$\lim_{x\to 0}\frac{a^x-b^x}{x}\left(\text{Form of }\frac{0}{0}\right)=\lim_{x\to 0}\frac{(a^x-b^x)'}{x'}=\lim_{x\to 0}\frac{a^x\log a-b^x\log b}{1}=a^0\log a-b^0\log b$$

$$=\log a-\log b$$

(8) $\lim_{x\to\infty}\left(1+\dfrac{1}{x}\right)^x$

$$\log\left(1+\frac{1}{x}\right)^x=x\log\left(1+\frac{1}{x}\right)=\frac{\log\left(1+\frac{1}{x}\right)}{\frac{1}{x}}$$

$$\lim_{x\to\infty}\log\left(1+\frac{1}{x}\right)^x=\lim_{x\to\infty}\frac{\log\left(1+\frac{1}{x}\right)}{\frac{1}{x}}\left(\text{Form of }\frac{0}{0}\right)=\lim_{x\to\infty}\frac{\{\log\left(1+\frac{1}{x}\right)\}'}{\left(\frac{1}{x}\right)'}=\lim_{x\to\infty}\frac{\frac{-\frac{1}{x^2}}{1+\frac{1}{x}}}{-\frac{1}{x^2}}=\lim_{x\to\infty}\frac{1}{1+\frac{1}{x}}=1$$

Since $\log\lim_{x\to\infty}\left(1+\frac{1}{x}\right)^x=\lim_{x\to\infty}\log\left(1+\frac{1}{x}\right)^x=1,\quad\lim_{x\to\infty}\left(1+\frac{1}{x}\right)^x=e^1=e$

(9) $\lim_{x\to 0}(\cos x)^{\frac{1}{x^2}}$

$$\log(\cos x)^{\frac{1}{x^2}}=\frac{1}{x^2}\log\cos x$$

$$\lim_{x\to 0}\log(\cos x)^{\frac{1}{x^2}}=\lim_{x\to 0}\frac{1}{x^2}\log\cos x=\lim_{x\to 0}\frac{\log\cos x}{x^2}\left(\text{Form of }\frac{0}{0}\right)=\lim_{x\to 0}\frac{(\log\cos x)'}{(x^2)'}$$

$$= \lim_{x \to 0} \frac{\frac{-\sin x}{\cos x}}{2x} = \lim_{x \to 0} \frac{-\sin x}{2x \cos x} = \lim_{x \to 0} \frac{-\tan x}{2x} \left(\text{Form of } \frac{0}{0}\right) = \lim_{x \to 0} \frac{(-\tan x)'}{(2x)'} = \lim_{x \to 0} \frac{-\sec^2 x}{2} = -\frac{1}{2}$$

$$\text{Since } \log \lim_{x \to 0} (\cos x)^{\frac{1}{x^2}} = \lim_{x \to 0} \log(\cos x)^{\frac{1}{x^2}} = -\frac{1}{2}, \quad \lim_{x \to 0} (\cos x)^{\frac{1}{x^2}} = e^{-\frac{1}{2}} = \frac{1}{\sqrt{e}}$$

#11 Show that each of the following is increasing on $(-\infty, \infty)$.

(1) $f(x) = x^3 - 6x^2 + 15x + 10$

$f'(x) = 3x^2 - 12x + 15 = 3(x^2 - 4x) + 15 = 3(x-2)^2 - 12 + 15 = 3(x-2)^2 + 3 > 0$

Since $f'(x) > 0$, $f(x)$ is increasing on $(-\infty, \infty)$.

(2) $f(x) = x^3 - 6x^2 + 12x - 8$

$f'(x) = 3x^2 - 12x + 12 = 3(x^2 - 4x + 4) = 3(x-2)^2 \geq 0$

$\therefore f'(x) \geq 0 \quad (f'(x) = 0 \text{ when } x = 2.)$

$\therefore f(x)$ is increasing on $(-\infty, \infty)$.

(3) $f(x) = x - \sin x$

$f'(x) = 1 - \cos x$

Since $-1 \leq \cos x \leq 1$, $-1 \leq -\cos x \leq 1$

$\therefore 0 \leq 1 - \cos x \leq 2$

$\therefore f'(x) \geq 0 \quad (f'(x) = 0 \text{ when } x = 2n\pi, \ n = 0, \pm 1, \pm 2, \cdots \cdots)$

$\therefore f(x)$ is increasing on $(-\infty, \infty)$.

#12 Prove the statement.

(1) A function $y = e^x - x$ is increasing on $[0, \infty)$.

$y = e^x - x \Rightarrow y' = e^x - 1$

Since $x \geq 0$, $e^x \geq 1$

$\therefore y' \geq 0 \quad (y' = 0 \text{ when } x = 0)$

Therefore, $y = e^x - x$ is increasing on $[0, \infty)$.

(2) A function $y = x^{\frac{1}{x}}$ is decreasing on $[3, \infty)$.

$y = x^{\frac{1}{x}} \Rightarrow \log y = \log x^{\frac{1}{x}} = \frac{1}{x} \log x$

$\Rightarrow \frac{d}{dx}(\log y) = \frac{d}{dx}\left(\frac{1}{x} \log x\right)$

$\therefore \frac{1}{y} \cdot \frac{dy}{dx} = -\frac{1}{x^2} \log x + \frac{1}{x} \cdot \frac{1}{x} = -\frac{1}{x^2} \log x + \frac{1}{x^2} = -\frac{1}{x^2}(\log x - 1)$

$\therefore \frac{dy}{dx} = y\left\{-\frac{1}{x^2}(\log x - 1)\right\} = x^{\frac{1}{x}}\left\{-\frac{1}{x^2}(\log x - 1)\right\}$

Since $x \geq 3$, $\left(\dfrac{1}{x^2} > 0,\ \log x - 1 > 0,\ x^{\frac{1}{x}} > 0\right)$

$\therefore\ y' = \dfrac{dy}{dx} < 0$

Therefore, $y = x^{\frac{1}{x}}$ is decreasing on $[3, \infty)$.

#13 Find the value of the real number a so that the following statement is true.

(1) A function $f(x) = x^3 + ax^2 + ax + 10$ is increasing on $(-\infty, \infty)$.

$f'(x) = 3x^2 + 2ax + a$

Let D be the discriminant of the equation $3x^2 + 2ax + a = 0$.

Since $f'(x) \geq 0$ for any real number x, $f'(x)$ does not have two different real number solutions.

$\therefore\ \dfrac{D}{4} \leq 0$

$\therefore\ \dfrac{D}{4} = a^2 - 3a = a(a - 3) \leq 0$

$\therefore\ 0 \leq a \leq 3$

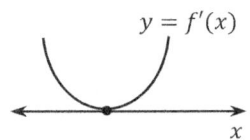
$y = f'(x)$

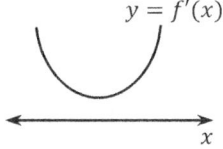
$y = f'(x)$

(2) A function $f(x) = x^3 - 6x^2 + ax - 2$ is decreasing on $(0, 3)$.

$f'(x) = 3x^2 - 12x + a = 3(x - 2)^2 - 12 + a$

Since $f'(x) \leq 0$ $(0 < x < 3)$, $f'(0) = a \leq 0$ and $f'(3) = -9 + a \leq 0$; $a \leq 9$

$\therefore\ a \leq 0$

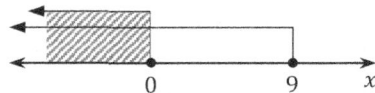
$0 \qquad 9 \qquad x$

(3) A function $f(x) = 2x^3 - 3ax^2 + 6ax - 6x - 1$ is decreasing on $[1, 3]$.

$f'(x) = 6x^2 - 6ax + 6a - 6 \leq 0$

$\therefore\ x^2 - ax + a - 1 \leq 0$ on $[1, 3]$ $\cdots\cdots$ ①

Since $1 \leq x \leq 3$, $(x - 1)(x - 3) \leq 0$ $\qquad \therefore\ x^2 - 4x + 3 \leq 0$ $\cdots\cdots$ ②

By ① and ②, $a = 4$

(4) A function $f(x) = ax + \sin x$ is increasing on $(-\infty, \infty)$.

$f'(x) = a + \cos x \geq 0$ $\qquad \therefore\ a \geq -\cos x$

Since $-1 \leq \cos x \leq 1$, $-1 \leq -\cos x \leq 1$

$\therefore\ a \geq 1$

(5) A function $f(x) = (ax^2 + 1)e^x$ is increasing on $(-\infty, \infty)$.

$f'(x) = (2ax)e^x + (ax^2 + 1)e^x = e^x(ax^2 + 2ax + 1) \geq 0$

Since $e^x > 0$, $ax^2 + 2ax + 1 \geq 0$

Let D be the discriminant of the equation $ax^2 + 2ax + 1 = 0$.

Then, $\dfrac{D}{4} \le 0$

$\therefore \ a^2 - a = a(a-1) \le 0$ $\qquad\qquad \therefore \ 0 \le a \le 1$

(6) A function $f(x) = (a-x)e^{x^2}$ is decreasing on $(-\infty, \infty)$.

$f'(x) = -e^{x^2} + 2x(a-x)e^{x^2} = e^{x^2}(-2x^2 + 2ax - 1)$

Since f is decreasing on $(-\infty, \infty)$, $f'(x) \le 0$ for all x.

Since $e^{x^2} > 0$, $-2x^2 + 2ax - 1 \le 0$; $2x^2 - 2ax + 1 \ge 0$

Let D be the discriminant of the equation $2x^2 - 2ax + 1 = 0$.

Then, $\dfrac{D}{4} \le 0$

$\therefore \ a^2 - 2 = \left(a - \sqrt{2}\right)\left(a + \sqrt{2}\right) \le 0$ $\qquad \therefore \ -\sqrt{2} \le a \le \sqrt{2}$

#14 Find all local extreme values of the graph of $f(x)$.

(1) $f(x) = x^3 - 3x - 2$

$f'(x) = 3x^2 - 3 = 3(x^2 - 1) = 3(x+1)(x-1)$

$f'(x) = 0 \ \Rightarrow \ x = 1 \text{ or } x = -1$

$f(1) = 1 - 3 - 2 = -4$

$f(-1) = -1 + 3 - 2 = 0$

$\therefore$ The local maximum value is 0 at $x = -1$ and local minimum value is -4 at $x = 1$.

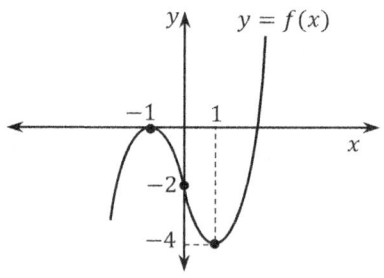

(2) $f(x) = -x^3 + 12x - 3$

$f'(x) = -3x^2 + 12 = -3(x^2 - 4) = -3(x+2)(x-2)$

$f'(x) = 0 \ \Rightarrow \ x = 2 \text{ or } x = -2$

$f(2) = -8 + 24 - 3 = 13$

$f(-2) = 8 - 24 - 3 = -19$

$\therefore$ The local maximum value is 13 at $x = 2$ and

local minimum value is -19 at $x = -2$.

(3) $f(x) = x^3 - 3x^2 + 3x + 1$

$f'(x) = 3x^2 - 6x + 3 = 3(x^2 - 2x + 1) = 3(x-1)^2 \ge 0$

Since f is increasing on $(-\infty, \infty)$, there is no local maximum value.

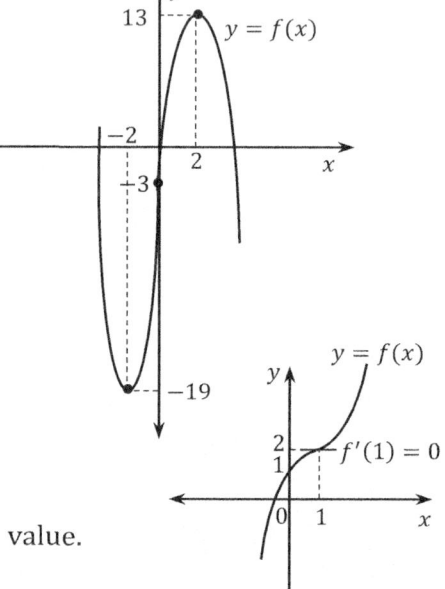

(4) $f(x) = 2x^3 + 3x^2 - 36x - 40$

$f'(x) = 6x^2 + 6x - 36 = 6(x^2 + x - 6) = 6(x + 3)(x - 2)$

$f'(x) = 0 \implies x = -3 \text{ or } x = 2$

$f(-3) = -2 \cdot 27 + 27 + 36 \cdot 3 - 40 = 41$

$f(2) = 16 + 12 - 72 - 40 = -84$

∴ The local maximum value is 41 at $x = -3$ and

local minimum value is -84 at $x = 2$.

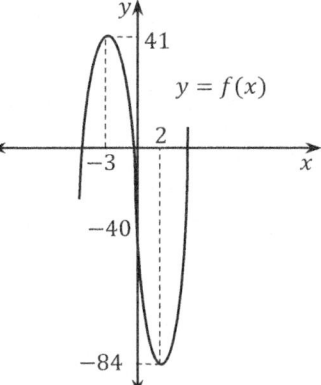

(5) $f(x) = -2x^3 + 9x^2 - 12x + 5$

$f'(x) = -6x^2 + 18x - 12 = -6(x^2 - 3x + 2) = -6(x - 1)(x - 2)$

$f'(x) = 0 \implies x = 1 \text{ or } x = 2$

$f(1) = -2 + 9 - 12 + 5 = 0$

$f(2) = -16 + 36 - 24 + 5 = 1$

∴ The local maximum value is 1 at $x = 2$ and

local minimum value is 0 at $x = 1$.

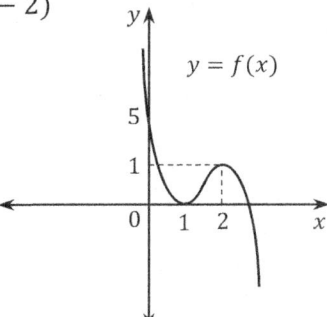

(6) $f(x) = \frac{1}{2}x^4 - \frac{4}{3}x^3 - x^2 + 4x + 5$

$f'(x) = 2x^3 - 4x^2 - 2x + 4 = 2x^2(x - 2) - 2(x - 2) = 2(x - 2)(x^2 - 1)$

$\qquad = 2(x - 2)(x - 1)(x + 1)$

$f'(x) = 0 \implies x = -1 \text{ or } x = 1 \text{ or } x = 2$

$f(-1) = \frac{1}{2} + \frac{4}{3} - 1 - 4 + 5 = \frac{11}{6}$

$f(1) = \frac{1}{2} - \frac{4}{3} - 1 + 4 + 5 = -\frac{5}{6} + 8 = \frac{43}{6}$

$f(2) = 8 - \frac{32}{3} - 4 + 8 + 5 = \frac{19}{3}$

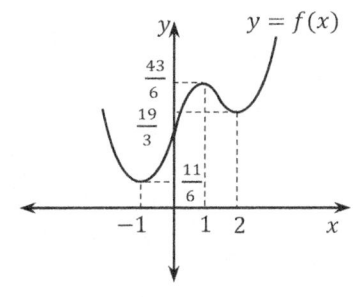

∴ The local maximum value is $\frac{43}{6}$ at $x = 1$ and local minimum value is $\frac{11}{6}$ at $x = -1$, and

local minimum value is $\frac{19}{3}$ at $x = 2$.

(7) $f(x) = |x - 1|$

$f(x) = |x - 1|$ is continuous on $(-\infty, \infty)$.

When $x > 1$, $f(x) = x - 1 \qquad$ ∴ $f'(x) = 1 > 0$

When $x < 1$, $f(x) = -x + 1 \qquad$ ∴ $f'(x) = -1 < 0$

At $x = 1$, the left and right sides of $f'(x)$ have different signs.

∴ $f(x)$ has a local minimum value at $x = 1$.

That is, $f(1) = |1 - 1| = 0$ is the local minimum value.

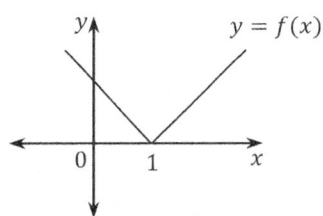

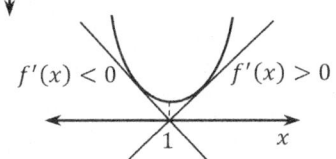

(8) $f(x) = \sqrt[3]{x^2}$

$f(x) = \sqrt[3]{x^2} \Rightarrow f(x) = x^{\frac{2}{3}}$

$\therefore f'(x) = \frac{2}{3}x^{-\frac{1}{3}} = \frac{2}{3\sqrt[3]{x}}$ $(x \neq 0)$

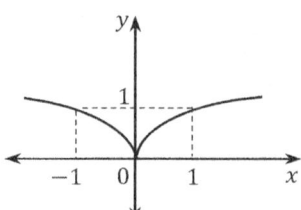

f is not differentiable at $x = 0$ but f is continuous on $(-\infty, \infty)$.

f has the local minimum value $f(0) = 0$ at $x = 0$.

x	$-\infty$	$\cdots$	0	$\cdots$	∞
$f'(x)$	$-$				$+$
$f(x)$		$\searrow$	Local Min.	$\nearrow$	

(9) $f(x) = \frac{2x-1}{x^2+6}$

$f'(x) = \frac{2(x^2+6)-(2x-1)(2x)}{(x^2+6)^2} = \frac{-2x^2+2x+12}{(x^2+6)^2} = \frac{-2(x^2-x-6)}{(x^2+6)^2} = \frac{-2(x-3)(x+2)}{(x^2+6)^2}$

$f'(x) = 0 \Rightarrow x = 3$ or $x = -2$

$f(3) = \frac{2 \cdot 3 - 1}{3^2+6} = \frac{5}{15} = \frac{1}{3}$

$f(-2) = \frac{2 \cdot (-2)-1}{(-2)^2+6} = \frac{-5}{10} = -\frac{1}{2}$

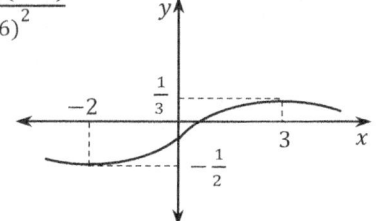

$\therefore$ The local maximum value is $\frac{1}{3}$ at $x = 3$ and local minimum value is $-\frac{1}{2}$ at $x = -2$.

(10) $f(x) = \frac{2-\cos x}{\sin x}$ $(0 < x < \frac{\pi}{2})$

$f'(x) = \frac{(\sin x)\sin x - (2-\cos x)\cos x}{\sin^2 x} = \frac{\sin^2 x + \cos^2 x - 2\cos x}{\sin^2 x} = \frac{1-2\cos x}{\sin^2 x}$

$f'(x) = 0 \Rightarrow 1 - 2\cos x = 0$; $\cos x = \frac{1}{2}$ $\therefore x = \frac{\pi}{3}$ $(\because 0 < x < \frac{\pi}{2})$

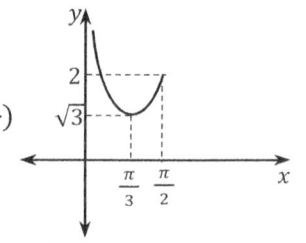

$x = \frac{\pi}{2} \Rightarrow f(x) = 2$; $x = \frac{\pi}{3} \Rightarrow f(x) = \frac{2-\frac{1}{2}}{\frac{\sqrt{3}}{2}} = \frac{3}{\sqrt{3}} = \sqrt{3}$

$\therefore$ The local minimum value is $\sqrt{3}$ at $x = \frac{\pi}{3}$.

(11) $f(x) = \frac{\log x}{x^4}$

$f'(x) = \frac{\frac{1}{x} \cdot x^4 - \log x \cdot (4x^3)}{x^8} = \frac{x^3 - 4x^3 \log x}{x^8} = \frac{1 - 4\log x}{x^5}$

$f'(x) = 0 \Rightarrow 1 - 4\log x = 0$; $\log x = \frac{1}{4}$ $\therefore x = e^{\frac{1}{4}}$

$\therefore f\left(e^{\frac{1}{4}}\right) = \frac{\log\left(e^{\frac{1}{4}}\right)}{\left(e^{\frac{1}{4}}\right)^4} = \frac{\frac{1}{4}\log e}{e} = \frac{1}{4e}$

$\therefore$ The local maximum value is $\frac{1}{4e}$ at $x = e^{\frac{1}{4}}$.

#15 State if the following statement is true or false.

(1) For a function $f(x) = xe^x$,

    1) $f(x)$ has a local minimum value at $x = 1$.

    2) $f(x)$ is decreasing on $x < -1$

    3) $f(x)$ is concave upward on the interval $(-\infty, -2)$.

1) $f(x) = xe^x \Rightarrow f'(x) = e^x + xe^x = e^x(1+x)$

    $f''(x) = e^x(1+x) + e^x = e^x(2+x)$

    $f'(x) = 0 \Rightarrow x = -1 \ (\because e^x > 0)$

    Since $f''(-1) = e^{-1} = \frac{1}{e} > 0$, $f(x)$ has a local minimum value at $x = -1$. (False)

2) $x < -1 \Rightarrow x + 1 < 0$

    $\therefore \ f'(x) = e^x(1+x) < 0$

    $\therefore \ f(x)$ is decreasing on $x < -1$ (True)

3) $x < -2 \Rightarrow x + 2 < 0$

    $\therefore \ f''(x) = e^x(2+x) < 0$

    $\therefore \ f(x)$ is concave downward on the interval $(-\infty, -2)$.   (False)

(2) For a function $f(x) = x^{\sqrt{x}} \ (x > 0)$,

    1) $f(x)$ is increasing on $x \geq 1$.

    2) $f(x) \geq 1$ for all $x \geq 1$.

    3) The equation of the function $f(x) = 2$ has two different real number solutions when $x \geq 1$.

    4) $f'(x) \geq 0$ for all $x > 0$.

1) $f(x) = x^{\sqrt{x}} \Rightarrow \log f(x) = \log x^{\sqrt{x}} = \sqrt{x} \log x$

    $\therefore \ \frac{d}{dx} \log f(x) = \frac{d}{dx} \sqrt{x} \log x$

    $\therefore \ \frac{f'(x)}{f(x)} = \frac{1}{2\sqrt{x}} \log x + \sqrt{x} \cdot \frac{1}{x} = \frac{\log x + 2}{2\sqrt{x}}$

    $\therefore \ f'(x) = \left(\frac{\log x + 2}{2\sqrt{x}}\right) f(x) = \left(\frac{\log x + 2}{2\sqrt{x}}\right) x^{\sqrt{x}}$

    Since $f'(x) > 0$ for $x \geq 1$, $f(x)$ is increasing. (True)

2) $f(x) = 1$ and $f(x)$ is increasing for $x \geq 1$, $f(x) \geq 1$ for all $x \geq 1$ (True)

3) Note that $f'(x) > 0$ for $x \geq 1$, $f(1) = 1$ and $\lim\limits_{x \to \infty} f(x) = \infty$

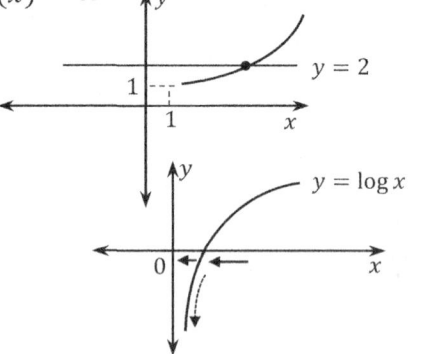

Since the graph of $y = f(x)$ and the line $y = 2$

have exactly one intersection point,

the equation $f(x) = 2$ has only one real number

solution when $x \geq 1$.　　(False)

4) As $x \to 0$, $\log x + 2 < 0$

∴ $f'(x) < 0$ as x approaches to 0.　　(False)

(3) For a function defined on an open interval $(0, 12)$,

the derivative $y = f'(x)$ is shown as the Figure.

1) $f(x)$ is increasing on $(4, 12)$.

2) $f(x)$ has a local maximum value at $x = 4$.

3) $f(x)$ has a minimum value at $x = 2$ in the open interval $(0, 4)$

4) $f(x)$ is not differentiable at $x = 10$.

1) Since $f'(x) > 0$ on $(4, 12)$, $f(x)$ is increasing.　(True)

2) $f'(4) = 0$ and $f'(x)$ has two different signs of the left $(-)$ and right $(+)$ sides of $x = 4$.

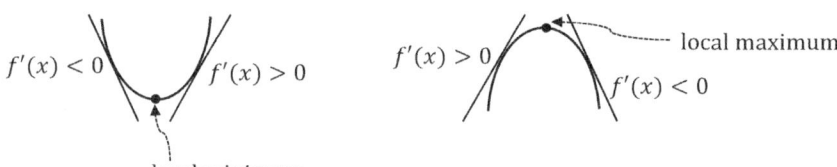

∴ $f(4)$ is the local minimum value.　　(False)

3) On the open interval $(0, 4)$, $f'(x) < 0$

∴ $f(x)$ is decreasing on $(0, 4)$.

That is, $f(0) > f(1) > f(2) > f(3) > f(4)$　　(False)

4) Since $f'(10)$ exists, $f(x)$ is differentiable at $x = 10$.　(False)

(4) When a function $f(x)$ with second derivative satisfies $f(-x) = -f(x)$ for all real

number x,

1) $f'(-x) = -f'(x)$

2) $\lim\limits_{x \to 0} f'(x) = -1$

3) If $f'(x)$ has a local maximum value at $x = a$ $(a \neq 0)$, then $f'(x)$ has a local

minimum value at $x = -a$.

1) $f(-x) = -f(x) \Rightarrow -f'(-x) = -f'(x)$　　∴ $f'(-x) = f'(x)$　　(False)

2) Since $f(-x) = -f(x)$, $f(x)$ is an odd function.

    Let $f(x) = x^3 - x$ Then, $f(-x) = (-x)^3 - (-x) = -x^3 + x = -(x^3 - x) = -f(x)$

    Since $f'(x) = 3x^2 - 1$ and $f''(x) = 6x$, $f''(x)$ exists.

    $\therefore \lim_{x \to 0} f'(x) = \lim_{x \to 0}(3x^2 - 1) = 0 - 1 = -1$ (True)

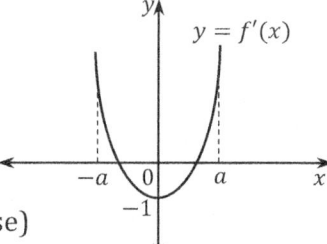

3) Since $f'(x) = 3x^2 - 1 = f'(-x)$, $f'(x)$ is an even function.

    $\therefore$ $f'(x)$ is symmetric with respect to the y-axis.

    $\therefore$ If $f'(x)$ has a local maximum value at $x = a$,

        then $f'(x)$ also has a local maximum value at $x = -a$.   (False)

(5) For a function $f(x) = x + \sin x$, let $g(x) = (f \circ f)(x)$.

    1) **The graph of $f(x)$ is concave upward on an open interval $(0, \pi)$.**

    2) **$g(x)$ is increasing on $(0, \pi)$.**

    3) **There is x $(0 < x < \pi)$ such that $g'(x) = 0$.**

1) $f'(x) = 1 + \cos x$ $f''(x) = -\sin x$

    Since $0 < x < \pi$, $0 < \sin x < 1$; $-1 < -\sin x < 0$ $\therefore -1 < f''(x) < 0$

    Since $f''(x) < 0$, $f(x)$ is concave downward.   (False)

2) $g(x) = (f \circ f)(x) = f(f(x))$

    $g'(x) = f'(f(x))f'(x) = \{1 + \cos f(x)\}(1 + \cos x)$

    Since $1 + \cos f(x) > 0$ and $1 + \cos x > 0$ on $(0, \pi)$, $g'(x) > 0$ on $(0, \pi)$.

    $\therefore$ $g(x)$ is increasing on $(0, \pi)$.   (True)

3) $g(0) = f(f(0)) = f(0) = 0$; $g(\pi) = f(f(\pi)) = f(\pi) = \pi$

    Since $g(x)$ is continuous on $[0, \pi]$ and differentiable on $(0, \pi)$, by Mean Value Theorem,

    there is at least one c such that $g'(c) = \frac{g(\pi) - g(0)}{\pi - 0}$.

    $\therefore$ $g'(c) = \frac{g(\pi) - g(0)}{\pi - 0} = \frac{\pi - 0}{\pi} = 1 \neq 0$ (False)

(6) For any polynomial function $f(x)$ with local maximum value at $x = 0$,

    1) **$|f(x)|$ has a local maximum value at $x = 0$.**

    2) **$f(|x|)$ has a local maximum value at $x = 0$.**

    3) **$f(x) - x^2|x|$ has a local maximum value at $x = 0$.**

1) Let $f(x) = -x^2$

    Then, $|f(x)| = x^2$; $|f(x)|$ is concave upward.

    $\therefore$ $|f(x)|$ has a local minimum value at $x = 0$. (False)

2) Since $f(|x|) = \begin{cases} f(x), & x \geq 0 \\ f(-x), & x < 0 \end{cases}$, $f'(|x|) = \begin{cases} f'(x), & x \geq 0 \\ -f'(-x), & x < 0 \end{cases}$

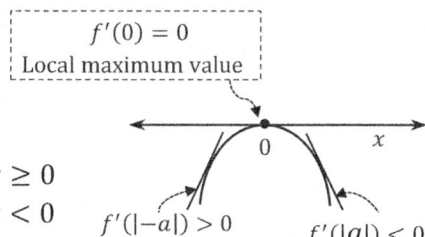

For any positive a, $f'(|-a|) = -f'(-(-a)) = -f'(a) > 0$ $\quad (\because a > 0 \Rightarrow f'(a) < 0)$

$$f'(|a|) = f'(a) < 0$$

$\therefore$ $f(|x|)$ has a local maximum value at $x = 0$. (True)

3) Let $g(x) = f(x) - x^2|x|$

Then, $g(x) = \begin{cases} f(x) - x^3, & x \geq 0 \\ f(x) + x^3, & x < 0 \end{cases}$, $\quad g'(x) = \begin{cases} f'(x) - 3x^2, & x \geq 0 \\ f'(x) + 3x^2, & x < 0 \end{cases}$

For any positive a, $g'(a) = f'(a) - 3a^2 < 0$ $\quad (\because a > 0 \Rightarrow f'(a) < 0)$

$$g'(-a) = f'(-a) + 3(-a)^2 > 0 \ (\because a > 0 \Rightarrow -a < 0 \quad \therefore f'(-a) > 0)$$

$\therefore$ $g(x)$ has a local maximum value at $x = 0$. (True)

#16 Find the local extreme values using the second derivatives.

(1) $f(x) = x^2 e^{-x}$

$f'(x) = (2x)e^{-x} + x^2 e^{-x}(-1) = e^{-x}(2x - x^2)$

$f''(x) = (-e^{-x})(2x - x^2) + e^{-x}(2 - 2x) = e^{-x}(x^2 - 4x + 2)$

$f'(x) = 0 \Rightarrow x = 0$ or $x = 2$

$\therefore$ $f(x)$ has local extreme values at $x = 0$ and $x = 2$

Since $f''(0) = 2 > 0$ and $f''(2) = -\dfrac{2}{e^2} < 0$,

$f(0)$ is a local minimum value and $f(2)$ is a local maximum value.

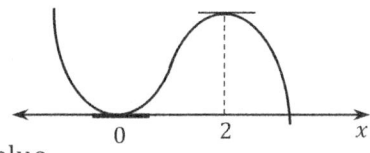

$\therefore$ The local minimum value is 0 at $x = 0$ and local maximum value is $\dfrac{4}{e^2}$ at $x = 2$.

(2) $f(x) = \sqrt{3} \sin x + \cos x \ \ (0 \leq x \leq 2\pi)$

$f'(x) = \sqrt{3} \cos x - \sin x$

$f''(x) = -\sqrt{3} \sin x - \cos x$

$f'(x) = 0 \Rightarrow \cos x = \dfrac{\sin x}{\sqrt{3}}$; $\dfrac{\sin x}{\cos x} = \sqrt{3}$; $\tan x = \sqrt{3}$; $x = \dfrac{\pi}{3}, x = \dfrac{4\pi}{3}$

$\therefore$ $f(x)$ has local extreme values at $x = \dfrac{\pi}{3}$ and $x = \dfrac{4\pi}{3}$

Since $f''\left(\dfrac{\pi}{3}\right) = -\sqrt{3} \sin \dfrac{\pi}{3} - \cos \dfrac{\pi}{3} = -\sqrt{3}\left(\dfrac{\sqrt{3}}{2}\right) - \dfrac{1}{2} = -2 < 0$ and

$f''\left(\dfrac{4\pi}{3}\right) = -\sqrt{3} \sin \dfrac{4\pi}{3} - \cos \dfrac{4\pi}{3} = -\sqrt{3}\left(-\dfrac{\sqrt{3}}{2}\right) - \left(-\dfrac{1}{2}\right) = 2 > 0$,

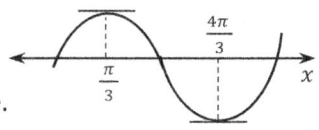

$f\left(\dfrac{\pi}{3}\right)$ is a local maximum value and $f\left(\dfrac{4\pi}{3}\right)$ is a local minimum value.

$f\left(\dfrac{\pi}{3}\right) = \sqrt{3} \sin \dfrac{\pi}{3} + \cos \dfrac{\pi}{3} = \sqrt{3}\left(\dfrac{\sqrt{3}}{2}\right) + \dfrac{1}{2} = 2$

$f\left(\dfrac{4\pi}{3}\right) = \sqrt{3} \sin \dfrac{4\pi}{3} + \cos \dfrac{4\pi}{3} = -\sqrt{3}\left(\dfrac{\sqrt{3}}{2}\right) - \dfrac{1}{2} = -2$

∴ The local maximum value is 2 at $x = \dfrac{\pi}{3}$ and local minimum value is -2 at $x = \dfrac{4\pi}{3}$.

(3) $f(x) = 2x - \tan x \quad \left(-\dfrac{\pi}{2} < x < \dfrac{\pi}{2}\right)$

$f'(x) = 2 - \sec^2 x$

$f''(x) = -2\sec x \, (\sec x \tan x) = -2\sec^2 x \tan x$

$f'(x) = 0 \;\Rightarrow\; \sec^2 x = 2 \;;\; \sec x = \pm\sqrt{2} \;;\; \cos x = \pm\dfrac{1}{\sqrt{2}} \;;\; x = \dfrac{\pi}{4}, \; x = -\dfrac{\pi}{4} \;\left(\because -\dfrac{\pi}{2} < x < \dfrac{\pi}{2}\right)$

∴ $f(x)$ has local extreme values at $x = \dfrac{\pi}{4}$ and $x = -\dfrac{\pi}{4}$

Since $f''\left(\dfrac{\pi}{4}\right) = -2\sec^2\dfrac{\pi}{4}\tan\dfrac{\pi}{4} = \dfrac{-2}{\left(\cos\frac{\pi}{4}\right)^2}\tan\dfrac{\pi}{4} = \dfrac{-2}{\frac{1}{2}}\cdot 1 = -4 < 0$ and

$f''\left(-\dfrac{\pi}{4}\right) = -2\sec^2\left(-\dfrac{\pi}{4}\right)\tan\left(-\dfrac{\pi}{4}\right) = \dfrac{-2}{\left(\cos\left(-\frac{\pi}{4}\right)\right)^2}\tan\left(-\dfrac{\pi}{4}\right)$

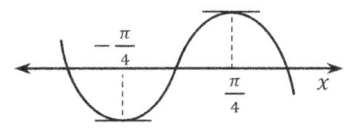

$\qquad = \dfrac{-2}{\frac{1}{2}}\cdot(-1) = 4 > 0,$

$f\left(\dfrac{\pi}{4}\right)$ is a local maximum value and $f\left(-\dfrac{\pi}{4}\right)$ is a local minimum value.

$f\left(\dfrac{\pi}{4}\right) = 2\left(\dfrac{\pi}{4}\right) - \tan\dfrac{\pi}{4} = \dfrac{\pi}{2} - 1$

$f\left(-\dfrac{\pi}{4}\right) = 2\left(-\dfrac{\pi}{4}\right) - \tan\left(-\dfrac{\pi}{4}\right) = -\dfrac{\pi}{2} - (-1) = -\dfrac{\pi}{2} + 1$

∴ The local maximum value is $\dfrac{\pi}{2} - 1$ at $x = \dfrac{\pi}{4}$ and local minimum value is $-\dfrac{\pi}{2} + 1$ at $x = -\dfrac{\pi}{4}$.

#17 Find the value.

(1) For a function $f(x) = x^3 + ax^2 + ax + 1$,

    1) Find the range of a so that $f(x)$ has local maximum and minimum values.

    2) Find the range of a so that $f(x)$ does not have local extreme values.

$f'(x) = 3x^2 + 2ax + a$

Let D be the discriminant of the equation $3x^2 + 2ax + a = 0$.

1) Since $f'(x) = 0$ has two different real number solutions, $\dfrac{D}{4} > 0$

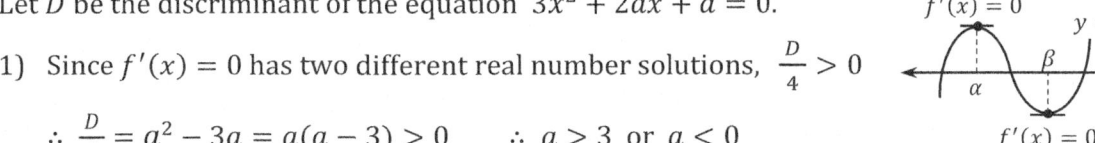

$\qquad \therefore \dfrac{D}{4} = a^2 - 3a = a(a - 3) > 0 \qquad \therefore a > 3 \text{ or } a < 0$

2) Since $f'(x)$ has a double root or no real number solution, $\dfrac{D}{4} \le 0$

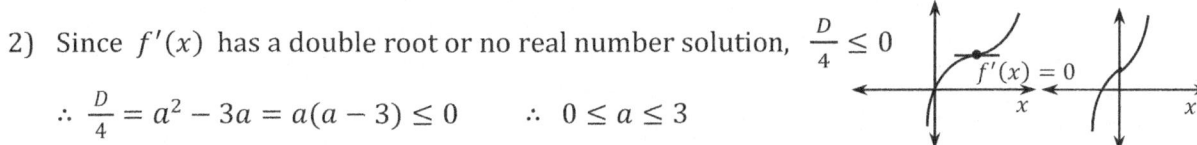

$\qquad \therefore \dfrac{D}{4} = a^2 - 3a = a(a - 3) \le 0 \qquad \therefore 0 \le a \le 3$

(2) When a function $f(x) = x^3 - 3x + a$ has a local maximum value 12,

find the value of a and local minimum value.

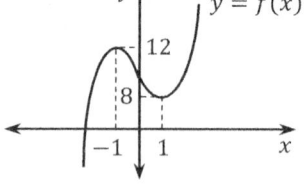

$f'(x) = 3x^2 - 3 = 3(x + 1)(x - 1)$

$f'(x) = 0 \Rightarrow x = -1$ or $x = 1$

Since $f(x)$ has a local maximum value 12 at $x = -1$, $f(-1) = 12$

$\therefore (-1)^3 - 3(-1) + a = 2 + a = 12$; $a = 10$

$\therefore$ Local minimum value is $f(1) = 1^3 - 3(1) + a = -2 + a = -2 + 10 = 8$

(3) When a function $f(x) = x^3 + ax^2 + bx + 3$ has a local minimum value 2 at $x = 1$,

find the values of a and b.

$f'(x) = 3x^2 + 2ax + b$

Since $f'(1) = 0$ and $f(1) = 2$, we have $3 \cdot 1^2 + 2a \cdot 1 + b = 0$ and $1^3 + a \cdot 1^2 + b \cdot 1 + 3 = 2$

$\therefore 2a + b = -3$, $a + b = -2$ $\qquad \therefore a = -1$, $b = -1$

(4) A function $f(x) = x^3 + 3ax^2 + bx + c$ has a local maximum value at $x = 1$ and a local

minimum value at $x = 3$.

1) Find the values of a and b.

2) Find the local extreme values.

3) When the local maximum value is 10, find the value of c and the local minimum

value.

1) $f(x) = x^3 + 3ax^2 + bx + c \Rightarrow f'(x) = 3x^2 + 6ax + b$

Since $f(x)$ has local extreme values at $x = 1$ and $x = 3$, $f'(1) = f'(3) = 0$

$\therefore 3 + 6a + b = 0$, $27 + 18a + b = 0$

$\therefore 12a + 24 = 0$; $a = -2$, $b = 9$

2) Local maximum value at $x = 1$ is $f(1) = 1 + 3a + b + c = 1 - 6 + 9 + c = 4 + c$.

Local minimum value at $x = 3$ is $f(3) = 27 + 27a + 3b + c = 27 - 54 + 27 + c = c$.

$\therefore$ Local extreme values are $4 + c$ and c.

3) Since $4 + c = 10$, $c = 6$

$\therefore$ Local minimum value is 6.

(5) Find the range of a so that a function $f(x) = x^4 - 4x^3 + 6ax^2$ has no local extreme

values.

$f(x) = x^4 - 4x^3 + 6ax^2 \Rightarrow f'(x) = 4x^3 - 12x^2 + 12ax = 4x(x^2 - 3x + 3a)$

Since $f(x)$ does not have local extreme values,

$f'(x) = 0$ has one real number solution and two complex number solutions, or

$f'(x) = 0$ has one real number solution and a double root.

i) When $f'(x) = 0$ has one real number solution and two complex number solutions,

   $x^2 - 3x + 3a = 0$ has two complex number solutions.

   Let D be the discriminant of the equation. Then, $D < 0$

   $\therefore$ $D = 9 - 12a < 0$; $a > \dfrac{3}{4}$

ii) When $f'(x) = 0$ has one real number solution and a double root,

   Case1: If $x = 0$ is the real number solution, then $0 - 0 + 3a = 0$. $\therefore$ $a = 0$

   Case2: If $x \neq 0$ is the double root, $D = 9 - 12a = 0$ $\therefore$ $a = \dfrac{3}{4}$

By i) and ii), $a = 0$ or $a \geq \dfrac{3}{4}$

(6) When a function $f(x) = x^3 + \left(a - \dfrac{1}{2}\right)x^2 - 2(a + 1)x + a$ has a local maximum value

at $x = 1$, find the range of the real number a.

$f'(x) = 3x^2 + 2\left(a - \dfrac{1}{2}\right)x - 2(a + 1) = (x - 1)(3x + 2a + 2)$

$$\begin{array}{r|ccc} & 3 & 2\left(a - \frac{1}{2}\right) & -2(a+1) \\ 1 & & 3 & 2a+2 \\ \hline & 3 & 2a+2 & 0 \end{array}$$

$f'(x) = 0$ $\Rightarrow$ $x = 1$ or $x = -\dfrac{2a+2}{3}$

Since $f(x)$ has a local maximum value at $x = 1$, $1 < -\dfrac{2a+2}{3}$

$\therefore$ $\dfrac{2a+2}{3} < -1$; $2a + 2 < -3$

$\therefore$ $a < -\dfrac{5}{2}$

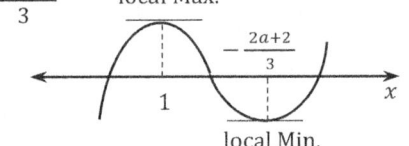

(7) When a function $f(x) = 2x^3 + 3ax^2$ has a local minimum value in the open interval

$(1, 2)$, find the range of the real number a.

$f'(x) = 6x^2 + 6ax = 6x(x + a)$

$f'(x) = 0$ $\Rightarrow$ $x = 0$ or $x = -a$

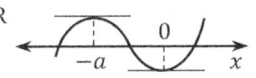

Since $f(x)$ has a local minimum value in $(1, 2)$, $1 < -a < 2$

$\therefore$ $-2 < a < -1$

(8) When a differentiable function $f(x)$ defined on real numbers satisfies:

   i) $f(x - y) = f(x) - f(y) + xy(x - y)$ for any real numbers, and

   ii) $f'(0) = 10$,

   $f(x)$ has a local maximum value at $x = a$ and local minimum value at $x = b$.

   Find the values of a and b.

Substituting $x = 0$ and $y = 0$ into i), $f(0) = f(0) - f(0) + 0 = 0$

By ii), $f'(0) = \lim\limits_{x \to 0} \dfrac{f(x) - f(0)}{x - 0} = 10$

$$f'(x) = \lim_{h \to 0} \frac{f(x+h)-f(x)}{h-0} = \lim_{h \to 0} \frac{f(x)-f(-h)-xh(x+h)-f(x)}{h} = \lim_{h \to 0} -\left\{\frac{f(-h)+xh(x+h)}{h}\right\}$$

$$= \lim_{h \to 0} -\left\{\frac{f(-h)}{h} + x(x+h)\right\} = 10 - x^2 = -(x - \sqrt{10})(x + \sqrt{10})$$

$f'(x) = 0 \Rightarrow x = \sqrt{10}$ or $x = -\sqrt{10}$

∴ $f(x)$ has a local maximum value at $x = \sqrt{10}$ and local minimum value at $x = -\sqrt{10}$.

Therefore, $a = \sqrt{10}$ and $b = -\sqrt{10}$

(9) When a cubic function $y = f(x)$ with integer coefficients satisfies:

   i) $f(-x) = -f(x)$ for all x,

   ii) $f(1) = 5$, and

   iii) $1 < f'(1) < 6$,

   Find the sum of the local extreme values.

By i), $y = f(x)$ is an odd function.

Let $f(x) = ax^3 + bx$, $(a, b;$ integers, $a \neq 0)$

By ii), $a + b = 5$; $b = 5 - a$

Since $f'(x) = 3ax^2 + b$, $f'(1) = 3a + b$

By iii), $1 < 3a + b < 6$; $1 < 3a + (5 - a) < 6$; $1 < 2a + 5 < 6$; $-2 < a < \frac{1}{2}$

Since a is an non-zero integer, $a = -1$ ∴ $b = 5 - a = 5 - (-1) = 6$

∴ $f(x) = -x^3 + 6x$; $f'(x) = -3x^2 + 6 = -3(x^2 - 2)$

$f'(x) = 0 \Rightarrow x = \pm\sqrt{2}$

$f(\sqrt{2}) = -2\sqrt{2} + 6\sqrt{2} = 4\sqrt{2}$; $f(-\sqrt{2}) = 2\sqrt{2} - 6\sqrt{2} = -4\sqrt{2}$

∴ The local maximum value is $4\sqrt{2}$ at $x = \sqrt{2}$ and local minimum value is $-4\sqrt{2}$ at $x = -\sqrt{2}$

Therefore, the sum is $4\sqrt{2} + (-4\sqrt{2}) = 0$

(10) When a function $f(x) = \log x + \frac{a}{x} - x$ has local extreme values, find the range of a.

$f'(x) = \frac{1}{x} - \frac{a}{x^2} - 1 = \frac{x - a - x^2}{x^2}$

$f'(x) = 0 \Rightarrow x^2 - x + a = 0$

Since $f(x)$ has local extreme values, $x^2 - x + a = 0$ has two different real number solutions.

$(x > 0)$

Let D be the discriminant of the equation. Then, $D > 0$

∴ $D = (-1)^2 - 4a = 1 - 4a > 0$; $a < \frac{1}{4}$

Since $x > 0$, the sum and product of the two roots are positive.   ∴ $a > 0$

Therefore, $0 < a < \frac{1}{4}$

(11) For a function $f(x) = x^3 - 2x^2 \sin \theta + x + 1$ $\left(0 \le \theta \le \frac{\pi}{2}\right)$,

    1) Find the range of θ at which $f(x)$ has local extreme values.

    2) Find the range of θ at which $f(x)$ has no local extreme values.

1) $f'(x) = 3x^2 - 4x \sin \theta + 1$ $\left(0 \le \theta \le \frac{\pi}{2}\right)$

Let D be the discriminant of the equation $3x^2 - 4x \sin \theta + 1 = 0$ $\cdots\cdots$ ①

Since $f(x)$ has local extreme values, the equation ① has two different real number solutions.

$\therefore \dfrac{D}{4} > 0$

$\dfrac{D}{4} = 4 \sin^2 \theta - 3 = (2 \sin \theta + \sqrt{3})(2 \sin \theta - \sqrt{3}) > 0$

$\therefore \sin \theta > \dfrac{\sqrt{3}}{2}$ or $\sin \theta < -\dfrac{\sqrt{3}}{2}$

Since $0 \le \theta \le \dfrac{\pi}{2}$, $\dfrac{\pi}{3} < \theta \le \dfrac{\pi}{2}$

2) Since the equation ① has no real number solution or has a double root, $\dfrac{D}{4} \le 0$

$\therefore -\dfrac{\sqrt{3}}{2} \le \sin \theta \le \dfrac{\sqrt{3}}{2}$

$\therefore 0 \le \theta \le \dfrac{\pi}{3}$

(12) When a function $f(x) = ax + 2 \sin x$ has local extreme values, find the range of a.

$f'(x) = a + 2 \cos x = 2\left(\cos x + \dfrac{a}{2}\right)$

To get a local extreme value, $f'(x)$ must have different signs for the left and right sides of x such that $f'(x) = 0$.

$\therefore \left|\dfrac{a}{2}\right| < 1$

Therefore, $|a| < 2$

(13) When a function $f(x) = e^{ax} \sin x$ has a local maximum value at $x = \dfrac{\pi}{4}$, find the value of a and local maximum value.

$f'(x) = ae^{ax} \sin x + e^{ax} \cos x = e^{ax}(a \sin x + \cos x)$

Since $f'\left(\dfrac{\pi}{4}\right) = e^{\frac{a\pi}{4}}\left(a \sin \dfrac{\pi}{4} + \cos \dfrac{\pi}{4}\right) = e^{\frac{a\pi}{4}}\left(\dfrac{a\sqrt{2}}{2} + \dfrac{\sqrt{2}}{2}\right) = \dfrac{\sqrt{2}}{2}e^{\frac{a\pi}{4}}(a+1) = 0$, $a = -1$

$f''(x) = ae^{ax}(a \sin x + \cos x) + e^{ax}(a \cos x - \sin x) = e^{ax}(a^2 \sin x + 2a \cos x - \sin x)$

$\qquad = e^{-x}(\sin x - 2 \cos x - \sin x) = -2e^{-x} \cos x$

$\therefore f''(x) < 0$

Since $f''\left(\dfrac{\pi}{4}\right) < 0$, $f\left(\dfrac{\pi}{4}\right)$ is a local maximum value.

$\therefore$ The local maximum value is $f\left(\dfrac{\pi}{4}\right) = e^{-\frac{\pi}{4}} \sin \dfrac{\pi}{4} = \dfrac{\sqrt{2}}{2}e^{-\frac{\pi}{4}}$.

#18 Find the cubic function such that:

i)   **a point $(0, 2)$ is the inflection point of the curve $y = f(x)$,**

ii) **the tangent line at $(0, 2)$ and a line $y = -2x$ are parallel to each other, and**

iii) **$f(x)$ has a local minimum value at $x = 1$.**

Let $y = f(x) = ax^3 + bx^2 + cx + d$ $(a \neq 0)$.

Then $f'(x) = 3ax^2 + 2bx + c$, $f''(x) = 6ax + 2b$

Since $(0, 2)$ is the inflection point, $f(0) = 2$ and $f''(0) = 0$

$\therefore$ $d = 2$, $b = 0$

By ii), $f'(0)$ is the slope of the tangent line.

$\therefore$ $f'(0) = -2$ ($\because$ Parallel to $y = -2x$.) ; $c = -2$

By iii), $f'(1) = 0$ $\therefore$ $3a + 0 - 2 = 0$; $a = \dfrac{2}{3}$

Therefore, $f(x) = \dfrac{2}{3}x^3 - 2x + 2$

#19 Find the inflection points for the following function.

(1) $f(x) = x^4 - 4x^3 + 4$

$f'(x) = 4x^3 - 12x^2 = 4x^2(x - 3)$

$f'(x) = 0 \Rightarrow x = 0$ or $x = 3$

$\therefore$ $f(x)$ has local extreme values at $x = 0$ and $x = 3$.

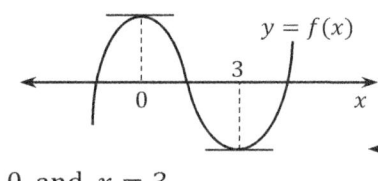
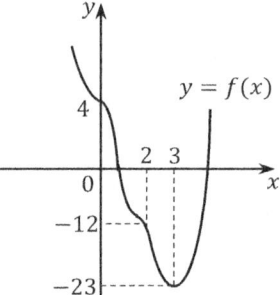

$f(0) = 4$, $f(3) = 3^4 - 4 \cdot 3^3 + 4 = 3^3(3 - 4) + 4 = -27 + 4 = -23$

$f''(x) = 12x^2 - 24x = 12x(x - 2)$

$f''(x) = 0 \Rightarrow x = 0$ or $x = 2$

$\therefore$ $f(x)$ has inflection points at $x = 0$ and $x = 2$.

$\therefore$ $(0, 4)$ and $(2, -12)$ are the inflection points.

(2) $y = \left(\ln\dfrac{1}{ax}\right)^2$

$y = \left(\ln\dfrac{1}{ax}\right)^2 = \{\ln(ax)^{-1}\}^2 = \{-\ln(ax)\}^2 = \{\ln(ax)\}^2$

$y' = 2\ln(ax) \cdot \dfrac{1}{ax} \cdot a = \dfrac{2\ln(ax)}{x}$

$y'' = \dfrac{2\frac{1}{ax} \cdot a \cdot x - 2\ln(ax) \cdot 1}{x^2} = \dfrac{2 - 2\ln(ax)}{x^2} = \dfrac{2\{1 - \ln(ax)\}}{x^2}$

$y'' = 0 \Rightarrow \ln(ax) = 1$; $ax = e$; $x = \dfrac{e}{a}$

When $x < \dfrac{e}{a}$, $1 - \ln(ax) > 0$ $\therefore$ $y'' > 0$

When $x > \dfrac{e}{a}$, $1 - \ln(ax) < 0$ $\therefore$ $y'' < 0$

$\therefore$ y'' has different signs at the left and right sides of $x = \dfrac{e}{a}$.

$\therefore$ $\left(\dfrac{e}{a}, 1\right)$ is the inflection point.

#20 Find the equation of the tangent line at the inflection point of the function.

(1) $y = \sin^2 x \quad \left(0 \le x \le \dfrac{\pi}{2}\right)$

$y' = 2\sin x \cdot \cos x = \sin 2x$

$y'' = 2\cos 2x$

$y'' = 0 \;\Rightarrow\; \cos 2x = 0 \;;\; x = \dfrac{\pi}{4}$

$x = \dfrac{\pi}{4} \;\Rightarrow\; y = \sin^2 \dfrac{\pi}{4} = \dfrac{1}{2} \qquad \therefore \left(\dfrac{\pi}{4}, \dfrac{1}{2}\right)$ is the inflection point.

The slope of the tangent line at $x = \dfrac{\pi}{4}$ is $\sin 2\left(\dfrac{\pi}{4}\right) = \sin\dfrac{\pi}{2} = 1$

$\therefore$ The equation of the tangent line at $\left(\dfrac{\pi}{4}, \dfrac{1}{2}\right)$ is $y - \dfrac{1}{2} = 1\left(x - \dfrac{\pi}{4}\right)$

That is, $y = x - \dfrac{\pi}{4} + \dfrac{1}{2}$

(2) $y = xe^{-x}$

$y' = e^{-x} + x(-e^{-x}) = e^{-x}(1 - x)$

$y'' = -e^{-x}(1 - x) + e^{-x}(-1) = e^{-x}(x - 2)$

$y'' = 0 \;\Rightarrow\; x - 2 = 0 \;;\; x = 2$

$x = 2 \;\Rightarrow\; y = 2e^{-2} = \dfrac{2}{e^2} \qquad \therefore \left(2, \dfrac{2}{e^2}\right)$ is the inflection point.

The slope of the tangent line at $x = 2$ is $e^{-2}(1 - 2) = -e^{-2} = -\dfrac{1}{e^2}$

$\therefore$ The equation of the tangent line at $\left(2, \dfrac{2}{e^2}\right)$ is $y - \dfrac{2}{e^2} = -\dfrac{1}{e^2}(x - 2)$

That is, $y = -\dfrac{1}{e^2}x + \dfrac{4}{e^2}$

#21 When the graph of $y = \log(ax^2 + b)$ has an inflection point $(1, \log 4)$, find the values of a and b.

$y = \log(ax^2 + b) \;\Rightarrow\; y' = \dfrac{2ax}{ax^2 + b}$

$$y'' = \dfrac{2a(ax^2 + b) - 2ax(2ax)}{(ax^2 + b)^2} = \dfrac{-2a^2x^2 + 2ab}{(ax^2 + b)^2} = \dfrac{-2a(ax^2 - b)}{(ax^2 + b)^2}$$

Since $(1, \log 4)$ is the inflection point, $y_{x=1} = \log 4$ and $y''_{x=1} = 0$

That is, $\log(a + b) = \log 4$ and $\dfrac{-2a(a - b)}{(a + b)^2} = 0$

$\therefore\ a + b = 4,\ -2a(a - b) = 0$

$\therefore\ a + b = 4\ \cdots\cdots$ ①

$\qquad a = 0\ $ or $\ a - b = 0\ \cdots\cdots$ ②

If $a = 0$, then $y = \log b$; There is no inflection point. $\qquad \therefore\ a \neq 0$

$\therefore\ a - b = 0$; $\ a = b$

By ①, $a = b = 2$

#22 Find the maximum value of the integer a at which a function

$\qquad f(x) = -x^3 + 3(a - 1)x^2 + 3(a - 3)x + 2\ $ **has no local extreme values.** $(x \geq 0)$

$f'(x) = -3x^2 + 6(a - 1)x + 3(a - 3) = -3x^2 + 6(a - 1)x + 3a - 9$

Since $f(x)$ has no local extreme values on $x \geq 0$, $f(x)$ has no local extreme values for all real

number x or $f(x)$ has local extreme values on $x < 0$.

Case1: When $f(x)$ has no local extreme value for all real number x,

$f'(x) = 0$ has a double root or has complex number solutions.

Let D be the discriminant of the equation $-3x^2 + 6(a - 1)x + 3a - 9 = 0$.

Then, $\dfrac{D}{4} \leq 0$

$\therefore\ \dfrac{D}{4} = 9(a - 1)^2 + 3(3a - 9) = 9a^2 - 9a - 18 = 9(a^2 - a - 2) = 9(a - 2)(a + 1) \leq 0$

$\therefore\ -1 \leq a \leq 2$

Case2: When $f(x)$ has local extreme values on $x < 0$,

$f'(x) = 0$ has two different negative real number solutions.

$\therefore\ \dfrac{D}{4} > 0\qquad$ That is, $\ \dfrac{D}{4} = 9(a - 2)(a + 1) > 0$

$\therefore\ a > 2\ $ or $\ a < -1\ \cdots\cdots$ ①

Since the sum of the two roots are negative, $\quad -\dfrac{6(a-1)}{-3} = 2(a - 1) < 0$; $\quad a < 1\ \cdots\cdots$ ②

Since the product of the two roots are positive, $\quad \dfrac{3a-9}{-3} = -a + 3 > 0\ $; $\quad a < 3\ \cdots\cdots$ ③

By ①, ②, and ③, $\ a < -1$

Therefore, by the cases 1 and 2, $\ a \leq 2$

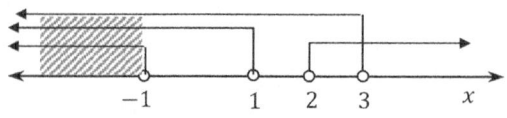

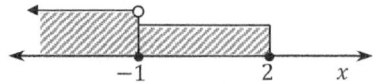

Hence, the maximum value of the integer a is 2.

#23 Find the maximum and minimum values for each function.

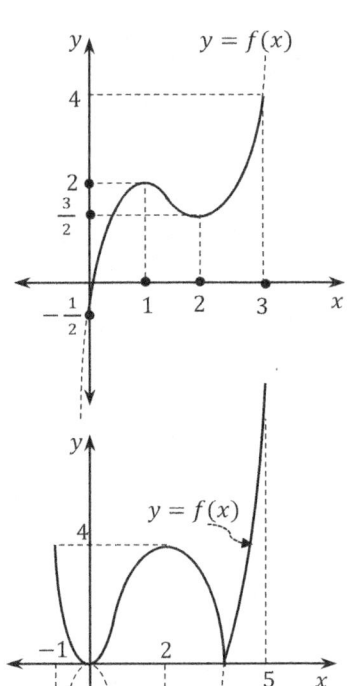

(1) $f(x) = x^3 - \frac{9}{2}x^2 + 6x - \frac{1}{2}$ $(0 \leq x \leq 3)$

$f'(x) = 3x^2 - 9x + 6 = 3(x^2 - 3x + 2) = 3(x-1)(x-2)$

$f'(x) = 0 \Rightarrow x = 1$ or $x = 2$

$f(1) = 1 - \frac{9}{2} + 6 - \frac{1}{2} = 2$; $f(2) = 8 - 18 + 12 - \frac{1}{2} = \frac{3}{2}$

$f(0) = 0 - \frac{1}{2} = -\frac{1}{2}$; $f(3) = 27 - \frac{81}{2} + 18 - \frac{1}{2} = 4$

∴ Maximum value: 4 at $x = 3$, Minimum value: $-\frac{1}{2}$ at $x = 0$

(2) $f(x) = |x^2(x-3)|$ $(-1 \leq x \leq 5)$

Let $g(x) = x^2(x-3)$

Then, $g'(x) = 2x(x-3) + x^2 = 3x(x-2)$

$g'(x) = 0 \Rightarrow x = 0$ or $x = 2$

$g(0) = 0$; $g(2) = -4$

$g(-1) = -4$; $g(5) = 50$

∴ Maximum value: 50 at $x = 5$, Minimum value: 0 at $x = 0$

(3) $f(x) = x + \sqrt{1 - 2x - x^2}$

Since $1 - 2x - x^2 \geq 0$, $x^2 + 2x - 1 \leq 0$ ⋯⋯ ①

$x^2 + 2x - 1 = 0 \Rightarrow x = \frac{-1 \pm \sqrt{1+1}}{1} = -1 \pm \sqrt{2}$

By ①, $-1 - \sqrt{2} \leq x \leq -1 + \sqrt{2}$

$f'(x) = 1 + \frac{-2-2x}{2\sqrt{1-2x-x^2}} = 1 - \frac{1+x}{\sqrt{1-2x-x^2}}$

$f'(x) = 0 \Rightarrow \frac{1+x}{\sqrt{1-2x-x^2}} = 1$; $(1+x)^2 = 1 - 2x - x^2$; $2x^2 + 4x = 0$; $2x(x+2) = 0$

$x = 0 \Rightarrow f'(x) = 1 - 1 = 0$

$x = -2 \Rightarrow f'(x) = 1 - \frac{1-2}{\sqrt{1+4-4}} = 1 + 1 = 2 \neq 0$

∴ $f(x)$ does not have a local extreme value at $x = -2$.

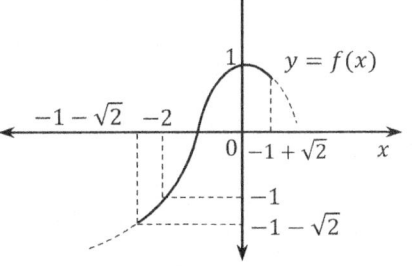

$f(0) = 1$; $f(-2) = -2 + 1 = -1$

$f(-1-\sqrt{2}) = -1 - \sqrt{2}$; $f(-1+\sqrt{2}) = -1 + \sqrt{2}$

∴ Maximum value: 1 at $x = 0$, Minimum value: $-1 - \sqrt{2}$ at $x = -1 - \sqrt{2}$

(4) $f(x) = x - 3 + \sqrt{9 - x^2}$

Since $9 - x^2 \geq 0$, $x^2 - 9 \leq 0$ $\quad \therefore -3 \leq x \leq 3$

$f'(x) = 1 + \dfrac{-2x}{2\sqrt{9-x^2}} = \dfrac{\sqrt{9-x^2}-x}{\sqrt{9-x^2}}$

$f'(x) = 0 \Rightarrow \sqrt{9 - x^2} - x = 0$; $\quad x = \sqrt{9 - x^2} \geq 0$

$\therefore x^2 = 9 - x^2$; $2x^2 = 9$ $\quad \therefore x = \pm\dfrac{3}{\sqrt{2}}$

Since $x \geq 0$, $x = \dfrac{3}{\sqrt{2}}$

$f(-3) = -6$; $\quad f(3) = 0$

$f\left(\dfrac{3}{\sqrt{2}}\right) = \dfrac{3}{\sqrt{2}} - 3 + \sqrt{9 - \dfrac{9}{2}} = \dfrac{3}{\sqrt{2}} - 3 + \dfrac{3}{\sqrt{2}} = 3\sqrt{2} - 3$

$\therefore$ Maximum value: $3\sqrt{2} - 3$ at $x = \dfrac{3}{\sqrt{2}}$, Minimum value: -6 at $x = -3$

(5) $f(x) = \dfrac{x}{x^2 - x + 1}$

$f'(x) = \dfrac{x^2 - x + 1 - x(2x-1)}{(x^2-x+1)^2} = \dfrac{-x^2+1}{(x^2-x+1)^2} = \dfrac{-(x+1)(x-1)}{(x^2-x+1)^2}$

$f'(x) = 0 \Rightarrow x = 1$ or $x = -1$

$f(1) = 1$, $\quad f(-1) = -\dfrac{1}{3}$

$\therefore$ Maximum value: 1 at $x = 1$, Minimum value: $-\dfrac{1}{3}$ at $x = -1$

(6) $f(x) = 2\sin^3 x + 3\cos^2 x \quad (0 \leq x \leq 2\pi)$

$f(x) = 2\sin^3 x + 3(1 - \sin^2 x)$

Let $\sin x = t$ $\quad$ Then, $-1 \leq t \leq 1$

Let $y = 2t^3 + 3(1 - t^2)$ $\quad$ Then, $y' = 6t^2 - 6t = 6t(t - 1)$

$y' = 0 \Rightarrow t = 0$ or $t = 1$

$y_{t=0} = 3$; $\quad y_{t=1} = 2$; $\quad y_{t=-1} = -2$

$\therefore$ Maximum value: 3 at $t = 0$ $(x = 0, \pi, 2\pi)$, Minimum value: -2 at $t = -1$ $\left(x = \dfrac{3\pi}{2}\right)$

(7) $f(x) = \sin x\,(1 + \cos x) \quad (0 \leq x \leq \pi)$

$f'(x) = \cos x\,(1 + \cos x) + \sin x\,(-\sin x) = \cos^2 x - \sin^2 x + \cos x$

$\qquad = \cos^2 x - (1 - \cos^2 x) + \cos x = 2\cos^2 x + \cos x - 1 = (2\cos x - 1)(\cos x + 1)$

$f'(x) = 0 \Rightarrow \cos x = \dfrac{1}{2}$ or $\cos x = -1$

Since $0 \leq x \leq \pi$, $x = \dfrac{\pi}{3}$ or $x = \pi$

$f\left(\dfrac{\pi}{3}\right) = \sin\dfrac{\pi}{3}\left(1 + \cos\dfrac{\pi}{3}\right) = \dfrac{\sqrt{3}}{2}\left(1 + \dfrac{1}{2}\right) = \dfrac{3\sqrt{3}}{4}$

$f(\pi) = \sin\pi\,(1 + \cos\pi) = 0$ $\qquad f(0) = 0$

$\therefore$ Maximum value: $\frac{3\sqrt{3}}{4}$ at $x = \frac{\pi}{3}$, Minimum value: 0 at $x = \pi$, $x = 0$

(8) $f(x) = \sin x + \sin 2x \cos x + \frac{3}{4}\cos 2x$ $\quad (0 \le x \le 2\pi)$

$f(x) = \sin x + \sin 2x \cos x + \frac{3}{4}\cos 2x = \sin x + (2\sin x \cos x)\cos x + \frac{3}{4}(1 - 2\sin^2 x)$

$\qquad = \sin x + 2\sin x \cos^2 x + \frac{3}{4}(1 - 2\sin^2 x)$

$\qquad = \sin x + 2\sin x (1 - \sin^2 x) + \frac{3}{4}(1 - 2\sin^2 x)$

$\qquad = -2\sin^3 x - \frac{3}{2}\sin^2 x + 3\sin x + \frac{3}{4}$

Let $\sin x = t$ $\quad (-1 \le t \le 1)$

Then, $f(x) = -2t^3 - \frac{3}{2}t^2 + 3t + \frac{3}{4}$

$f'(x) = -6t^2 - 3t + 3 = -3(2t^2 + t - 1) = -3(t + 1)(2t - 1)$

$f'(x) = 0 \Rightarrow t = -1$ or $t = \frac{1}{2}$

$f(-1) = 2 - \frac{3}{2} - 3 + \frac{3}{4} = \frac{8 - 6 - 12 + 3}{4} = -\frac{7}{4}$

$f\left(\frac{1}{2}\right) = -2\left(\frac{1}{8}\right) - \frac{3}{2}\left(\frac{1}{4}\right) + 3\left(\frac{1}{2}\right) + \frac{3}{4} = \frac{-2 - 3 + 12 + 6}{8} = \frac{13}{8}$

$f(1) = -2 - \frac{3}{2} + 3 + \frac{3}{4} = 1 - \frac{3}{4} = \frac{1}{4}$

$\therefore$ Maximum value: $\frac{13}{8}$ at $t = \frac{1}{2}$ $\left(x = \frac{\pi}{6}, \ x = \frac{5\pi}{6}\right)$, Minimum value: $-\frac{7}{4}$ at $t = -1$ $\left(x = \frac{3\pi}{2}\right)$

(9) $f(x) = \sin x + \sqrt{3}\cos x + x$ $\quad (0 \le x \le \pi)$

$f'(x) = \cos x - \sqrt{3}\sin x + 1 = -2\sin\left(x - \frac{\pi}{6}\right) + 1$

$f'(x) = 0 \Rightarrow \sin\left(x - \frac{\pi}{6}\right) = \frac{1}{2}$; $\quad x - \frac{\pi}{6} = \frac{\pi}{6}$ or $x - \frac{\pi}{6} = \frac{5\pi}{6}$ $\quad (\because 0 \le x \le \pi)$

$\therefore x = \frac{\pi}{3}$ or $x = \pi$

$f\left(\frac{\pi}{3}\right) = \sin\frac{\pi}{3} + \sqrt{3}\cos\frac{\pi}{3} + \frac{\pi}{3} = \frac{\sqrt{3}}{2} + \sqrt{3}\left(\frac{1}{2}\right) + \frac{\pi}{3} = \sqrt{3} + \frac{\pi}{3}$

$f(\pi) = \sin\pi + \sqrt{3}\cos\pi + \pi = -\sqrt{3} + \pi$ (≈ 1.41)

$f(0) = \sin 0 + \sqrt{3}\cos 0 + 0 = \sqrt{3}$ (≈ 1.732)

$\therefore$ Maximum value: $\sqrt{3} + \frac{\pi}{3}$ at $x = \frac{\pi}{3}$, Minimum value: $-\sqrt{3} + \pi$ at $x = \pi$

(10) $f(x) = \sin x + |\cos x|$ $\quad (0 \le x \le 2\pi)$

$|\cos x| = \sqrt{\cos^2 x} = \sqrt{1 - \sin^2 x}$

Let $\sin x = t$ $\quad (0 \le x \le 2\pi)$ $\quad$ Then, $-1 \le t \le 1$

Let $y = t + \sqrt{1 - t^2}$ $\quad$ Then, $y' = 1 + \frac{-2t}{2\sqrt{1 - t^2}} = \frac{\sqrt{1 - t^2} - t}{\sqrt{1 - t^2}}$

$y' = 0 \Rightarrow \sqrt{1 - t^2} - t = 0 ; \quad t = \sqrt{1 - t^2} \geq 0$

Since $t^2 = 1 - t^2, \ 2t^2 = 1 \quad \therefore \ t = \frac{1}{\sqrt{2}} \quad (\because t \geq 0)$

$y_{t = \frac{1}{\sqrt{2}}} = \frac{1}{\sqrt{2}} + \sqrt{1 - \frac{1}{2}} = \frac{1}{\sqrt{2}} + \frac{1}{\sqrt{2}} = \sqrt{2}$

$y_{t = 1} = 1 + 0 = 1$

$y_{t = -1} = -1 + 0 = -1$

$\therefore$ Maximum value: $\sqrt{2}$ at $t = \frac{1}{\sqrt{2}} \left(x = \frac{\pi}{4}, \ x = \frac{3\pi}{4} \right)$, Minimum value: -1 at $t = -1 \left(x = \frac{3\pi}{2} \right)$

(11) $f(x) = (\sin x + \cos x)^3 - 12 \sin x \cos x$

Let $\sin x + \cos x = t$

Then, $t = \sqrt{2} \sin \left(x + \frac{\pi}{4} \right)$

Since $-1 \leq \sin \left(x + \frac{\pi}{4} \right) \leq 1, \ -\sqrt{2} \leq \sqrt{2} \sin \left(x + \frac{\pi}{4} \right) \leq \sqrt{2} \qquad \therefore \ -\sqrt{2} \leq t \leq \sqrt{2}$

$\sin x + \cos x = t \ \Rightarrow \ (\sin x + \cos x)^2 = t^2$

$\therefore \sin^2 x + 2 \sin x \cos x + \cos^2 x = t^2 ; \quad 1 + 2 \sin x \cos x = t^2 ; \quad \sin x \cos x = \frac{t^2 - 1}{2}$

$\therefore \ y = t^3 - 12 \left(\frac{t^2 - 1}{2} \right) = t^3 - 6t^2 + 6$

$y' = 3t^2 - 12t = 3t(t - 4)$

$y' = 0 \ \Rightarrow \ t = 0 \ \text{or} \ t = 4$

Since $-\sqrt{2} \leq t \leq \sqrt{2}, \ t = 0$

$y_{t = 0} = 6$

$y_{t = -\sqrt{2}} = -2\sqrt{2} - 12 + 6 = -2\sqrt{2} - 6$

$y_{t = \sqrt{2}} = 2\sqrt{2} - 12 + 6 = 2\sqrt{2} - 6$

$\therefore$ Maximum value: 6 at $t = 0$, Minimum value: $-2\sqrt{2} - 6$ at $t = -\sqrt{2}$

(12) $f(x) = \frac{\sin x + \cos x}{\sin x + \cos x - 2}$

Let $\sin x + \cos x = t \qquad$ Then, $-\sqrt{2} \leq t \leq \sqrt{2}$

$y = \frac{t}{t - 2} = \frac{t - 2 + 2}{t - 2} = 1 + \frac{2}{t - 2}$

Since $-\sqrt{2} \leq t \leq \sqrt{2}, \ -\sqrt{2} - 2 \leq t - 2 \leq \sqrt{2} - 2 ; \quad \frac{2}{\sqrt{2} - 2} \leq \frac{2}{t - 2} \leq \frac{2}{-\sqrt{2} - 2}$

$\therefore \ \frac{\sqrt{2}}{\sqrt{2} - 2} \leq 1 + \frac{2}{t - 2} \leq \frac{-\sqrt{2}}{-\sqrt{2} - 2} ; \quad \frac{\sqrt{2}}{\sqrt{2} - 2} \leq 1 + \frac{2}{t - 2} \leq \frac{\sqrt{2}}{\sqrt{2} + 2} ; \quad \frac{\sqrt{2}(\sqrt{2} + 2)}{2 - 4} \leq 1 + \frac{2}{t - 2} \leq \frac{\sqrt{2}(\sqrt{2} - 2)}{2 - 4}$

$\therefore \ \frac{2 + 2\sqrt{2}}{-2} \leq 1 + \frac{2}{t - 2} \leq \frac{2 - 2\sqrt{2}}{-2} ; \quad -1 - \sqrt{2} \leq 1 + \frac{2}{t - 2} \leq -1 + \sqrt{2}$

$\therefore$ Maximum value: $-1 + \sqrt{2}$, Minimum value: $-1 - \sqrt{2}$

(13) $f(x) = (x^2 - 8)e^x \quad (-3 \le x \le 3)$

$f'(x) = 2xe^x + (x^2 - 8)e^x = e^x(x^2 + 2x - 8) = e^x(x + 4)(x - 2)$

$f'(x) = 0 \implies x = -4$ or $x = 2$

Since $-3 \le x \le 3, \ x = 2$

$f(2) = (4 - 8)e^2 = -4e^2 \ ; \quad f(-3) = e^{-3} \ ; \quad f(3) = e^3$

$\therefore$ Maximum value: e^3 at $x = 3$, Minimum value: $-4e^2$ at $x = 2$

(14) $f(x) = \sqrt{6 - x^2}\,e^x$

Since $6 - x^2 \ge 0, \ -\sqrt{6} \le x \le \sqrt{6}$

$f'(x) = \dfrac{-2x}{2\sqrt{6-x^2}}e^x + \sqrt{6 - x^2}\,e^x = e^x\left(\dfrac{-x}{\sqrt{6-x^2}} + \sqrt{6 - x^2}\right) = e^x\left(\dfrac{-x+6-x^2}{\sqrt{6-x^2}}\right) = -e^x\left(\dfrac{x^2+x-6}{\sqrt{6-x^2}}\right)$

$\qquad = -e^x\dfrac{(x+3)(x-2)}{\sqrt{6-x^2}}$

$f'(x) = 0 \implies x = -3$ or $x = 2$

Since $-\sqrt{6} \le x \le \sqrt{6}, \ x = 2$

$f(2) = \sqrt{6 - 4}\,e^2 = \sqrt{2}e^2 \ ; \quad f(-\sqrt{6}) = 0 \ ; \quad f(\sqrt{6}) = 0$

$\therefore$ Maximum value: $\sqrt{2}e^2$ at $x = 2$, Minimum value: 0 at $x = -\sqrt{6}, \ x = \sqrt{6}$

#24 Find the value.

(1) When a function $f(x) = ax^4 - 4ax^3 + b \ (a > 0)$ on a closed interval $[1, 4]$ has the maximum and minimum values 3 and -15, respectively, find the values of a and b.

$f'(x) = 4ax^3 - 12ax^2 = 4ax^2(x - 3)$

$f'(x) = 0 \implies x = 0$ or $x = 3$

Since $1 \le x \le 4, \ x = 3$

$f(3) = 81a - 108a + b = -27a + b$

$f(1) = a - 4a + b = -3a + b$

$f(4) = 4^4 a - 4^4 a + b = b$

Since $a > 0$, maximum value is b at $x = 4$ and minimum value is $-27a + b$ at $x = 3$.

$\therefore \ b = 3, \ -27a + b = -15$

Therefore, $a = \dfrac{2}{3}, \ b = 3$

(2) When a function $f(x) = \dfrac{x+1}{x^2+8} \ (0 \le x \le a)$ has maximum and minimum values $\dfrac{1}{4}$ and $\dfrac{1}{8}$, respectively, find the maximum value of a.

$f'(x) = \dfrac{x^2+8-(x+1)(2x)}{(x^2+8)^2} = \dfrac{-x^2-2x+8}{(x^2+8)^2} = \dfrac{-(x+4)(x-2)}{(x^2+8)^2}$

$f'(x) = 0 \Rightarrow x = -4$ or $x = 2$

Since $x \geq 0$, $x = 2$

$f(2) = \dfrac{3}{12} = \dfrac{1}{4}$; $f(0) = \dfrac{1}{8}$; $f(a) = \dfrac{a+1}{a^2+8}$

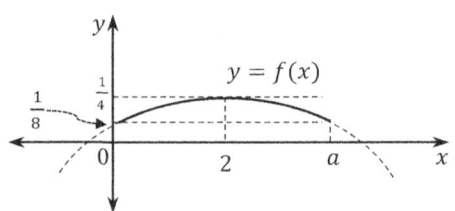

Note that $0 \leq x \leq a$ and maximum and minimum values are $\dfrac{1}{4}$ and $\dfrac{1}{8}$, respectively.

Thus, when $f(a) = \dfrac{a+1}{a^2+8} = \dfrac{1}{8}$, a will be the greatest number.

$\dfrac{a+1}{a^2+8} = \dfrac{1}{8} \Rightarrow a^2 + 8 = 8a + 8$; $a^2 - 8a = a(a-8) = 0$ $\quad \therefore a = 0$ or $a = 8$

$\therefore 2 \leq a \leq 8$ $\quad$ Therefore, the maximum value of a is 8.

(3) For a function $y = \sin^2 x$ $\left(0 \leq x \leq \dfrac{\pi}{2}\right)$, the function is concave downward in the

range $a < x \leq b$. Find the values of a and b.

$y = \sin^2 x$ $\left(0 \leq x \leq \dfrac{\pi}{2}\right) \Rightarrow y' = 2 \sin x \cos x = \sin 2x$

$$y'' = 2 \cos 2x$$

When $y'' < 0$, y is concave downward.

$\therefore 2 \cos 2x < 0$; $\cos 2x < 0$

Since $0 \leq x \leq \dfrac{\pi}{2}$, $0 \leq 2x \leq \pi$

Since $\cos 2x < 0$, $\dfrac{\pi}{2} < 2x \leq \pi$

$\therefore \dfrac{\pi}{4} < x \leq \dfrac{\pi}{2}$

Therefore, $a = \dfrac{\pi}{4}$, $b = \dfrac{\pi}{2}$

(4) For a function $f(x) = e^{2x}$ on a closed interval $[0, x]$, find the values of a and b so that

the inequality $a < \dfrac{1}{x} \ln \dfrac{e^{2x}-1}{2x} < b$ is always true.

Note that $f(x) = e^{2x}$ is continuous on the closed interval $[0, x]$ and differentiable on the open

interval $(0, x)$.

By Mean Value Theorem, there is at least one c $(0 < c < x)$ such that $\dfrac{f(x)-f(0)}{x-0} = f'(c)$

$\therefore f'(c) = \dfrac{e^{2x}-1}{x}$

Note that $f(x) = e^{2x} \Rightarrow f'(x) = 2e^{2x}$ $\quad \therefore f'(c) = 2e^{2c}$

Thus, $\dfrac{e^{2x}-1}{x} = 2e^{2c}$; $\dfrac{e^{2x}-1}{2x} = e^{2c}$

$\therefore \log\dfrac{e^{2x}-1}{2x} = \log e^{2c}$; $\log\dfrac{e^{2x}-1}{2x} = 2c$

Since $0 < c < x$, $0 < 2c < 2x$

$$\therefore \ 0 < \log \frac{e^{2x}-1}{2x} < 2x \ ; \ 0 < \frac{1}{x}\log \frac{e^{2x}-1}{2x} < 2$$

Therefore, $a = 0$, $b = 2$

(5) When a function $f(x) = x \log x + 3x + a$ has minimum value 1, find the value of a.

$$f'(x) = \log x + x \cdot \frac{1}{x} + 3 = \log x + 4$$

$$f'(x) = 0 \ \Rightarrow \ \log x = -4 \quad \therefore \ x = e^{-4}$$

$\therefore \ f(x)$ has a local extreme value at $x = e^{-4}$.

Since $f(x)$ has minimum value 1, $f(e^{-4}) = 1$

$$\therefore \ f(e^{-4}) = e^{-4}\log e^{-4} + 3e^{-4} + a = -4e^{-4} + 3e^{-4} + a = -e^{-4} + a = 1$$

That is, $a = 1 + e^{-4}$

#25 Find the range.

(1) For an equation $2x^3 - 6x + a = 0$, find the range of a so that:

   1) The equation has three different real number solutions.

   2) The equation has a double root and a real number solution.

   3) The equation has a real number solution and two complex number solutions.

1)   Let $f(x) = 2x^3 - 6x + a$

     Then, $f'(x) = 6x^2 - 6 = 6(x+1)(x-1)$

     $f(-1) = -2 + 6 + a = 4 + a > 0$

     $f(1) = 2 - 6 + a = -4 + a < 0$

     $\therefore \ -4 < a < 4$

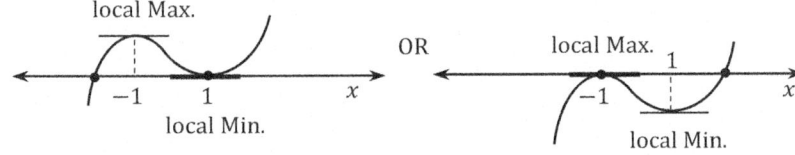

2)

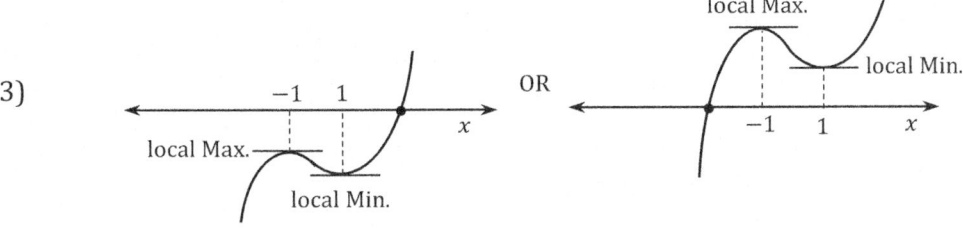

     $f(1) = -4 + a = 0 \ \ \text{or} \ \ f(-1) = 4 + a = 0$

     $\therefore \ a = 4 \ \text{or} \ a = -4$

3)

     Case 1: $f(-1) < 0 \ $ and $ \ f(1) < 0$

$$\therefore (4+a)(-4+a) > 0 ; \quad a > 4 \text{ or } a < -4$$

Case 2: $f(-1) > 0$ and $f(1) > 0$

$$\therefore (4+a)(-4+a) > 0 ; \quad a > 4 \text{ or } a < -4$$

Therefore, $a > 4$ or $a < -4$

(2) When an equation $x^3 - 3x^2 - 24x - a = 0$ has two different positive number solutions and a negative number solution, find the range of a.

Let $f(x) = x^3 - 3x^2 - 24x - a$

Then, $f'(x) = 3x^2 - 6x - 24 = 3(x^2 - 2x - 8) = 3(x-4)(x+2)$

$f'(x) = 0 \Rightarrow x = 4$ or $x = -2$

$f(4) = 4^3 - 3 \cdot 4^2 - 24 \cdot 4 - a = 4^2(4 - 3 - 6) - a = 4^2(-5) - a = -80 - a < 0$

$f(-2) = -8 - 12 + 48 - a = 28 - a > 0$

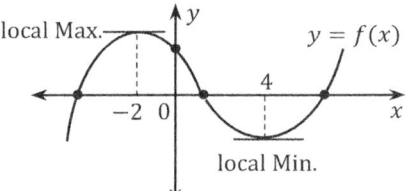

$\therefore a > -80$ and $a < 28$; $-80 < a < 28$ $\cdots\cdots$ ①

Since y-intercept is positive, $-a > 0$; $a < 0$ $\cdots\cdots$ ②

By ① and ②, $-80 < a < 0$

(3) Find the range of a so that $-x^3 + 3x^2 - a = 0$ has a solution which is greater than 5.

$-x^3 + 3x^2 - a = 0 \iff -x^3 + 3x^2 = a$

Let $y = -x^3 + 3x^2$ $\cdots\cdots$ ①

$\quad\ y = a$ $\cdots\cdots$ ②

Then, the solutions of the equation $-x^3 + 3x^2 - a = 0$ are the x-coordinates of the

intersection points of ① and ②.

By ①, $y' = -3x^2 + 6x = -3x(x-2)$

$y' = 0 \Rightarrow x = 0$ or $x = 2$

That is, $y = -x^3 + 3x^2$ has local extreme values

at $x = 0$ and $x = 2$.

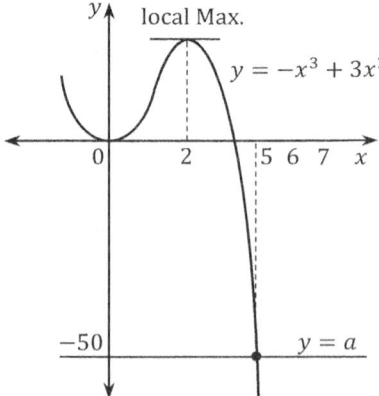

Since $y_{x=5} = -5^3 + 3 \cdot 5^2 = 5^2(-5 + 3) = 5^2(-2) = -50$,

$a < -50$

(4) When an equation $3x^4 - 4x^3 - 12x^2 - a = 0$ has

exactly two different negative real number solutions,

find the range of a.

$3x^4 - 4x^3 - 12x^2 - a = 0 \iff 3x^4 - 4x^3 - 12x^2 = a$

Let $f(x) = 3x^4 - 4x^3 - 12x^2$ and $g(x) = a$

Then, $f'(x) = 12x^3 - 12x^2 - 24x = 12x(x^2 - x - 2) = 12x(x-2)(x+1)$

$f'(x) = 0 \Rightarrow x = 0$ or $x = 2$ or $x = -1$

$f(0) = 0$; local maximum value

$f(2) = -32$; local minimum value

$f(-1) = -5$; local minimum value

From the graph, $-32 < a < -5$

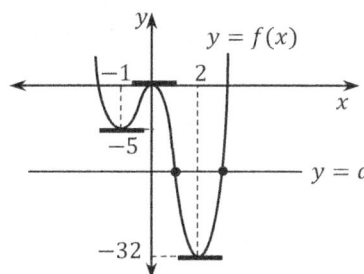

(5) When an equation $x^4 - 4a^3x + 27 = 0$ has no real number solution,

find the range of a.

Let $f(x) = x^4 - 4a^3x + 27$

Then, $f'(x) = 4x^3 - 4a^3 = 4(x - a)(x^2 + ax + a^2)$

$f'(x) = 0 \Rightarrow x = a$ $\left(\because x^2 + ax + a^2 = \left(x + \frac{a}{2}\right)^2 - \frac{a^2}{4} + a^2 = \left(x + \frac{a}{2}\right)^2 + \frac{3a^2}{4} > 0\right)$

$f(a) = a^4 - 4a^4 + 27 = -3a^4 + 27$

$\therefore$ $f(a)$ is the local minimum value and also the minimum value.

Since $f(x) = 0$ has no real number solution, $f(a) > 0$

That is, $-3a^4 + 27 > 0$; $a^4 - 9 < 0$; $(a^2 - 3)(a^2 + 3) < 0$; $(a - \sqrt{3})(a + \sqrt{3})(a^2 + 3) < 0$

Since $a^2 + 3 > 0$, $(a - \sqrt{3})(a + \sqrt{3}) < 0$

$\therefore$ $-\sqrt{3} < a < \sqrt{3}$

(6) When a function $f(x) = x^3 - 3ax - 2a$ has local extreme values and an equation

$f(x) = 0$ has only one real number solution, find the range of a.

$f'(x) = 3x^2 - 3a = 3(x^2 - a) = 3(x - \sqrt{a})(x + \sqrt{a})$

Since $f(x)$ has local extreme values, $f'(x) = 0$ has two different real number solutions.

$\therefore$ $a > 0$ ······ ①

$f(-\sqrt{a}) = -a\sqrt{a} + 3a\sqrt{a} - 2a = 2a\sqrt{a} - 2a = 2a(\sqrt{a} - 1)$; local maximum value

$f(\sqrt{a}) = a\sqrt{a} - 3a\sqrt{a} - 2a = -2a\sqrt{a} - 2a = -2a(\sqrt{a} + 1)$; local minimum value

Since $f(x) = 0$ has only one real number solution, $f(-\sqrt{a}) \cdot f(\sqrt{a}) > 0$

$\therefore$ $2a(\sqrt{a} - 1) \cdot -2a(\sqrt{a} + 1) = -4a^2(a - 1) > 0$

$\therefore$ $a^2(a - 1) < 0$ ······ ②

By ① and ②, $0 < a < 1$

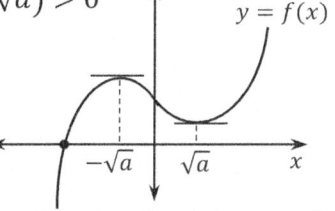

(7) Find the range of a so that a function $y = \frac{1}{2}x^4 - 3x^2 + 2ax$ has local extreme values.

$y' = 2x^3 - 6x + 2a$

Since y has local extreme values, $y' = 0$ has three different real number solutions.

Let $f(x) = 2x^3 - 6x + 2a$

Then, $f'(x) = 6x^2 - 6 = 6(x+1)(x-1)$

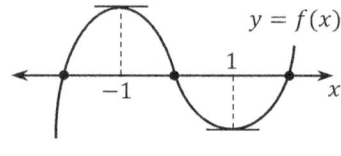

$f'(x) = 0 \Rightarrow x = -1$ or $x = 1$

$\therefore f(-1) > 0$ and $f(1) < 0$

Since $f(-1) = -2 + 6 + 2a = 4 + 2a > 0$ and $f(1) = 2 - 6 + 2a = -4 + 2a < 0$,

$a > -2$ and $a < 2$

$\therefore -2 < a < 2$

(8) When the curve $y = e^{ax}$ and a line $y = x$ intersect at two different points,

   find the range of a.

$y = e^{ax} \Rightarrow \log y = \log e^{ax} = ax \log e = ax$

$\therefore$ The number of intersection points of the curve $\log y = ax$ and a line $y = x$ is the number of

solutions of the equation $\log y = ax$.

The equation of the tangent line at a point $(t, \log t)$ on the curve $y = \log x$ is

$y - \log t = \dfrac{1}{t}(x - t) \cdots\cdots \text{①}$

Since ① passes through the origin $(0,0)$, $\ -\log t = -1$

$\therefore\ t = e$

① $\Leftrightarrow y - \log e = \dfrac{1}{e}(x - e)$; $y - 1 = \dfrac{1}{e}(x - e)$; $y = \dfrac{1}{e}x$ (Tangent line)

Since the curve $y = \log x$ and a line $y = ax$ intersect at two different points, $\ 0 < a < \dfrac{1}{e}$

(9) For a curve $y = 2\sin x \ (0 \le x \le \pi)$ and a line $y = x + a$, find the range of a so that

   there is at least one intersection point.

$2\sin x = x + a$; $\ 2\sin x - x = a$

Let $f(x) = 2\sin x - x$ and $g(x) = a$

Then, $f'(x) = 2\cos x - 1$

$f'(x) = 0 \Rightarrow \cos x = \dfrac{1}{2}$

Since $0 \le x \le \pi$, $\ x = \dfrac{\pi}{3}$

$f\left(\dfrac{\pi}{3}\right) = 2\sin\dfrac{\pi}{3} - \dfrac{\pi}{3} = \sqrt{3} - \dfrac{\pi}{3}$; $f(0) = 0$; $f(\pi) = 2\sin\pi - \pi = -\pi$

$\therefore\ f\left(\dfrac{\pi}{3}\right)$ is the local maximum value and also maximum value.

$\therefore\ -\pi \le a \le \sqrt{3} - \dfrac{\pi}{3}$

(10) When a curve $y = x^3 - 3x$ has exactly one tangent line at $(-1, a)$,

   find the range of a.

$y = x^3 - 3x \Rightarrow y' = 3x^2 - 3 = 3(x+1)(x-1)$

Let $P(t,\ t^3 - 3t)$ be a point on the curve.

Then, the equation of the tangent line at P is $y - (t^3 - 3t) = (3t^2 - 3)(x - t)$ $\cdots\cdots$ ①

Since $(-1, a)$ lies on the line ①, $a - (t^3 - 3t) = (3t^2 - 3)(-1 - t)$

∴ $a - t^3 + 3t = -3t^2 + 3 - 3t^3 + 3t$; $2t^3 + 3t^2 + a - 3 = 0$ $\cdots\cdots$ ②

Since the equation ② has only one real number solution, local extreme values are both positive or negative.

Let $f(t) = 2t^3 + 3t^2 + a - 3$

Then, $f'(t) = 6t^2 + 6t = 6t(t + 1)$

$f'(t) = 0 \Rightarrow t = 0$ or $t = -1$

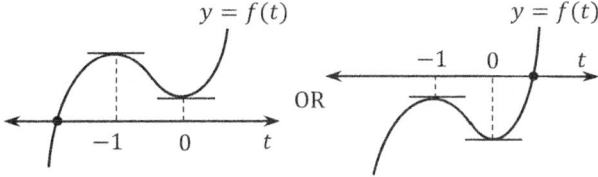

∴ ($f(0) > 0$ and $f(-1) > 0$) or ($f(0) < 0$ and $f(-1) < 0$) That is, $f(0)f(-1) > 0$

Since $f(0) = a - 3$ and $f(-1) = -2 + 3 + a - 3 = a - 2$, $(a - 3)(a - 2) > 0$

∴ $a > 3$ or $a < 2$

(11) For a curve $y = x^3 - x + 1$ and a line $y = 11x + a$, find the range of a so that:

    1) The curve and the line intersect at three different points.

    2) The curve and the line intersect at two points. At one of the points, the line is the tangent of the curve.

    3) The curve and the line intersect only at a point.

1) $x^3 - x + 1 = 11x + a \Rightarrow x^3 - 12x + 1 - a = 0$

    Let $f(x) = x^3 - 12x + 1 - a$

    Then, $f'(x) = 3x^2 - 12 = 3(x^2 - 4) = 3(x + 2)(x - 2)$

    $f'(x) = 0 \Rightarrow x = -2$ or $x = 2$

    $f(-2) > 0$ and $f(2) < 0$

    $f(-2) = -8 + 24 + 1 - a = 17 - a > 0$ ∴ $a < 17$

    $f(2) = 8 - 24 + 1 - a = -15 - a < 0$ ∴ $a > -15$

    Therefore, $-15 < a < 17$

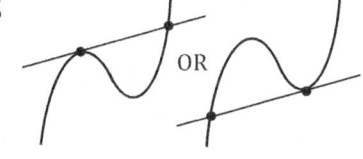

2) $f(-2) = 0$ or $f(2) = 0$

    ∴ $a = 17$ or $x = -15$

3) ($f(-2) > 0$ and $f(2) > 0$) or ($f(-2) < 0$ and $f(2) < 0$)

    ∴ $f(-2)f(2) > 0$; $(17 - a)(-15 - a) > 0$

    ∴ $a > 17$ or $a < -15$

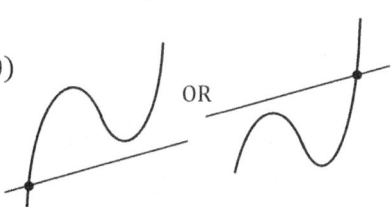

(12) When an inequality $x^4 - 4x - a^2 + 4a > 0$ is always true for all real number x, find the range of a.

Let $f(x) = x^4 - 4x - a^2 + 4a$

Then, $f'(x) = 4x^3 - 4 = 4(x^3 - 1) = 4(x - 1)(x^2 + x + 1)$

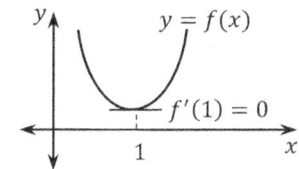

$f'(x) = 0 \Rightarrow x = 1 \ (\because x^2 + x + 1 > 0)$

$f(1) = 1 - 4 - a^2 + 4a = -a^2 + 4a - 3$; local minimum value and minimum value

Since $f(x) > 0$ for all x, $f(1) > 0$

$\therefore f(1) = -(a^2 - 4a + 3) = -(a-1)(a-3) > 0$; $(a-1)(a-3) < 0$

$\therefore 1 < a < 3$

(13) For any real number x such that $x > -1$, the inequality $\frac{4}{3}x^3 - x^2 - 2x + 1 - a > 0$

is always true. Find the range of a.

Let $f(x) = \frac{4}{3}x^3 - x^2 - 2x + 1 - a$

Then, $f'(x) = 4x^2 - 2x - 2 = 2(2x^2 - x - 1) = 2(2x+1)(x-1)$

$f'(x) = 0 \Rightarrow x = -\frac{1}{2}$ or $x = 1$

Since $f(x) > 0$ when $x > -1$, $f(-1) \geq 0$ and $f(1) > 0$

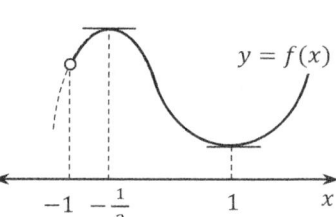

$f(-1) = -\frac{4}{3} - 1 + 2 + 1 - a = \frac{2}{3} - a \geq 0$; $a \leq \frac{2}{3}$

$f(1) = \frac{4}{3} - 1 - 2 + 1 - a = -\frac{2}{3} - a > 0$; $a < -\frac{2}{3}$

Therefore, $a < -\frac{2}{3}$

(14) For any positive number a, the inequality $\sqrt{x} > a \log x$ is always true.

Find the range of a.

$\sqrt{x} > a \log x \Rightarrow \sqrt{x} - a \log x > 0$

Let $f(x) = \sqrt{x} - a \log x$

Then, $f'(x) = \frac{1}{2\sqrt{x}} - \frac{a}{x} = \frac{\sqrt{x} - 2a}{2x}$

$\therefore f'(x) = 0 \Rightarrow \sqrt{x} - 2a = 0$; $x = 4a^2$

$f(4a^2) = \sqrt{4a^2} - a \log(4a^2) = 2a - a \log(4a^2)$; local extreme value

Since $f(x) > 0$ for all x, $2a - a \log(4a^2) > 0$; $\log(4a^2) < 2$

$\therefore 0 < 4a^2 < e^2$; $0 < 2a < e$

$\therefore 0 < a < \frac{e}{2}$

(15) For any real number x such that $0 \leq x \leq \frac{\pi}{2}$, $\cos x - a = a \sin x$ is always true.

Find the range of a.

$\cos x - a = a \sin x \Rightarrow \cos x = a(1 + \sin x)$

$\therefore \frac{\cos x}{1 + \sin x} = a$

Let $f(x) = \frac{\cos x}{1 + \sin x}$

Then, $f'(x) = \dfrac{-\sin x(1+\sin x)-\cos x(\cos x)}{(1+\sin x)^2} = \dfrac{-(\sin^2 x+\cos^2 x)-\sin x}{(1+\sin x)^2} = \dfrac{-1-\sin x}{(1+\sin x)^2} = \dfrac{-1}{1+\sin x}$

$\therefore\ f'(x) < 0\ \ \left(\because\ 0 \le x \le \dfrac{\pi}{2}\right)$

$\therefore\ f(x)$ is decreasing on $\left[0,\ \dfrac{\pi}{2}\right]$.

$\therefore\ f(0)$ is the maximum value and $f\left(\dfrac{\pi}{2}\right)$ is the minimum value.

Since $f(0) = 1$ and $f\left(\dfrac{\pi}{2}\right) = 0,\ \ 0 \le a \le 1$

#26 Find the value.

(1) When an equation $\log x = x + a$ has two different real number solutions, find the value of a.

$\log x = x + a\ \ \Rightarrow\ \ \log x - x = a\ \ \cdots\cdots\ ①$

Let $y = \log x - x\ \ \cdots\cdots\ ②$

$\qquad y = a\ \ \cdots\cdots\ ③$

The roots of the equation ① are the x-coordinates of the intersection points of ② and ③.

By ②, $y' = \dfrac{1}{x} - 1 = \dfrac{1-x}{x}$

$y' = 0\ \ \Rightarrow\ \ 1 - x = 0\ ;\ x = 1$

$\therefore\ y_{x=1} = \log 1 - 1 = -1$

$\therefore\ a < -1$

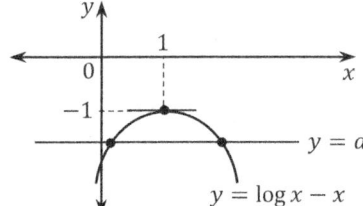

(2) When an equation $x \log x - 3x - a = 0$ has exactly one real number solution, find the minimum value of a.

Let $f(x) = x \log x - 3x$ and $g(x) = a$

Then, $f(x)$ and $g(x)$ intersect only at one point.

$f'(x) = \log x + x \cdot \dfrac{1}{x} - 3 = \log x - 2$

$f'(x) = 0\ \ \Rightarrow\ \ \log x = 2\ \ \therefore\ x = e^2$

$f(e^2) = e^2 \log e^2 - 3e^2 = 2e^2 - 3e^2 = -e^2$

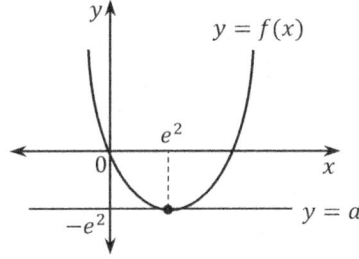

Since $f(x)$ and $g(x)$ have only one intersection point, $g(x) = a$ is the tangent line at $x = e^2$.

$\therefore\ a = -e^2$

(3) When an inequality $ax \ge \log x$ is always true for all $x > 0$, find the minimum value of a positive number a.

The equation of the tangent line at a point $(t, \log t)$ on the graph of $y = \log x$ is

$y - \log t = \dfrac{1}{t}(x - t)\ ;\ \text{i.e.,}\ \ y = \dfrac{1}{t}x - 1 + \log t\ \ \cdots\cdots\ ①$

Since the tangent line ① passes through the origin $(0, 0)$, $0 = 0 - 1 + \log t$

$\therefore \ \log t = 1 \ ; \ t = e$

$\therefore \ y = \dfrac{1}{e}x - 1 + \log e = \dfrac{1}{e}x$

$\therefore \ ax \geq \dfrac{1}{e}x \ ; \quad a \geq \dfrac{1}{e}$ $\therefore$ The minimum value of a is $\dfrac{1}{e}$.

(4) When an inequality $3x^4 + 4x^3 - 12x^2 \geq a - 2\sin\left(\dfrac{\pi}{2}x\right)$ is always true for any real

   number x, find the maximum value of a.

Let $f(x) = 3x^4 + 4x^3 - 12x^2$

      $g(x) = a - 2\sin\left(\dfrac{\pi}{2}x\right)$

Then, $f'(x) = 12x^3 + 12x^2 - 24x = 12x(x^2 + x - 2) = 12x(x + 2)(x - 1)$

$f'(x) = 0 \ \Rightarrow x = 0$ or $x = -2$ or $x = 1$

$f(0) = 0$; Local maximum

$f(-2) = 2^2(12 - 8 - 12) = -32$; Local minimum

$f(1) = 3 + 4 - 12 = -5$; Local minimum

Since $-1 \leq \sin\left(\dfrac{\pi}{2}x\right) \leq 1, \ -2 \leq -2\sin\left(\dfrac{\pi}{2}x\right) \leq 2$

$\therefore \ a - 2 \leq a - 2\sin\left(\dfrac{\pi}{2}x\right) \leq a + 2$

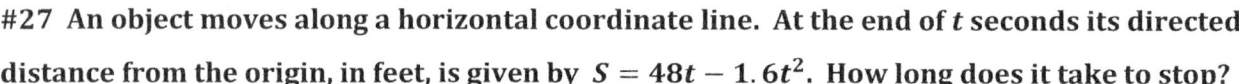

Since $f(x) \geq g(x)$, (The minimum value of $f(x)$) $\geq$ (The maximum value of $g(x)$)

$\therefore \ -32 \geq a + 2 \ ; \ a \leq -34$ $\therefore$ The maximum value of a is -34.

#27 An object moves along a horizontal coordinate line. At the end of t seconds its directed distance from the origin, in feet, is given by $S = 48t - 1.6t^2$. How long does it take to stop?

When the velocity is zero, the object will stop.

$\therefore \ V = S' = 48 - 3.2t = 0 \ ; \ t = \dfrac{48}{3.2} = \dfrac{30}{2} = 15$ (seconds)

#28 An object, projected vertically upward with an initial velocity of 100 feet/sec, moves according to the law $S = 100t - 8t^2$, where S is the distance from the starting point.

   (1) Find the velocity and acceleration when $t = 6$ and when $t = 7$.

   (2) Find the greatest height reached.

   (3) When will its height be 200 feet?

(1) $V(t) = \dfrac{dS}{dt} = 100 - 16t$ $A = V' = -16$

   At $t = 6$, $V(t) = 100 - 16 \cdot 6 = 4$ and $A = -16$

∴ The object is rising at 4 feet/sec.

At $t = 7$, $V(t) = 100 - 16 \cdot 7 = -12$ and $A = -16$

∴ The object is falling at 12 feet/sec.

(2) At the highest point of the motion, $V = 0$.

Solving $V = 0 = 100 - 16t$, $t = 6.25$

At this time, $S = 100 \cdot (6.25) - 8 \cdot (6.25)^2 = 6.25(100 - 8(6.25)) = 6.25(100 - 50)$

$$= 312.5 \text{ feet}$$

(3) $200 = 100t - 8t^2$

∴ $t^2 - 12.5t + 25 = 0$; $(t - 2.5)(t - 10) = 0$

∴ $t = 2.5$ or $t = 10$

At the end of 2.5 sec of motion, the object at a height of 200 feet is rising since $V > 0$.

At the end of 10 sec of motion, it is at the same height but is falling since $V < 0$.

#29 An object moves on a horizontal coordinate line. Its directed distance S from the origin at the end of t seconds is $S = -t^3 + at^2 + bt + 1$ feet. When $t = 1$, $S = -3$. When is the object moving to the left?

Let $f(t) = S = -t^3 + at^2 + bt + 1$

Then, the velocity of the object is $\frac{dS}{dt} = f'(t) = -3t^2 + 2at + b$

Since $f(1) = -3$ and $f'(1) = 0$,

$-1 + a + b + 1 = a + b = -3$ ⋯⋯ ①

$-3 + 2a + b = 0$ ⋯⋯ ②

By ① and ②, $a = 6$ and $b = -9$

∴ $f'(t) = -3t^2 + 2at + b = -3t^2 + 12t - 9 = -3(t^2 - 4t + 3) = -3(t - 3)(t - 1)$

$f'(t) = 0 \Rightarrow t = 3$ or $t = 1$

After $t = 1$, the object will change its direction at $t = 3$.

#30 The positions of two objects P and Q, on a coordinate line at the end of t seconds are given by $S_1 = t^2(t^2 - 6t + 12)$ and $S_2 = mt$, respectively. Find the value of m so that the objects have the same velocity three times.

$S_1 = t^2(t^2 - 6t + 12) = t^4 - 6t^3 + 12t^2$

∴ The velocity of P is $S_1' = 4t^3 - 18t^2 + 24t$ and the velocity of Q is $S_2' = m$

Since the objects have the same velocity three times,

the equation $4t^3 - 18t^2 + 24t = m$ has three different real number solutions.

Let $f(t) = 4t^3 - 18t^2 + 24t - m$

Then, $f'(t) = 12t^2 - 36t + 24 = 12(t^2 - 3t + 2) = 12(t - 1)(t - 2)$

∴ $f(1)f(2) < 0$

$f(1) = 4 - 18 + 24 - m = 10 - m$

$f(2) = 32 - 72 + 48 - m = 8 - m$

∴ $(10 - m)(8 - m) < 0$ That is, $8 < m < 10$

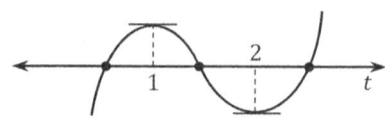

#31 Ships A and B start from the origin at the same time. Ship A travels due east at a rate of 20 miles per hour and ship B travels due north at the rate of 40 miles per hour. Find the speed at the intersection point of the line segment $\overline{AB}$ and a line $y = 3x$.

After t hours, the positions of A and B are $A(20t, 0)$ and $B(0, 40t)$, respectively.

∴ The equation of the line segment $\overline{AB}$ is $\dfrac{x}{20t} + \dfrac{y}{40t} = 1$ ······①

Substituting $y = 3x$ into ①, $\dfrac{x}{20t} + \dfrac{3x}{40t} = 1$; $\dfrac{2x + 3x}{40t} = 1$; $5x = 40t$ ∴ $x = 8t, \ y = 24t$

∴ The intersection point C is $C(8t, 24t)$.

Thus, the distance from the origin to C is

$$\sqrt{(8t)^2 + (24t)^2} = \sqrt{(8^2 + 24^2)t^2} = \sqrt{8^2(1 + 3^2)t^2} = 8\sqrt{10}\, t$$

Therefore, the speed at C is $\left(8\sqrt{10}\, t\right)' = 8\sqrt{10}$ (mile/hour)

#32 A ship A is sailing due south at 8 miles/hour and ship B, 16 miles south of A, is sailing due east at 6 miles/hour.

(1) At what rate are they approaching or separating at the end of 1 hour?

(2) At the end of 2 hours?

(3) When do they cease to approach each other and how far apart are they at the time?

Let A and B be the initial positions of the two ships, and A_t and B_t their positions t hour later.

Let D be the distance between them t hour later.

Then, $D^2 = (16 - 8t)^2 + (6t)^2$

$\dfrac{d}{dt}(D^2) = \dfrac{d}{dt}\{(16 - 8t)^2 + (6t)^2\}$

∴ $2D \cdot \dfrac{dD}{dt} = 2(16 - 8t) \cdot (-8) + 2(6t) \cdot 6$

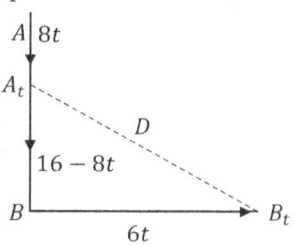

$\therefore \; D \cdot \dfrac{dD}{dt} = (16 - 8t) \cdot (-8) + (6t) \cdot 6 = -128 + 64t + 36t = 100t - 128$

$\therefore \; \dfrac{dD}{dt} = \dfrac{100t - 128}{D}$

(1) When $t = 1$, $D^2 = (16 - 8)^2 + 6^2 = 64 + 36 = 100$ $\therefore D = 10$

$\quad \dfrac{dD}{dt} = \dfrac{100 - 128}{10} = -\dfrac{28}{10} = -2.8$

$\quad \therefore$ They are approaching at 2.8 miles/hour

(2) When $t = 2$, $D^2 = 12^2$ $\therefore D = 12$

$\quad \dfrac{dD}{dt} = \dfrac{200 - 128}{12} = \dfrac{72}{12} = 6$

$\quad \therefore$ They are separating at 6 miles/hour

(3) They will cease to approach each other when $\dfrac{dD}{dt} = 0$; i.e., when $t = \dfrac{128}{100} = 1.28$ hour.

$\quad$ At which time they are $D = \sqrt{\{16 - 8(1.28)\}^2 + \{6(1.28)\}^2} = \sqrt{(5.76)^2 + (7.68)^2}$

$$= \sqrt{92.16} = 9.6 \text{ miles apart.}$$

#33 An object travels along the x-axis so that its position S is $S = \dfrac{1}{3}t^3 - 4t^2 + 12t$ meters after t seconds. What is its maximum speed on the interval $2 \leq t \leq 7$.

Let $S = f(t)$

Then, $f'(t) = t^2 - 8t + 12 = (t - 2)(t - 6)$

$f'(t) = 0 \;\Rightarrow\; t = 2$ or $t = 6$

$f'(t) = t^2 - 8t + 12 = (t - 4)^2 - 16 + 12 = (t - 4)^2 - 4$

From the graph, the maximum speed is $|f'(7)| = 5$.

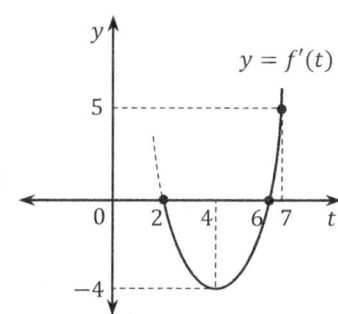

Note

(1) Minimum velocity of the object is -4 and the maximum velocity of it is 5 on $[2, 7]$.

(2) The speed of an object is the absolute value of its velocity.

#34 An object moves in the plane so that its position is $(e^t \cos t, \; e^t \sin t)$.
Find the value of t when the speed of the object is $\sqrt{2}e^2$.

$x = e^t \cos t, \; y = e^t \sin t$

$\dfrac{d}{dt}(x) = \dfrac{d}{dt}(e^t \cos t);\quad \dfrac{dx}{dt} = e^t \cos t - e^t \sin t = e^t(\cos t - \sin t)$

$\dfrac{d}{dt}(y) = \dfrac{d}{dt}(e^t \sin t);\quad \dfrac{dy}{dt} = e^t \sin t + e^t \cos t = e^t(\sin t + \cos t)$

$V(t) = \langle \dfrac{dx}{dt}, \dfrac{dy}{dt} \rangle$

$$|V(t)| = \sqrt{\left(\frac{dx}{dt}\right)^2 + \left(\frac{dy}{dt}\right)^2} = \sqrt{\{e^t(\cos t - \sin t)\}^2 + \{e^t(\sin t + \cos t)\}^2}$$

$$= \sqrt{e^{2t}(\cos t - \sin t)^2 + e^{2t}(\sin t + \cos t)^2}$$

$$= \sqrt{e^{2t}(\cos^2 t - 2\sin t \cos t + \sin^2 t) + e^{2t}(\sin^2 t + 2\sin t \cos t + \cos^2 t)}$$

$$= \sqrt{e^{2t}(1 - 2\sin t \cos t) + e^{2t}(1 + 2\sin t \cos t)} = \sqrt{2e^{2t}} = \sqrt{2}e^t$$

$\therefore \ \sqrt{2}e^t = \sqrt{2}e^2$

Therefore, $t = 2$

#35 A point P is moving in the plane so that its coordinates after t seconds are $(t - \sin t,$ $\cos t)$, measured in feet.

  (1) Find the maximum speed of the point P.

  (2) Find the angle between the velocity vector of P at $t = \dfrac{\pi}{3}$ and x-axis.

(1) $x = t - \sin t$, $y = \cos t$

$$\frac{d}{dt}(x) = \frac{d}{dt}(t - \sin t) \ ; \quad \frac{dx}{dt} = 1 - \cos t$$

$$\frac{d}{dt}(y) = \frac{d}{dt}(\cos t) \ ; \quad \frac{dy}{dt} = -\sin t$$

Let V be the velocity vector.

Then, $V = \langle 1 - \cos t, -\sin t \rangle$

$\therefore$ The speed is $|V| = \sqrt{(1 - \cos t)^2 + (-\sin t)^2} = \sqrt{(1 - 2\cos t + \cos^2 t) + \sin^2 t}$

$$= \sqrt{2 - 2\cos t} = \sqrt{2(1 - \cos t)}$$

Therefore, we have the maximum speed $\sqrt{2 \cdot 2} = 2$ when $\cos t = -1$.

(2) $t = \dfrac{\pi}{3} \ \Rightarrow \ \dfrac{dx}{dt} = 1 - \cos\dfrac{\pi}{3} = 1 - \dfrac{1}{2} = \dfrac{1}{2}$

$$\frac{dy}{dt} = -\sin\frac{\pi}{3} = -\frac{\sqrt{3}}{2}$$

Let θ be the angle between the velocity vector of P at $t = \dfrac{\pi}{3}$ and x-axis.

Then, $\tan\theta = \dfrac{\frac{dy}{dt}}{\frac{dx}{dt}} = \dfrac{-\frac{\sqrt{3}}{2}}{\frac{1}{2}} = -\sqrt{3}$

$\therefore \ \theta = -\dfrac{\pi}{3}$

#36 A particle rotates counterclockwise from rest according to the law $\theta = \dfrac{t^3}{81} - \dfrac{t}{3}$**, where**

θ **is in radians and** t **in seconds. Find the angular displacement** θ**, the angular velocity** V**, and the angular accelation** A **at the end of 9 seconds.**

$$\theta = \frac{t^3}{81} - \frac{t}{3} = \frac{9^3}{81} - \frac{9}{3} = 9 - 3 = 6$$

$$V = \frac{d\theta}{dt} = \frac{3t^2}{81} - \frac{1}{3} = \frac{3 \cdot 9^2}{81} - \frac{1}{3} = 3 - \frac{1}{3} = \frac{8}{3} \text{ rad/sec}$$

$$A = \frac{dV}{dt} = \frac{6t}{81} = \frac{6 \cdot 9}{81} = \frac{2}{3} \text{ rad/sec}^2$$

#37 Suppose that a point P **moves around a circle with center** $(0,0)$ **and radius** r **at a constant angular speed of** ω **radians per second. When its initial position is** $(r, 0)$**,**

(1) Express x **and** y **at** t **as** t**.**

$P(x, y) = (r \cos \omega t,\ r \sin \omega t)$

That is, $x = r \cos \omega t,\ y = r \sin \omega t$

(2) Find expressions for the velocity $V(t)$ **and its acceleration** $A(t)$**.**

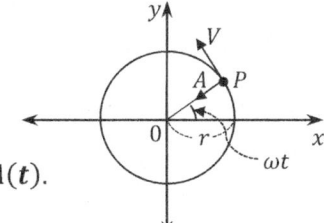

$\dfrac{dx}{dt} = -r\omega \sin \omega t$

$\dfrac{dy}{dt} = r\omega \cos \omega t$

$\therefore\ V(t) = \langle \dfrac{dx}{dt}, \dfrac{dy}{dt} \rangle = \langle -r\omega \sin \omega t,\ r\omega \cos \omega t \rangle$

$\dfrac{d^2x}{dt^2} = -r\omega^2 \cos \omega t,\quad \dfrac{d^2y}{dt^2} = -r\omega^2 \sin \omega t$

$\therefore\ A(t) = \langle \dfrac{d^2x}{dt^2}, \dfrac{d^2y}{dt^2} \rangle = \langle -r\omega^2 \cos \omega t,\ -r\omega^2 \sin \omega t \rangle$

(3) Find the speed $|V(t)|$**.**

$$|V(t)| = \sqrt{(-r\omega \sin \omega t)^2 + (r\omega \cos \omega t)^2} = \sqrt{(r\omega)^2(\sin^2 \omega t + \cos^2 \omega t)} = \sqrt{(r\omega)^2} = r\omega$$

(4) Show that the velocity and acceleration vectors are always perpendicular to each other.

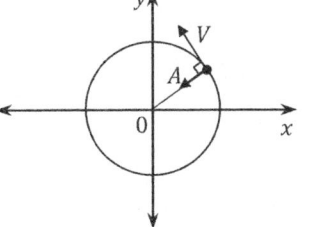

$V(t) \cdot A(t) = \langle -r\omega \sin \omega t,\ r\omega \cos \omega t \rangle \cdot \langle -r\omega^2 \cos \omega t,\ -r\omega^2 \sin \omega t \rangle$

$\qquad = \langle -\omega y, \omega x \rangle \cdot \langle -\omega^2 x, -\omega^2 y \rangle$

$\qquad = (-\omega y)(-\omega^2 x) + (\omega x)(-\omega^2 y)$

$\qquad = \omega^3 xy - \omega^3 xy = 0$

$\therefore\ V(t) \perp A(t)$

#38 Water is running into a conical reservoir, 6 feet in diameter and 12 feet tall (with vertex down) at a rate of 1 cubic feet per minute.

    (1) At what rate is the water level rising when the water is 4 feet deep?

    (2) At what rate is the area of the water surface increasing when the water is 5 feet deep?

    (3) At what rate is the wetted surface of the reservoir increasing when the water is 8 feet deep?

Let r be the radius of the surface of water and h be the depth at time t.

Then, $\dfrac{r}{3} = \dfrac{h}{12}$; $r = \dfrac{h}{4}$

The volume of cone of water is $V = \dfrac{1}{3}\pi r^2 h = \dfrac{1}{3}\pi \left(\dfrac{h}{4}\right)^2 h = \dfrac{1}{48}\pi h^3$

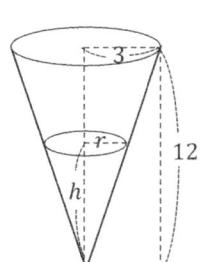

$\therefore \ \dfrac{dV}{dt} = \dfrac{3}{48}\pi h^2 \dfrac{dh}{dt} = \dfrac{1}{16}\pi h^2 \dfrac{dh}{dt}$

Since $\dfrac{dV}{dt} = 1$, $\dfrac{dh}{dt} = \dfrac{16}{\pi h^2}$

(1) When $h = 4$, $\dfrac{dh}{dt} = \dfrac{16}{\pi 4^2} = \dfrac{1}{\pi}$

    $\therefore$ The water level is increasing at the rate $\dfrac{1}{\pi}$ feet/min .

(2) Area of water surface is $A = \pi r^2 = \pi \left(\dfrac{h}{4}\right)^2 = \dfrac{1}{16}\pi h^2$

    $\therefore \ \dfrac{dA}{dt} = \dfrac{2\pi h}{16}\dfrac{dh}{dt} = \dfrac{1}{8}\pi h \dfrac{dh}{dt} = \dfrac{1}{8}\pi h \dfrac{16}{\pi h^2} = \dfrac{2}{h}$

    When $h = 5$, $\dfrac{dA}{dt} = \dfrac{2}{5}$ feet^2/min

(3) The wetted surface is a cone whose lateral area is π (radius) (slant height).

$$S = \pi r \sqrt{h^2 + r^2} = \pi \cdot \dfrac{h}{4} \cdot \sqrt{h^2 + \left(\dfrac{h}{4}\right)^2} = \pi \cdot \dfrac{h}{4} \cdot \sqrt{\dfrac{17h^2}{16}} = \dfrac{\sqrt{17}}{16}\pi h^2$$

$$\dfrac{dS}{dt} = 2\dfrac{\sqrt{17}}{16}\pi h \dfrac{dh}{dt} = \dfrac{\sqrt{17}}{8}\pi h \dfrac{16}{\pi h^2} = 2\sqrt{17}\dfrac{1}{h}$$

When $h = 8$, $\dfrac{dS}{dt} = 2\sqrt{17}\dfrac{1}{8} = \dfrac{\sqrt{17}}{4}$ feet^2/min

Chapter 5. The Indefinite Integral

#1 Find the values of the constants a, b, and c.

(1) $\int(3x^2 + ax - 4)dx = bx^3 + 2x^2 + cx + 3$

$3x^2 + ax - 4 = (bx^3 + 2x^2 + cx + 3)' = 3bx^2 + 4x + c$

$\therefore \ 3 = 3b, \ a = 4, \ -4 = c$

$\therefore \ a = 4, \ b = 1, \ c = -4$

(2) $\frac{d}{dx}[\int(ax^2 + 3x + 2)dx] = 8x^2 + bx + c$

$\left(\frac{a}{3}x^3 + \frac{3}{2}x^2 + 2x\right)' = ax^2 + 3x + 2$

$\therefore \ \int(ax^2 + 3x + 2)dx = \left(\frac{a}{3}x^3 + \frac{3}{2}x^2 + 2x\right) + C$

$\therefore \ \frac{d}{dx}[\int(ax^2 + 3x + 2)dx] = \frac{d}{dx}\left[\left(\frac{a}{3}x^3 + \frac{3}{2}x^2 + 2x\right) + C\right] = ax^2 + 3x + 2$

$\therefore \ ax^2 + 3x + 2 = 8x^2 + bx + c$

$\therefore \ a = 8, \ b = 3, \ c = 2$

#2 Integrate the expression.

(1) $\int x^5 dx$

Add 1 to the exponent ($5 + 1 = 6$) and divide the expression by the new exponent.

$\int x^5 dx = \frac{1}{5+1}x^{5+1} + C = \frac{1}{6}x^6 + C$

(2) $\int(3x^2 + 6x)dx$

The integral of a sum or difference is equal to the sum or difference of the individual integrals.

$\int(3x^2 + 6x)dx = \int 3x^2 dx + \int 6x dx = 3\int x^2 dx + 6\int x dx$

$= 3\left(\frac{1}{2+1}x^{2+1} + C_1\right) + 6\left(\frac{1}{1+1}x^{1+1} + C_2\right) = 3\left(\frac{1}{3}x^3 + C_1\right) + 6\left(\frac{1}{2}x^2 + C_2\right)$

$= x^3 + 3C_1 + 3x^2 + 6C_2 = x^3 + 3x^2 + (3C_1 + 6C_2) = x^3 + 3x^2 + C$

(Two arbitrary constants C_1 and C_2 add up to some other unknown constant, C.)

(3) $\int(2x^2 + 3x - 4)dx$

$\int(2x^2 + 3x - 4)dx = \int 2x^2 dx + \int 3x dx - \int 4 dx$

$= \left(\frac{2}{3}x^3 + C_1\right) + \left(\frac{3}{2}x^2 + C_2\right) - (4x + C_3) = \frac{2}{3}x^3 + \frac{3}{2}x^2 - 4x + (C_1 + C_2 - C_3)$

$= \frac{2}{3}x^3 + \frac{3}{2}x^2 - 4x + C$

(4) $\int \left(\frac{2}{x} - \frac{3}{x^2} \right) dx$

$\int \left(\frac{2}{x} - \frac{3}{x^2} \right) dx = \int \left(\frac{2}{x} - 3x^{-2} \right) dx = 2 \int \frac{1}{x} dx - 3 \int x^{-2} dx$

$= 2(\log|x| + C_1) - 3 \left(\frac{1}{-2+1} x^{-2+1} + C_2 \right) = 2\log|x| + 3x^{-1} + C = 2\log|x| + \frac{3}{x} + C$

(5) $\int \left(\frac{2x^3 - x^4}{x^4} \right) dx$

$\int \left(\frac{2x^3 - x^4}{x^4} \right) dx = \int \frac{2x^3}{x^4} dx - \int \frac{x^4}{x^4} dx = 2 \int \frac{1}{x} dx - \int 1 \, dx = (2\log|x| + C_1) - (x + C_2)$

$= 2\log|x| - x + C$

(6) $\int x^2 (3 - \sqrt{x}) dx$

$\int x^2 (3 - \sqrt{x}) dx = \int 3x^2 dx - \int x^2 \sqrt{x} \, dx = 3 \int x^2 dx - \int x^{\frac{5}{2}} dx$

$= 3 \left(\frac{1}{3} x^3 + C_1 \right) - \left(\frac{1}{\frac{5}{2}+1} x^{\frac{5}{2}+1} + C_2 \right) = x^3 - \frac{2}{7} x^{\frac{7}{2}} + C$

(7) $\int x(x-1)(x+2) dx$

$\int x(x-1)(x+2) dx = \int x(x^2 + x - 2) dx = \int (x^3 + x^2 - 2x) dx$

$= \frac{1}{4} x^4 + \frac{1}{3} x^3 - \frac{2}{2} x^2 + C = \frac{1}{4} x^4 + \frac{1}{3} x^3 - x^2 + C$

(8) $\int (\sin \theta + \cos \theta)^2 d\theta + \int (\sin \theta - \cos \theta)^2 d\theta$

$\int (\sin \theta + \cos \theta)^2 d\theta + \int (\sin \theta - \cos \theta)^2 d\theta$

$= \int \{ (\sin \theta + \cos \theta)^2 + (\sin \theta - \cos \theta)^2 \} d\theta$

$= \int \{ (\sin^2 \theta + 2 \sin \theta \cos \theta + \cos^2 \theta) + (\sin^2 \theta - 2 \sin \theta \cos \theta + \cos^2 \theta) \} d\theta$

$= \int \{ (1 + 2\sin \theta \cos \theta) + (1 - 2\sin \theta \cos \theta) \} d\theta = \int 2 d\theta = 2 \int d\theta = 2(\theta + C_1) = 2\theta + C$

(9) $\int \left(\frac{\sin^2 x}{\cos^2 x} - \frac{1}{\cos^2 x} \right) dx$

$\int \left(\frac{\sin^2 x}{\cos^2 x} - \frac{1}{\cos^2 x} \right) dx = \int \left(\frac{\sin^2 x - 1}{\cos^2 x} \right) dx = \int \left(\frac{-\cos^2 x}{\cos^2 x} \right) dx = \int (-1) dx = -x + C$

(10) $\int \frac{x^3}{x+1} dx + \int \frac{1}{x+1} dx$

$\int \left(\frac{x^3}{x+1} + \frac{1}{x+1} \right) dx = \int \left(\frac{x^3+1}{x+1} \right) dx = \int \left\{ \frac{(x+1)(x^2-x+1)}{x+1} \right\} dx = \int (x^2 - x + 1) dx$

$= \frac{1}{3} x^3 - \frac{1}{2} x^2 + x + C$

(11) $\int \left(\frac{x^2 - 2x - 3}{\sqrt{x}} \right) dx$

$\int \left(\frac{x^2 - 2x - 3}{\sqrt{x}} \right) dx = \int \left(\frac{x^2 - 2x - 3}{x^{\frac{1}{2}}} \right) dx = \int \left(x^{\frac{3}{2}} - 2x^{\frac{1}{2}} - 3x^{-\frac{1}{2}} \right) dx$

$= \frac{2}{5} x^{\frac{5}{2}} - 2 \cdot \frac{2}{3} x^{\frac{3}{2}} - 3 \cdot 2x^{\frac{1}{2}} + C = \frac{2}{5} x^2 \sqrt{x} - \frac{4}{3} x\sqrt{x} - 6\sqrt{x} + C$

(12) $\int \left(x - \frac{1}{x}\right)^3 dx$

$\int \left(x - \frac{1}{x}\right)^3 dx = \int \left\{x^3 - 3x^2\frac{1}{x} + 3x\left(\frac{1}{x}\right)^2 - \left(\frac{1}{x}\right)^3\right\} dx = \int \left(x^3 - 3x + \frac{3}{x} - x^{-3}\right) dx$

$\qquad = \frac{1}{4}x^4 - \frac{3}{2}x^2 + 3\log|x| - \frac{1}{-2}x^{-2} + C = \frac{1}{4}x^4 - \frac{3}{2}x^2 + 3\log|x| + \frac{1}{2x^2} + C$

(13) $\int \frac{(\sqrt{x}-2)^2}{x} dx$

$\int \frac{(\sqrt{x}-2)^2}{x} dx = \int \left(\frac{x - 4\sqrt{x} + 4}{x}\right) dx = \int \left(1 - 4x^{-\frac{1}{2}} + 4\frac{1}{x}\right) dx$

$\qquad = x - 4 \cdot 2x^{\frac{1}{2}} + 4\log|x| + C = x - 8\sqrt{x} + 4\log|x| + C$

(14) $\int \left(\frac{x^4+x^2+1}{x^2-x+1}\right) dx$

$\int \left(\frac{x^4+x^2+1}{x^2-x+1}\right) dx = \int \left(\frac{(x^2-x+1)(x^2+x+1)}{x^2-x+1}\right) dx = \int (x^2 + x + 1)dx = \frac{1}{3}x^3 + \frac{1}{2}x^2 + x + C$

#3 **For two functions $f(x)$ and $g(x)$ satisfying $\frac{d}{dx}\{f(x) + g(x)\} = 4$ and $\frac{d}{dx}\{f(x)g(x)\} = 8x$,**

$f(0) = -1$ **and** $g(0) = 1$. **Find the following functions:**

(1) $f(x) + g(x)$

$\frac{d}{dx}\{f(x) + g(x)\} = 4 \Rightarrow f(x) + g(x) = 4x + C$

Since $f(0) = -1$ and $g(0) = 1$, $f(0) + g(0) = 4 \cdot 0 + C = (-1) + 1$

$\therefore\ C = 0$

$\therefore\ f(x) + g(x) = 4x$

(2) $f(x)g(x)$

$\frac{d}{dx}\{f(x)g(x)\} = 8x \Rightarrow f(x)g(x) = 4x^2 + C$

Since $f(0) = -1$ and $g(0) = 1$, $f(0)g(0) = 4 \cdot 0 + C = (-1) \cdot 1 = -1$

$\therefore\ C = -1$

$\therefore\ f(x)g(x) = 4x^2 - 1$

(3) $f(x),\ g(x)$

Since $f(x) + g(x) = 4x$ and $f(x)g(x) = 4x^2 - 1 = (2x + 1)(2x - 1)$,

$\begin{cases} f(x) = 2x + 1 \\ g(x) = 2x - 1 \end{cases}$ or $\begin{cases} f(x) = 2x - 1 \\ g(x) = 2x + 1 \end{cases}$

Since $f(0) = -1$ and $g(0) = 1$, $\begin{cases} f(x) = 2x - 1 \\ g(x) = 2x + 1 \end{cases}$

#4 Answer the question.

(1) When $f'(x) = 2x^2 - 3x + 1$ and $f(1) = 2$, find the function $f(x)$.

$f(x) = \int f'(x)dx = \int (2x^2 - 3x + 1)dx = \frac{2}{3}x^3 - \frac{3}{2}x^2 + x + C$

Since $f(1) = 2$, $\frac{2}{3} - \frac{3}{2} + 1 + C = 2$; $C = \frac{11}{6}$

$\therefore \; f(x) = \frac{2}{3}x^3 - \frac{3}{2}x^2 + x + \frac{11}{6}$

(2) For the problem to integrate $f(x)$, you misunderstand to differentiate it and obtain

   $3x + 4$. Integrate $f(x)$ and find the coefficient of x^2 in the integral.

Since $f'(x) = 3x + 4$, $f(x) = \int (3x + 4)dx = \frac{3}{2}x^2 + 4x + C_1$

$\therefore \; \int f(x)dx = \int \left(\frac{3}{2}x^2 + 4x + C_1\right)dx = \frac{3}{2} \cdot \frac{1}{3}x^3 + 4 \cdot \frac{1}{2}x^2 + C_1 x + C_2$

$\qquad = \frac{1}{2}x^3 + 2x^2 + C_1 x + C_2$

$\therefore$ The coefficient of x^2 is 2.

(3) When $f(x)$ is the antiderivative of $x^2\sqrt{x} - x + 3$ such that $f(0) = 4$, find $f(x)$.

$f(x) = \int (x^2\sqrt{x} - x + 3)dx = \int (x^{\frac{5}{2}} - x + 3)dx = \frac{2}{7}x^{\frac{7}{2}} - \frac{1}{2}x^2 + 3x + C$

Since $f(0) = 4$, $C = 4$

$\therefore \; f(x) = \frac{2}{7}x^{\frac{7}{2}} - \frac{1}{2}x^2 + 3x + 4$

(4) Find the solution of the equation $\log_x \left\{\frac{d}{dx}(\int x^2 dx)\right\} = 2x^3 + 3x^2 - 8x + 5$.

$\frac{d}{dx}(\int x^2 dx) = x^2$

$\therefore \; \log_x \left\{\frac{d}{dx}(\int x^2 dx)\right\} = \log_x x^2 = 2 \;\; (x > 0, \; x \neq 1)$

Thus, the given equation is $2 = 2x^3 + 3x^2 - 8x + 5$; $2x^3 + 3x^2 - 8x + 3 = 0$

$2x^3 + 3x^2 - 8x + 3 = (x - 1)(2x^2 + 5x - 3) = (x - 1)(2x - 1)(x + 3) = 0$

$\therefore \; x = 1$ or $x = \frac{1}{2}$ or $x = -3$

Since $x > 0$ and $x \neq 1$, $x = \frac{1}{2}$

(5) For the graph of a function $y = f(x)$ passing through the two points $(1, 0)$ and $(-1, 1)$,

   $f''(x) = 2x - 4$. Find the sum of all solutions of $f(x) = 0$.

Since $f''(x) = 2x - 4$, $f'(x) = \int (2x - 4)dx = x^2 - 4x + C_1$

$\therefore \; f(x) = \int (x^2 - 4x + C_1)dx = \frac{1}{3}x^3 - 4 \cdot \frac{1}{2}x^2 + C_1 x + C_2 = \frac{1}{3}x^3 - 2x^2 + C_1 x + C_2$

Since $f(x)$ passes through the two points $(1, 0)$ and $(-1, 1)$,

$f(1) = \frac{1}{3} - 2 + C_1 + C_2 = 0$ and $f(-1) = -\frac{1}{3} - 2 - C_1 + C_2 = 1$

$\therefore$ $C_1 + C_2 = \frac{5}{3}$ and $-C_1 + C_2 = \frac{10}{3}$

$\therefore$ $2C_2 = \frac{15}{3} = 5$; $C_2 = \frac{5}{2}$, $C_1 = \frac{5}{3} - \frac{5}{2} = -\frac{5}{6}$

$\therefore$ $f(x) = \frac{1}{3}x^3 - 2x^2 - \frac{5}{6}x + \frac{5}{2}$

$\therefore$ By the relationship between the roots and coefficients,

the sum of all solutions of $f(x) = 0$ is $-\frac{-2}{\frac{1}{3}} = 6$

(6) When the slope of the tangent line at (a, b) on the graph of $y = f(x)$ is $\frac{1}{a}$ $(a > 0)$,

find the function passing through the point $(1, 3)$.

Since $f'(a) = \frac{1}{a}$ $(a > 0)$, $f(x) = \int f'(x)dx = \int \frac{1}{x}dx = \log|x| + C = \log x + C$ $(\because x > 0)$

Since $f(1) = 3$, $\log 1 + C = 3$ $\quad \therefore C = 3$

$\therefore$ $f(x) = \log x + 3$

(7) For a continuous function $f(x)$ satisfying $f'(x) = |x|$ for any real number x,

$f(1) = 2$. Find the value of $f(-1)$.

Since $f'(x) = \begin{cases} x, & x \geq 0 \\ -x, & x < 0 \end{cases}$, $f(x) = \begin{cases} \frac{1}{2}x^2 + C_1, & x \geq 0 \\ -\frac{1}{2}x^2 + C_2, & x < 0 \end{cases}$

Since $f(x)$ is continuous for any real number x, $f(x)$ is continuous at $x = 0$.

$\therefore$ $\lim\limits_{x \to 0} f(x) = f(0)$ $\quad \therefore C_1 = C_2$

Since $f(1) = 2$, $\frac{1}{2} + C_1 = 2$; $C_1 = \frac{3}{2}$ $\quad \therefore C_1 = C_2 = \frac{3}{2}$

$\therefore$ $f(-1) = -\frac{1}{2}(-1)^2 + \frac{3}{2} = 1$

(8) For a continuous function $f(x)$, the derivative of it is $f'(x) = \begin{cases} 2x, & x < 1 \\ 1, & x \geq 1 \end{cases}$.

When the graph of the function $y = f(x)$ passes through a point $(0, 1)$, find the value

of $f(10) + f(-10)$.

$f'(x) = \begin{cases} 2x, & x < 1 \\ 1, & x \geq 1 \end{cases}$ $\Rightarrow$ $f(x) = \begin{cases} x^2 + C_1, & x < 1 \\ x + C_2, & x \geq 1 \end{cases}$

Since $(0, 1)$ lies on the graph of $y = f(x)$, $f(0) = 1$ $\quad \therefore 0^2 + C_1 = 1$; $C_1 = 1$

Since $f(x)$ is a continuous function, $\lim\limits_{x \to 1^-} f(x) = \lim\limits_{x \to 1^+} f(x) = f(1)$

$\therefore$ $\lim\limits_{x \to 1^-}(x^2 + 1) = \lim\limits_{x \to 1^+}(x + C_2) = 1 + C_2$; $1^2 + 1 = 1 + C_2$; $C_2 = 1$

$\therefore$ $f(x) = \begin{cases} x^2 + 1, & x < 1 \\ x + 1, & x \geq 1 \end{cases}$

Therefore, $f(10) + f(-10) = 10 + 1 + ((-10)^2 + 1) = 11 + 101 = 112$

(9) When $\int(1 - f(x))dx = \frac{1}{2}x^2(4 - x^2) + C$, find the local extreme values of $f(x)$.

$\frac{d}{dx}\left[\int(1 - f(x))dx\right] = \frac{d}{dx}\left[\frac{1}{2}x^2(4 - x^2) + C\right]$

$\therefore\ 1 - f(x) = \frac{1}{2}\{2x(4 - x^2) + x^2(-2x)\} = \frac{1}{2}(-4x^3 + 8x) = -2x^3 + 4x$

$\therefore\ f(x) = 2x^3 - 4x + 1$

$\therefore\ f'(x) = 6x^2 - 4 = 2(3x^2 - 2)$

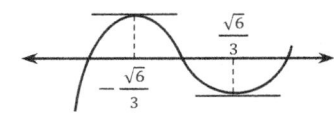

$f'(x) = 0\ \Rightarrow\ 3x^2 - 2 = 0\ ;\ x^2 = \frac{2}{3}\qquad \therefore\ x = \pm\sqrt{\frac{2}{3}} = \pm\frac{\sqrt{6}}{3}$

$\therefore$ The local maximum value is $f\left(-\frac{\sqrt{6}}{3}\right) = 2\left(-\frac{\sqrt{6}}{3}\right)^3 - 4\left(-\frac{\sqrt{6}}{3}\right) + 1 = -\frac{12\sqrt{6}}{27} + \frac{4\sqrt{6}}{3} + 1$

$$= \frac{-4\sqrt{6}+12\sqrt{6}}{9} + 1 = \frac{8\sqrt{6}}{9} + 1$$

and the local minimum value is $f\left(\frac{\sqrt{6}}{3}\right) = 2\left(\frac{\sqrt{6}}{3}\right)^3 - 4\left(\frac{\sqrt{6}}{3}\right) + 1 = \frac{12\sqrt{6}}{27} - \frac{4\sqrt{6}}{3} + 1$

$$= \frac{4\sqrt{6}-12\sqrt{6}}{9} + 1 = -\frac{8\sqrt{6}}{9} + 1$$

(10) For a function $y = f(x)$ with a local extreme value 4 at $x = 1$,

$f'(x) = 3x^2 - 8x + a$. Find the values of a and the other local extreme value.

Since $f'(x) = 3x^2 - 8x + a$, $f(x) = \int(3x^2 - 8x + a)dx = x^3 - 4x^2 + ax + C$

Since $f(x)$ has a local extreme value 4 at $x = 1$, $f'(1) = 0$ and $f(1) = 4$

$\therefore\ f'(1) = 3 - 8 + a = 0\ ;\ a = 5$

$\quad f(1) = 1 - 4 + 5 + C = 4\ ;\ C = 2$

$\therefore\ f(x) = x^3 - 4x^2 + 5x + 2$ and $f'(x) = 3x^2 - 8x + 5 = (x - 1)(3x - 5)$

Since $f'\left(\frac{5}{3}\right) = 0$, $f\left(\frac{5}{3}\right)$ is an another extreme value.

$f\left(\frac{5}{3}\right) = \left(\frac{5}{3}\right)^3 - 4\left(\frac{5}{3}\right)^2 + 5\left(\frac{5}{3}\right) + 2 = \left(\frac{5}{3}\right)^2\left(\frac{5}{3} - 4\right) + \frac{25}{3} + 2$

$= \frac{25}{9}\left(-\frac{7}{3}\right) + \frac{25}{3} + 2 = \frac{25}{3}\left(-\frac{7}{9} + 1\right) + 2 = \frac{25}{3}\left(\frac{2}{9}\right) + 2 = \frac{50}{27} + 2 = \frac{104}{27}$

(11) When $f'(x) = x^2 - 4x + 3$ and the local maximum value of $f(x)$ is 5,

find the function $f(x)$ and the local minimum value of $f(x)$.

$f(x) = \int f'(x)dx = \int(x^2 - 4x + 3)dx = \frac{1}{3}x^3 - 2x^2 + 3x + C$

Since $f'(x) = x^2 - 4x + 3 = (x - 1)(x - 3)$, $f(x)$ has local extreme values at $x = 1$ and $x = 3$.

Since $f(1) = 5$, $\frac{1}{3} - 2 + 3 + C = 5$ $\qquad \therefore\ C = \frac{11}{3}$

$\therefore\ f(x) = \frac{1}{3}x^3 - 2x^2 + 3x + \frac{11}{3}$

The local minimum value is $f(3) = \frac{1}{3}\cdot 3^3 - 2\cdot 3^2 + 3\cdot 3 + \frac{11}{3} = 9 - 18 + 9 + \frac{11}{3} = \frac{11}{3}$

(12) For a function $f(x)$ satisfying $f'(x) = x^2 - x - 6$, find the difference between the local extreme values.

$f(x) = \int f'(x)dx = \int (x^2 - x - 6)dx = \frac{1}{3}x^3 - \frac{1}{2}x^2 - 6x + C$

Since $f'(x) = x^2 - x - 6 = (x - 3)(x + 2)$,

$f(x)$ has a local maximum value at $x = -2$ and a local minimum value at $x = 3$.

∴ The difference between the local extreme values is $f(-2) - f(3) = \frac{125}{6}$

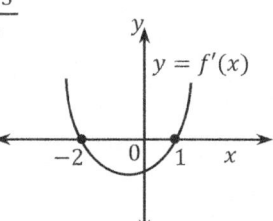

(13) For a function $f(x)$, the derivative of $f(x)$ is shown as the Figure. When the local maximum value of $f(x)$ is 4 and local minimum value of $f(x)$ is -2, find the value of $f(0)$.

From the graph, $f'(-2) = 0$ and $f'(1) = 0$

∴ $f'(x) = a(x + 2)(x - 1)$, $a > 0$

∴ $f'(x) = a(x^2 + x - 2)$

∴ $f(x) = \int f'(x)dx = \int a(x^2 + x - 2)dx = \frac{1}{3}ax^3 + \frac{1}{2}ax^2 - 2ax + C$

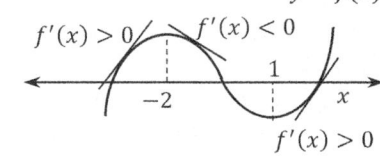

Form the graph, $\quad f'(x) > 0$ when $x < -2$

$\qquad\qquad\qquad f'(x) < 0$ when $-2 < x < 1$

$\qquad\qquad\qquad f'(x) > 0$ when $x > 1$

∴ $f(x)$ has local maximum value at $x = -2$ and local minimum value at $x = 1$.

∴ $f(-2) = 4$ and $f(1) = -2$

$-\frac{8}{3}a + 2a + 4a + C = 4$; $\frac{10}{3}a + C = 4$

$\frac{1}{3}a + \frac{1}{2}a - 2a + C = -2$; $-\frac{7}{6}a + C = -2$

∴ $a = \frac{4}{3}$, $C = -\frac{4}{9}$

∴ $f(x) = \frac{4}{9}x^3 + \frac{2}{3}x^2 - \frac{8}{3}x - \frac{4}{9}$

Therefore, $f(0) = -\frac{4}{9}$

(14) For a function $f(x)$, the derivative of $f(x)$ is shown as the Figure. When $f(0) = 1$, find the range of k so that the equation $f(x) = kx + 1$ of x has three different real number solutions.

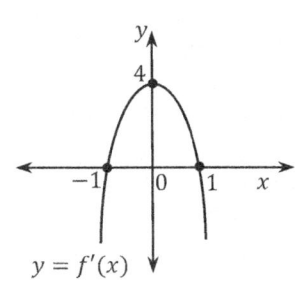

From the graph, $f'(-1) = 0$ and $f'(1) = 0$

∴ $f'(x) = a(x + 1)(x - 1)$, $a < 0$

∴ $f'(x) = a(x^2 - 1)$

Since the point $(0, 4)$ lies on the graph of $y = f'(x)$, $f'(0) = 4$

$\therefore\ a(0 - 1) = 4$; $a = -4$

$\therefore\ f'(x) = -4(x^2 - 1) = -4x^2 + 4$

$\therefore\ f(x) = \int f'(x)dx = \int(-4x^2 + 4)dx = -\dfrac{4}{3}x^3 + 4x + C$

Since $f(0) = 1$, $C = 1$

$\therefore\ f(x) = -\dfrac{4}{3}x^3 + 4x + 1$

Since $f(x) = kx + 1$, $-\dfrac{4}{3}x^3 + 4x + 1 = kx + 1$; $-\dfrac{4}{3}x^3 + (4 - k)x = 0$

$\therefore\ x\left(-\dfrac{4}{3}x^2 + 4 - k\right) = 0$; $x = 0$ or $-\dfrac{4}{3}x^2 + 4 - k = 0$

Since $f(x) = kx + 1$ has three different real number solutions, the equation $-\dfrac{4}{3}x^2 + 4 - k = 0$

has two different nonzero real number solutions.

Let D be the discriminant of the equation $-\dfrac{4}{3}x^2 + 4 - k = 0$. Then, $D > 0$

$\therefore\ D = 0^2 - 4 \cdot \left(-\dfrac{4}{3}\right) \cdot (4 - k) = \dfrac{16}{3}(4 - k) > 0$

Therefore, $k < 4$

(15) When $f(x) = \int(x \log x + e^x + x)dx$, find the limit $\lim\limits_{h \to 0} \dfrac{f(1-2h)-f(1)}{h}$.

$\lim\limits_{h \to 0} \dfrac{f(1-2h)-f(1)}{h} = \lim\limits_{h \to 0} \dfrac{f(1-2h)-f(1)}{-2h}(-2) = f'(1)(-2) = -2f'(1)$

Since $f(x) = \int(x \log x + e^x + x)dx$, $f'(x) = x \log x + e^x + x$

$\therefore\ f'(1) = \log 1 + e^1 + 1 = e + 1$

$\therefore\ \lim\limits_{h \to 0} \dfrac{f(1-2h)-f(1)}{h} = -2f'(1) = -2(e + 1) = -2e - 2$

(16) For a cubic function $f(x)$, the antiderivative of $f(x)$ is $F(x) = xf(x) - 2x^3(x - 1)$.

When $f(-1) = 0$, find the value of $f(1)$.

$\dfrac{d}{dx}\{F(x)\} = \dfrac{d}{dx}\{xf(x) - 2x^3(x - 1)\}$

$\therefore\ f(x) = f(x) + xf'(x) - 2\{3x^2(x - 1) + x^3\}$

$\therefore\ xf'(x) = 2\{3x^2(x - 1) + x^3\}$; $f'(x) = 2(4x^2 - 3x) = 8x^2 - 6x$

$\therefore\ f(x) = \int(8x^2 - 6x)dx = \dfrac{8}{3}x^3 - 3x^2 + C$

Since $f(-1) = 0$, $-\dfrac{8}{3} - 3 + C = 0$ $\therefore\ C = \dfrac{17}{3}$

$\therefore\ f(x) = \dfrac{8}{3}x^3 - 3x^2 + \dfrac{17}{3}$

$\therefore\ f(1) = \dfrac{8}{3} - 3 + \dfrac{17}{3} = \dfrac{16}{3}$

(17) For a function $f(x) = x\log x - x$, $g(x)$ is the inverse of $f'(x)$.

When the antiderivative of $g(x)$ at $x = 1$ is $3e$, find the antiderivative of $g(x)$.

$f'(x) = \log x + x \cdot \dfrac{1}{x} - 1 = \log x$

Let $y = \log x$

Then, $x = e^y$

Changing x and y, $y = e^x$

$\therefore\ g(x) = e^x$

Let $G(x) = \int g(x)dx = \int e^x\,dx = e^x + C$

Since $G(1) = 3e$, $e^1 + C = 3e$; $C = 2e$

$\therefore\ \int g(x)dx = e^x + 2e$

(18) For two polynomial functions $f(x)$ and $g(x)$,

(i) $f(x) + g(x) = 2x^3 - 2x^2 + 2x + 4$ and

(ii) $f'(x) - g'(x) = 6x^2 - 4$.

When the two graphs of the functions intersect at $x = 1$,

find the value of $f(1) - g(-1)$.

Since the two graphs of the functions intersect at $x = 1$, $f(1) = g(1)$.

$\therefore\ f(1) - g(1) = 0$

Since $f'(x) - g'(x) = 6x^2 - 4$,

$f(x) - g(x) = \int\{f'(x) - g'(x)\}dx = \int(6x^2 - 4)dx = 2x^3 - 4x + C$

$\therefore\ f(1) - g(1) = 2 - 4 + C = 0$; $C = 2$

$\therefore\ f(x) - g(x) = 2x^3 - 4x + 2$ $\cdots\cdots$ ①

Note that $f(x) + g(x) = 2x^3 - 2x^2 + 2x + 4$ $\cdots\cdots$ ②

By ① and ②, $2f(x) = 4x^3 - 2x^2 - 2x + 6$

$\therefore\ f(x) = 2x^3 - x^2 - x + 3$ and $g(x) = 2x^3 - x^2 - x + 3 - (2x^3 - 4x + 2) = -x^2 + 3x + 1$

$\therefore\ f(1) = 2 - 1 - 1 + 3 = 3$ and $g(-1) = -1 - 3 + 1 = -3$

Therefore, $f(1) - g(-1) = 3 - (-3) = 6$

(19) For a function $f(x) = 1 + 2x + 3x^2 + \cdots\cdots + nx^{n-1}$, the indefinite integral $F(x)$

satisfies $\lim\limits_{n\to\infty} F\left(\dfrac{1}{2}\right) = 10$. Find the value of $F(0)$.

$F(x) = \int f(x)dx = \int(1 + 2x + 3x^2 + \cdots\cdots + nx^{n-1})dx = x + x^2 + x^3 + \cdots\cdots + x^n + C$

$\lim\limits_{n\to\infty} F\left(\dfrac{1}{2}\right) = \lim\limits_{n\to\infty}\left\{\dfrac{1}{2} + \left(\dfrac{1}{2}\right)^2 + \left(\dfrac{1}{2}\right)^3 + \cdots\cdots + \left(\dfrac{1}{2}\right)^n + C\right\} = \lim\limits_{n\to\infty}\left[\dfrac{\left(\frac{1}{2}\right)\left\{1 - \left(\frac{1}{2}\right)^n\right\}}{1 - \left(\frac{1}{2}\right)} + C\right]$

$$= \lim_{n \to \infty} \left[\left\{ 1 - \left(\frac{1}{2} \right)^n \right\} + C \right] = 1 + C = 10 \qquad \therefore \ C = 9$$

$\therefore \ F(x) = x + x^2 + x^3 + \cdots\cdots + x^n + 9$

Therefore, $F(0) = 9$

#5 Find the indefinite integrals.

(1) $\int (3 - \sin x) dx$

$= \int 3 dx - \int \sin x \, dx = 3x - (-\cos x) + C = 3x + \cos x + C$

(2) $\int \cot^2 x \, dx$

$= \int (\csc^2 x - 1) dx = -\cot x - x + C$

(3) $\int \frac{1}{1 - \sin^2 x} dx$

$= \int \frac{1}{\cos^2 x} dx = \int \sec^2 x \, dx = \tan x + C$

(4) $\int \frac{\sin^2 x}{1 - \cos x} dx$

$= \int \frac{1 - \cos^2 x}{1 - \cos x} dx = \int (1 + \cos x) dx = x + \sin x + C$

(5) $\int \frac{\sin x}{\cos^2 x} dx$

$= \int \left(\frac{1}{\cos x} \cdot \frac{\sin x}{\cos x} \right) dx = \int (\sec x \cdot \tan x) dx = \sec x + C$

(6) $\int \frac{\cos x}{\sin^2 x} dx$

$= \int \left(\frac{1}{\sin x} \cdot \frac{\cos x}{\sin x} \right) dx = \int (\csc x \cdot \cot x) dx = -\csc x + C$

(7) $\int \frac{\sin x + \cos x}{\sin x \cdot \cos x} dx$

$= \int \frac{\sin x}{\sin x \cdot \cos x} dx + \int \frac{\cos x}{\sin x \cdot \cos x} dx = \int \frac{1}{\cos x} dx + \int \frac{1}{\sin x} dx = \int \sec x \, dx + \int \csc x \, dx$

$= (\log|\sec x + \tan x| + C_1) + (-\log|\csc x + \cot x| + C_2)$

$= \log \left| \frac{\sec x + \tan x}{\csc x + \cot x} \right| + C$ (By the logarithmic property, $\log a - \log b = \log \frac{a}{b}$)

(8) $\int \frac{4 + \sin x}{\cos x} dx$

$= \int \frac{4}{\cos x} dx + \int \frac{\sin x}{\cos x} dx = 4 \int \sec x \, dx + \int \tan x \, dx$

$= 4(\log|\sec x + \tan x| + C_1) - \log|\cos x| + C_2$

$= 4 \log|\sec x + \tan x| - \log|\cos x| + C$

$= \log \frac{(\sec x + \tan x)^4}{|\cos x|} + C$

(9) $\int \sin^2 \frac{x}{2} dx$

$= \int \left(\frac{1-\cos x}{2}\right) dx = \frac{1}{2} \int (1 - \cos x) dx = \frac{1}{2}(x - \sin x) + C = \frac{1}{2}x - \frac{1}{2}\sin x + C$

(10) $\int (\tan x + \cot x)^2 dx$

$= \int (\tan^2 x + 2\tan x \cot x + \cot^2 x) dx = \int (\tan^2 x + 2 + \cot^2 x) dx$

$= \int \{(\sec^2 x - 1) + 2 + (\csc^2 x - 1)\} dx = \int (\sec^2 x + \csc^2 x) dx = \int \sec^2 x \, dx + \int \csc^2 x \, dx$

$= \tan x - \cot x + C$

(11) $\int \frac{2e^x \sin^2 x - 3}{\sin^2 x} dx$

$= \int \left(2e^x - \frac{3}{\sin^2 x}\right) dx = \int (2e^x - 3\csc^2 x) dx = 2e^x - 3(-\cot x) + C = 2e^x + 3\cot x + C$

(12) $\int \frac{8^x + 1}{2^x + 1} dx$

$= \int \frac{(2^x)^3 + 1}{2^x + 1} dx = \int \frac{(2^x + 1)\left((2^x)^2 - 2^x + 1\right)}{2^x + 1} dx = \int (4^x - 2^x + 1) dx = \frac{4^x}{\log 4} - \frac{2^x}{\log 2} + x + C$

(13) $\int \frac{x - \cos^2 x}{x \cos^2 x} dx$

$= \int \left(\frac{x}{x \cos^2 x} - \frac{\cos^2 x}{x \cos^2 x}\right) dx = \int \left(\sec^2 x - \frac{1}{x}\right) dx = \tan x - \log|x| + C$

#6 Answer the question.

(1) When $f'(x) = \dfrac{1}{\tan \frac{x}{2} + \cot \frac{x}{2}}$ **and** $f\left(\frac{\pi}{2}\right) = -1$, **find the function** $f(x)$. $(0 < x < \pi)$

$\tan \frac{x}{2} + \cot \frac{x}{2} = \frac{\sin \frac{x}{2}}{\cos \frac{x}{2}} + \frac{\cos \frac{x}{2}}{\sin \frac{x}{2}} = \frac{\sin^2\left(\frac{x}{2}\right) + \cos^2\left(\frac{x}{2}\right)}{\sin \frac{x}{2} \cos \frac{x}{2}} = \frac{1}{\frac{1}{2}\sin x} = \frac{2}{\sin x}$ $\quad \therefore \ f'(x) = \frac{2}{\sin x}$

$\therefore \ f(x) = \int f'(x) dx = \int \frac{\sin x}{2} dx = \frac{1}{2} \int \sin x \, dx = -\frac{1}{2}\cos x + C$

Since $f\left(\frac{\pi}{2}\right) = -1$, $-\frac{1}{2}\cos\left(\frac{\pi}{2}\right) + C = -1$ $\quad \therefore \ C = -1$

$\therefore \ f(x) = -\frac{1}{2}\cos x - 1$

(2) When the slope of the tangent line at (x, y) **on the graph of** $y = f(x)$ **is at a rate proportional to** $e^x + x$ **and the equation of the tangent line at** $x = 0$ **on the graph is** $y = 2x + 1$, **find the value of** $f(1)$.

For a constant k, $f'(x) = k(e^x + x)$

Since $f'(0) = 2$, $k = 2$

$\therefore \ f(x) = \int f'(x) dx = \int 2(e^x + x) dx = 2e^x + x^2 + C$

Since $y = f(x)$ passes through $(0, 1)$, $f(0) = 2e^0 + 0 + C = 1$ $\quad \therefore \ C = -1$

$\therefore \ f(x) = 2e^x + x^2 - 1$ $\quad$ Therefore, $f(1) = 2e^1 + 1 - 1 = 2e$

(3) At every point of a curve y, $y'' = x^2 - 1$. When the equation of the tangent line at $(1, -1)$ on the curve y is $x + 2y = 3$, find the equation of the curve y.

$\dfrac{d^2 y}{dx^2} = \dfrac{d}{dx} y' = x^2 - 1$

$\therefore \int \dfrac{d}{dx}(y')dx = \int (x^2 - 1)dx = \dfrac{1}{3}x^3 - x + C_1$

$x + 2y = 3 \implies y = -\dfrac{1}{2}x + \dfrac{3}{2}$

Since the slope of the curve at $(1, -1)$ is equal to $-\dfrac{1}{2}$, $\quad \dfrac{1}{3} - 1 + C_1 = -\dfrac{1}{2}$

$\therefore \ C_1 = -\dfrac{1}{2} - \dfrac{1}{3} + 1 = \dfrac{1}{6}$

$\therefore \ y' = \dfrac{dy}{dx} = \dfrac{1}{3}x^3 - x + \dfrac{1}{6}$

$\therefore \ y = \int y' dx = \int \left(\dfrac{1}{3}x^3 - x + \dfrac{1}{6}\right) dx = \dfrac{1}{3} \cdot \dfrac{1}{4}x^4 - \dfrac{1}{2}x^2 + \dfrac{1}{6}x + C = \dfrac{1}{12}x^4 - \dfrac{1}{2}x^2 + \dfrac{1}{6}x + C$

At $(1, -1)$, $\quad -1 = \dfrac{1}{12} - \dfrac{1}{2} + \dfrac{1}{6} + C \qquad \therefore \ C = -\dfrac{3}{4}$

Therefore, the equation of the curve is $y = \dfrac{1}{12}x^4 - \dfrac{1}{2}x^2 + \dfrac{1}{6}x - \dfrac{3}{4}$.

(4) Find the equation of the curve passing through the point $(0, 2)$. The slope of the curve at any point $P(x, y)$ is $= 3x^2 y$.

Since $m = \dfrac{dy}{dx} = 3x^2 y$, $\ dy = 3x^2 y \cdot dx$

$\therefore \ \dfrac{dy}{y} = 3x^2 dx$

$\therefore \ \int \dfrac{1}{y} dy = \int 3x^2 dx \qquad \therefore \ \log y = x^3 + C$

Since C is a constant, let $C = \log c$

Then, $y = e^{x^3 + C} = e^{x^3} e^C = e^{x^3} e^{\log c} = ce^{x^3}$

At $(0, 2)$, $\ 2 = ce^0 \qquad \therefore c = 2$

$\therefore$ The equation of the curve is $y = 2e^{x^3}$.

#7 A continuous function $f(x)$ on $(-\infty, \infty)$ satisfies $f(0) = 1$ and $f'(x) = -f(x) + e^{-x} \sin x$.

(1) Find the derivative of $e^x f(x)$.

Let $g(x) = e^x f(x)$

Then, $g'(x) = e^x f(x) + e^x f'(x) = e^x f(x) + e^x(-f(x) + e^{-x} \sin x) = e^x e^{-x} \sin x = \sin x$

(2) Find the function $f(x)$.

Since $g'(x) = \sin x$, $\ g(x) = \int g'(x)dx = \int \sin x \, dx = -\cos x + C$

Since $f(0) = 1$, $g(0) = e^0 f(0) = e^0 = 1$

$\therefore \quad g(0) = -\cos 0 + C = -1 + C = 1 \; ; \quad C = 2$

$\therefore \quad g(x) = -\cos x + 2$

By (1), $f(x) = e^{-x}g(x) = e^{-x}(-\cos x + 2)$

#8 Find the antiderivative by performing the variable substitution.

(1) $\int (x^3 + 2)^2 \cdot 3x^2 dx$

Let $x^3 + 2 = t$ Then, $3x^2 \frac{dx}{dt} = 1 \; ; \quad 3x^2 dx = dt$

$\therefore \quad \int (x^3 + 2)^2 \cdot 3x^2 dx = \int t^2 dt = \frac{1}{3} t^3 + C = \frac{1}{3}(x^3 + 2)^3 + C$

(2) $\int (x^3 + 2)^{\frac{1}{2}} \cdot x^2 dx$

Let $x^3 + 2 = t$ Then, $3x^2 \frac{dx}{dt} = 1 \; ; \quad 3x^2 dx = dt \; ; \quad x^2 dx = \frac{1}{3} dt$

$\int (x^3 + 2)^{\frac{1}{2}} \cdot x^2 dx = \int t^{\frac{1}{2}} \cdot \frac{1}{3} dt = \frac{1}{3} \int t^{\frac{1}{2}} dt = \frac{1}{3} \cdot \frac{2}{3} t^{\frac{3}{2}} + C = \frac{2}{9} t^{\frac{3}{2}} + C = \frac{2}{9}(x^3 + 2)^{\frac{3}{2}} + C$

(3) $\int \left(x^2 + 2x + 1\right)^3 (x + 1) dx$

Let $x^2 + 2x + 1 = t$ Then, $(2x + 2)\frac{dx}{dt} = 1 \; ; \quad 2(x + 1)dx = dt$

$\int (x^2 + 2x + 1)^3 (x + 1)dx = \int t^3 \cdot \frac{1}{2} dt = \frac{1}{2} \int t^3 \, dt = \frac{1}{2} \cdot \frac{1}{4} t^4 + C = \frac{1}{8}(x^2 + 2x + 1)^4 + C$

(4) $\int \frac{8x^2}{(x^3+4)^3} dx$

$\int \frac{8x^2}{(x^3+4)^3} dx = 8 \int (x^3 + 4)^{-3} x^2 dx = 8 \int (x^3 + 4)^{-3} \cdot 3x^2 \cdot \frac{1}{3} dx = \frac{8}{3} \int (x^3 + 4)^{-3} \cdot 3x^2 dx$

$\qquad = \frac{8}{3} \int t^{-3} dt = \frac{8}{3} \cdot \frac{1}{-2} t^{-2} + C = -\frac{4}{3}(x^3 + 4)^{-2} + C = -\frac{4}{3(x^3+4)^2} + C$

(5) $\int \frac{x^2}{\sqrt[4]{x^3+1}} dx$

$\int \frac{x^2}{\sqrt[4]{x^3+1}} dx = \int (x^3 + 1)^{-\frac{1}{4}} x^2 dx = \int (x^3 + 1)^{-\frac{1}{4}} \cdot 3x^2 \cdot \frac{1}{3} dx = \frac{1}{3} \int (x^3 + 1)^{-\frac{1}{4}} \cdot 3x^2 dx$

$\qquad = \frac{1}{3} \int t^{-\frac{1}{4}} dt = \frac{1}{3} \cdot \frac{4}{3} t^{\frac{3}{4}} + C = \frac{4}{9}(x^3 + 1)^{\frac{3}{4}} + C$

(6) $\int 2x\sqrt{1 - 3x^2}\,dx$

Let $1 - 3x^2 = t$ Then, $-6x \frac{dx}{dt} = 1; \quad -6xdx = dt$

$\int 2x\sqrt{1 - 3x^2}dx = -\frac{1}{3} \int \sqrt{1 - 3x^2} \cdot (-6x) \cdot dx = -\frac{1}{3} \int \sqrt{t} dt = -\frac{1}{3} \int t^{\frac{1}{2}} dt$

$\qquad = -\frac{1}{3} \cdot \frac{2}{3} t^{\frac{3}{2}} + C = -\frac{2}{9}(1 - 3x^2)^{\frac{3}{2}} + C$

(7) $\int x\sqrt{x^2 + 1}\,dx$

Let $x^2 + 1 = t$; $2x\,dx = dt$

$\int x\sqrt{x^2 + 1}\,dx = \int \sqrt{t}\,\frac{1}{2}\,dt = \frac{1}{2}\int t^{\frac{1}{2}}\,dt = \frac{1}{2}\cdot\frac{2}{3}t^{\frac{3}{2}} + C = \frac{1}{3}(x^2 + 1)^{\frac{3}{2}} + C$

$\qquad\qquad = \frac{1}{3}(x^2 + 1)\sqrt{x^2 + 1} + C$

(8) $\int \frac{x-1}{\sqrt{x+1}}\,dx$

Let $\sqrt{x+1} = t$; $x + 1 = t^2$; $dx = 2t\,dt$

$\int \frac{x-1}{\sqrt{x+1}}\,dx = \int \frac{t^2 - 1 - 1}{t}(2t\,dt) = 2\int(t^2 - 2)\,dt = 2\left(\frac{1}{3}t^3 - 2t + C_1\right) = \frac{2}{3}t^3 - 4t + 2C_1$

$\qquad\qquad = \frac{2}{3}\left(\sqrt{x+1}\right)^3 - 4\left(\sqrt{x+1}\right) + C = \frac{2}{3}(x+1)\sqrt{x+1} - 4\left(\sqrt{x+1}\right) + C$

$\qquad\qquad = \frac{2}{3}\sqrt{x+1}(x + 1 - 6) + C = \frac{2}{3}(x - 5)\sqrt{x+1} + C$

(9) $\int \frac{x+4}{\sqrt[3]{x^2+8x}}\,dx$

$\int \frac{x+4}{\sqrt[3]{x^2+8x}}\,dx = \int (x^2 + 8x)^{-\frac{1}{3}}(x + 4)\,dx = \frac{1}{2}\int(x^2 + 8x)^{-\frac{1}{3}}(2x + 8)\,dx$

$\qquad\qquad = \frac{1}{2}\int t^{-\frac{1}{3}}\,dt = \frac{1}{2}\cdot\frac{3}{2}t^{\frac{2}{3}} + C = \frac{3}{4}(x^2 + 8x)^{\frac{2}{3}} + C$

(10) $\int \sin x^\circ\,dx$

$\int \sin x^\circ\,dx = \int \sin\left(\frac{\pi}{180}x\right)dx = -\frac{180}{\pi}\cos\left(\frac{\pi}{180}x\right) + C = -\frac{180}{\pi}\cos x^\circ + C$

(11) $\int \sin x\cos x\,dx$

$\int \sin x\cos x\,dx = \int \sin x\,(\cos x\,dx) = \int t\,dt = \frac{1}{2}t^2 + C = \frac{1}{2}\sin^2 x + C$

Another approach:

$\int \sin x\cos x\,dx = -\int \cos x\,(-\sin x\,dx) = -\int t\,dt = -\frac{1}{2}t^2 + C = -\frac{1}{2}\cos^2 x + C$

For arbitrary constants C_1 and C_2,

$\frac{1}{2}\sin^2 x + C_1 = -\frac{1}{2}\cos^2 x + C_2$

$\left(\because -\frac{1}{2}\cos^2 x + C_2 = -\frac{1}{2}(1 - \sin^2 x) + C_2 = \frac{1}{2}\sin^2 x - \frac{1}{2} + C_2 = \frac{1}{2}\sin^2 x + C\right)$

(12) $\int \sin x\cos 2x\,dx$

$\int \sin x\cos 2x\,dx = \int \sin x\,(2\cos^2 x - 1)\,dx$

Let $\cos x = t$ Then, $-\sin x\,dx = dt$

$\therefore \int \sin x\cos 2x\,dx = -\int(2\cos^2 x - 1)(-\sin x\,dx) = -\int(2t^2 - 1)\,dt = -\frac{2}{3}t^3 + t + C$

$\qquad\qquad = -\frac{2}{3}\cos^3 x + \cos x + C$

(13) $\int \sin^2 x \cos x \, dx$

$\int \sin^2 x \cos x \, dx = \int \sin^2 x \, (\cos x \, dx) = \int t^2 dt = \frac{1}{3}t^3 + C = \frac{1}{3}\sin^3 x + C$

(14) $\int (\sin x + \cos x)^2 dx$

$\int (\sin x + \cos x)^2 dx = \int (\sin^2 x + 2 \sin x \cos x + \cos^2 x) dx = \int (1 + 2 \sin x \cos x) dx$

$$= \int 1 dx + \int \sin 2x \, dx = x - \frac{1}{2}\cos 2x + C$$

(15) $\int \cos^2 x \, dx$

$\int \cos^2 x \, dx = \int \frac{1+\cos 2x}{2} dx = \frac{1}{2}\int 1 dx + \frac{1}{2}\int \cos 2x \, dx = \frac{1}{2}x + \frac{1}{2}\cdot\frac{1}{2}\sin 2x + C = \frac{1}{2}x + \frac{1}{4}\sin 2x + C$

(16) $\int \sin^2 x \, dx$

$\int \sin^2 x \, dx = \int \frac{1-\cos 2x}{2} dx = \frac{1}{2}\int 1 dx - \frac{1}{2}\int \cos 2x \, dx = \frac{1}{2}x - \frac{1}{2}\cdot\frac{1}{2}\sin 2x + C = \frac{1}{2}x - \frac{1}{4}\sin 2x + C$

(17) $\int \sin 2x \cos 3x \, dx$

$\int \sin 2x \cos 3x \, dx = \int \frac{1}{2}\{\sin(2x + 3x) + \sin(2x - 3x)\} dx$

$$= \int \frac{1}{2}(\sin 5x - \sin x) dx$$

$$= \frac{1}{2}\int \sin 5x \, dx - \frac{1}{2}\int \sin x \, dx$$

$$= \frac{1}{2}\cdot\frac{1}{5}(-\cos 5x) - \frac{1}{2}(-\cos x) + C = -\frac{1}{10}\cos 5x + \frac{1}{2}\cos x + C$$

(18) $\int \cos 4x \cos 2x \, dx$

$\int \cos 4x \cos 2x \, dx = \int \frac{1}{2}\{\cos(4x + 2x) + \cos(4x - 2x)\} dx$

$$= \int \frac{1}{2}\{\cos 6x + \cos 2x\} dx = \frac{1}{2}\int \cos 6x \, dx + \frac{1}{2}\int \cos 2x \, dx$$

$$= \frac{1}{2}\cdot\frac{1}{6}(\sin 6x) + \frac{1}{2}\cdot\frac{1}{2}(\sin 2x) + C$$

$$= \frac{1}{12}\sin 6x + \frac{1}{4}\sin 2x + C$$

(19) $\int (1 + \cos x)^4 \sin x \, dx$

Let $1 + \cos x = t$ Then, $-\sin x \, dx = dt$

$\int (1 + \cos x)^4 \sin x \, dx = -\int t^4 dt = -\frac{1}{5}t^5 + C = -\frac{1}{5}(1 + \cos x)^5 + C$

(20) $\int \frac{\sin x}{\cos x} dx$

$\int \frac{\sin x}{\cos x} dx = -\int \frac{(-\sin x)}{\cos x} dx = -\int \frac{1}{t} dt = -\log|t| + C = -\log|\cos x| + C$

(Or, $-\log|t| + C = \log|t|^{-1} + C = \log\frac{1}{|t|} + C = \log\frac{1}{|\cos x|} + C = \log|\sec x| + C$)

(21) $\int \frac{\sec^2 x}{\tan x} dx$

$\int \frac{\sec^2 x}{\tan x} dx = \int \frac{(\tan x)'}{\tan x} dx = \log|\tan x| + C$

(22) $\int \frac{1}{1+\cos x} dx$

$$\int \frac{1}{1+\cos x} dx = \int \frac{1-\cos x}{1-\cos^2 x} dx = \int \frac{1-\cos x}{\sin^2 x} dx = \int \frac{1}{\sin^2 x} dx - \int \frac{\cos x}{\sin^2 x} dx$$

$$= \int \csc^2 x \, dx - \int \csc x \cot x \, dx = -\cot x + \csc x + C$$

(23) $\int \frac{1}{\sin x} dx$

$$\int \frac{1}{\sin x} dx = \int \frac{1}{2\sin\frac{x}{2}\cos\frac{x}{2}} dx = \int \frac{\frac{1}{\cos^2\frac{x}{2}}}{\frac{2\sin\frac{x}{2}\cos\frac{x}{2}}{\cos^2\frac{x}{2}}} dx = \int \frac{\sec^2\frac{x}{2}}{2\tan\frac{x}{2}} dx = \int \frac{\frac{1}{2}\sec^2\frac{x}{2}}{\tan\frac{x}{2}} dx$$

$$= \int \frac{1}{t} dt = \log|t| + C = \log\left|\tan\frac{x}{2}\right| + C$$

(24) $\int \frac{\sin\sqrt{x}}{\sqrt{x}} dx$

Let $\sqrt{x} = t$ Then, $\frac{1}{2}x^{-\frac{1}{2}}dx = dt$; $\frac{1}{2\sqrt{x}} dx = dt$; $\frac{1}{\sqrt{x}} dx = 2dt$

$$\int \frac{\sin\sqrt{x}}{\sqrt{x}} dx = \int \sin t \cdot 2dt = 2\int \sin t \, dt = -2\cos t + C = -2\cos\sqrt{x} + C$$

(25) $\int (e^x + 1)^4 \cdot e^x dx$

$$\int (e^x + 1)^4 \cdot e^x dx = \int (e^x + 1)^4 (e^x dx) = \int t^4 \cdot dt = \frac{1}{5}t^5 + C = \frac{1}{5}(e^x + 1)^5 + C$$

(26) $\int e^{-4x} dx$

Let $-4x = t$ Then, $-4dx = dt$

$$\int e^{-4x} dx = \int e^t \left(-\frac{1}{4}dt\right) = -\frac{1}{4}\int e^t dt = -\frac{1}{4}e^t + C = -\frac{1}{4}e^{-4x} + C$$

(27) $\int xe^{x^2} dx$

Let $x^2 = t$ Then, $2xdx = dt$

$$\int xe^{x^2} dx = \frac{1}{2}\int e^{x^2}(2xdx) = \frac{1}{2}\int e^t dt = \frac{1}{2}e^t + C = \frac{1}{2}e^{x^2} + C$$

(28) $\int \frac{e^x - 1}{e^x + 1} dx$

Let $e^x = t$ Then, $e^x dx = dt$; $tdx = dt$

$$\int \frac{e^x - 1}{e^x + 1} dx = \int \frac{t-1}{t+1} \cdot \frac{1}{t} dt$$

$$\frac{t-1}{t+1} \cdot \frac{1}{t} = \frac{a}{t+1} + \frac{b}{t} = \frac{at + b(t+1)}{(t+1)t} = \frac{(a+b)t+b}{(t+1)t} \qquad \therefore a+b=1,\ b=-1 \qquad \therefore a=2,\ b=-1$$

$$\therefore \int \frac{e^x-1}{e^x+1} dx = \int \frac{t-1}{t+1} \cdot \frac{1}{t} dt = \int \left(\frac{2}{t+1} + \frac{-1}{t}\right) dt = 2\int \frac{1}{t+1} dt - \int \frac{1}{t} dt = 2(\log|t+1|) - \log|t| + C$$

$$= 2(\log|e^x + 1|) - \log|e^x| + C = 2\log(e^x + 1) - \log e^x + C = 2\log(e^x + 1) - x + C$$

(29) $\int e^x \cos e^x \, dx$

$$\int e^x \cos e^x \, dx = \int \cos e^x (e^x dx) = \int \cos t \, dt = \sin t + C = \sin e^x + C$$

(30) $\int e^{3\cos 2x}\sin 2x\, dx$

$\int e^{3\cos 2x}\sin 2x\, dx = -\frac{1}{6}\int e^{3\cos 2x}(-6\sin 2x)dx = -\frac{1}{6}\int e^t dt = -\frac{1}{6}e^t + C = -\frac{1}{6}e^{3\cos 2x} + C$

(31) $\int a^{2x}dx$

$\int a^{2x}dx = \frac{1}{2\log a}\int a^{2x}\cdot \log a \cdot 2dx = \frac{a^{2x}}{2\log a} + C$

(32) $\int 5^{4x+3}dx$

$\int 5^{4x+3}dx = \frac{1}{4}\cdot \frac{5^{4x+3}}{\log 5} + C = \frac{5^{4x+3}}{4\log 5} + C$

#9 Find the integrals.

(1) $\int \frac{1}{x+2}\,dx = \int \frac{1}{x+2}\,dx = \int \frac{(x+2)'}{x+2}\,dx = \log|x+2| + C$

(2) $\int \frac{x+2}{x+1}\,dx$

$\int \frac{x+2}{x+1}\,dx = \int \frac{x+1+1}{x+1}\,dx = \int 1dx + \int \frac{1}{x+1}\,dx = \int 1dx + \int \frac{(x+1)'}{x+1} = x + \log|x+1| + C$

(3) $\int \frac{x^2+1}{x-1}\,dx$

$\int \frac{x^2+1}{x-1}\,dx = \int \left\{(x+1) + \frac{2}{x-1}\right\}dx = \int (x+1)dx + \int \frac{2}{x-1}\,dx$

$= \frac{1}{2}x^2 + x + 2\log|x-1| + C$

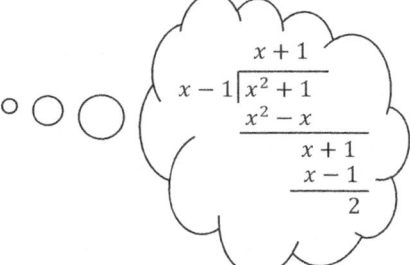

(4) $\int \frac{x}{x^2-1}\,dx$

$\int \frac{x}{x^2-1}\,dx = \frac{1}{2}\int \frac{2x}{x^2-1}\,dx = \frac{1}{2}\int \frac{(x^2-1)'}{x^2-1}\,dx = \frac{1}{2}\log|x^2-1| + C$

(5) $\int \frac{x^2}{1-2x^3}\,dx$

$\int \frac{x^2}{1-2x^3}\,dx = -\frac{1}{6}\int \frac{(1-2x^3)'}{1-2x^3}\,dx = -\frac{1}{6}\log|1-2x^3| + C$

(6) $\int \frac{2x+1}{x^2+x+1}\,dx$

$\int \frac{2x+1}{x^2+x+1}\,dx = \int \frac{(x^2+x+1)'}{x^2+x+1}\,dx = \log|x^2+x+1| + C$

(7) $\int \frac{3}{2x^2+x-1}\,dx$

$\frac{3}{2x^2+x-1} = \frac{3}{(2x-1)(x+1)} = \frac{a}{2x-1} + \frac{b}{x+1} = \frac{a(x+1)+b(2x-1)}{(2x-1)(x+1)} = \frac{(a+2b)x+a-b}{(2x-1)(x+1)}$

$\therefore\ a+2b = 0,\ a - b = 3\ ;\ a = 2,\ b = -1$

$\int \frac{3}{2x^2+x-1}\,dx = \int \left(\frac{2}{2x-1} + \frac{-1}{x+1}\right)dx = \int \frac{2}{2x-1}\,dx - \int \frac{1}{x+1}\,dx = \int \frac{(2x-1)'}{2x-1}\,dx - \int \frac{1}{x+1}\,dx$

$= \log|2x-1| - \log|x+1| + C$

(8) $\int \frac{1}{25-16y^2} dy$

$$\frac{1}{25-16y^2} = \frac{1}{(5-4y)(5+4y)} = \frac{a}{5-4y} + \frac{b}{5+4y} = \frac{a(5+4y)+b(5-4y)}{(5-4y)(5+4y)} = \frac{5(a+b)+4(a-b)y}{(5-4y)(5+4y)}$$

$$\therefore a+b = \frac{1}{5}, \ a-b = 0 \ ; \ 2a = \frac{1}{5} \ ; \ a = \frac{1}{10}, \ b = \frac{1}{10}$$

$$\int \frac{1}{25-16y^2} dy = \int \left(\frac{\frac{1}{10}}{5-4y} + \frac{\frac{1}{10}}{5+4y} \right) dy = \frac{1}{10} \int \frac{1}{5-4y} dy + \frac{1}{10} \int \frac{1}{5+4y} dy$$

$$= \frac{1}{10} \cdot \frac{1}{-4} \int \frac{-4}{5-4y} dy + \frac{1}{10} \cdot \frac{1}{4} \int \frac{4}{5+4y} dy = -\frac{1}{40} \log|5-4y| + \frac{1}{40} \log|5+4y| + C$$

$$= \frac{1}{40} \log \left| \frac{5+4y}{5-4y} \right| + C$$

(9) $\int \frac{1}{9x^2-16} dx$

$$\frac{1}{9x^2-16} = \frac{1}{(3x+4)(3x-4)} = \frac{a}{3x+4} + \frac{b}{3x-4} = \frac{a(3x-4)+b(3x+4)}{(3x+4)(3x-4)} = \frac{3(a+b)x-4(a-b)}{(3x+4)(3x-4)}$$

$$\therefore a+b = 0, \ a-b = -\frac{1}{4} \ ; \ 2a = -\frac{1}{4} \ ; \ a = -\frac{1}{8}, \ b = \frac{1}{8}$$

$$\int \frac{1}{9x^2-16} dx = \int \left(\frac{-\frac{1}{8}}{3x+4} + \frac{\frac{1}{8}}{3x-4} \right) dx = -\frac{1}{8} \int \frac{1}{3x+4} dx + \frac{1}{8} \int \frac{1}{3x-4} dx$$

$$= -\frac{1}{8} \cdot \frac{1}{3} \int \frac{3}{3x+4} dx + \frac{1}{8} \cdot \frac{1}{3} \int \frac{3}{3x-4} dx = -\frac{1}{24} \log|3x+4| + \frac{1}{24} \log|3x-4| + C$$

$$= \frac{1}{24} \log \left| \frac{3x-4}{3x+4} \right| + C$$

(10) $\int \frac{x+1}{(x-1)(x-2)} dx$

$$\frac{x+1}{(x-1)(x-2)} = \frac{a}{x-1} + \frac{b}{x-2} = \frac{a(x-2)+b(x-1)}{(x-1)(x-2)} = \frac{(a+b)x-2a-b}{(x-1)(x-2)} \qquad \therefore a+b = 1, \ -2a-b = 1$$

$$\therefore a = -2, \ b = 3$$

$$\int \frac{x+1}{(x-1)(x-2)} dx = \int \left(\frac{-2}{x-1} + \frac{3}{x-2} \right) dx = -2 \int \frac{1}{x-1} dx + 3 \int \frac{1}{x-2} dx$$

$$= -2 \log|x-1| + 3 \log|x-2| + C$$

(11) $\int \frac{1}{\sqrt{x+1}+\sqrt{x}} dx$

$$\int \frac{1}{\sqrt{x+1}+\sqrt{x}} dx = \int \frac{\sqrt{x+1}-\sqrt{x}}{(\sqrt{x+1}+\sqrt{x})(\sqrt{x+1}-\sqrt{x})} dx = \int \frac{\sqrt{x+1}-\sqrt{x}}{x+1-x} dx = \int (\sqrt{x+1}-\sqrt{x}) dx$$

$$= \int \left\{ (x+1)^{\frac{1}{2}} - x^{\frac{1}{2}} \right\} dx = \int (x+1)^{\frac{1}{2}} dx - \int x^{\frac{1}{2}} dx = \frac{2}{3} (x+1)^{\frac{3}{2}} - \frac{2}{3} x^{\frac{3}{2}} + C$$

$$= \frac{2}{3} (x+1)\sqrt{x+1} - \frac{2}{3} x\sqrt{x} + C$$

(12) $\int \frac{1}{\sqrt[3]{2x+1}} dx$

$$\int \frac{1}{\sqrt[3]{2x+1}} dx = \int (2x+1)^{-\frac{1}{3}} dx = \frac{1}{2} \cdot \frac{3}{2} (2x+1)^{\frac{2}{3}} + C = \frac{3}{4} (2x+1)^{\frac{2}{3}} + C = \frac{3}{4} \sqrt[3]{(2x+1)^2} + C$$

(13) $\int \tan x \, dx$

$\int \tan x \, dx = \int \frac{\sin x}{\cos x} dx = -\int \frac{(\cos x)'}{\cos x} dx = -\log|\cos x| + C$

(14) $\int \frac{1+\cos x}{x+\sin x} dx$

$\int \frac{1+\cos x}{x+\sin x} dx = \int \frac{(x+\sin x)'}{x+\sin x} dx = \log|x + \sin x| + C$

(15) $\int \frac{1}{x\log x} dx = \int \frac{1}{x\log x} dx = \int \frac{\frac{1}{x}}{\log x} dx = \int \frac{(\log x)'}{\log x} dx = \log|\log x| + C$

(16) $\int \frac{1}{e^x+1} dx$

$\int \frac{1}{e^x+1} dx = \int \frac{e^{-x}}{(e^x+1)e^{-x}} dx = \int \frac{e^{-x}}{1+e^{-x}} dx = -\int \frac{-e^{-x}}{1+e^{-x}} dx = -\int \frac{(1+e^{-x})'}{1+e^{-x}} dx = -\log|1 + e^{-x}| + C$

$= \log|1 + e^{-x}|^{-1} + C = \log\frac{1}{|1+e^{-x}|} + C = \log\frac{e^x}{|e^x+1|} + C = \log\frac{e^x}{e^x+1} + C$

$= \log e^x - \log(e^x + 1) + C = x - \log(e^x + 1) + C$

#10 Use integration by parts to perform the indicated integration.

(1) $\int xe^{3x} dx$

Let $u = x, \ v' = e^{3x} dx$

Then, $u' = dx, \ v = \frac{1}{3} e^{3x}$

$\therefore \ \int xe^{3x} dx = \frac{1}{3} e^{3x} \cdot x - \int \frac{1}{3} e^{3x} \cdot dx = \frac{1}{3} xe^{3x} - \frac{1}{3} \int e^{3x} dx$

$= \frac{1}{3} xe^{3x} - \frac{1}{3} \cdot \frac{1}{3} e^{3x} + C = \frac{1}{3} xe^{3x} - \frac{1}{9} e^{3x} + C$

(2) $\int x^2 e^{2x} dx$

Let $u = x^2, \ v' = e^{2x} dx$

Then, $u' = 2xdx, \ v = \frac{1}{2} e^{2x}$

$\therefore \ \int x^2 e^{2x} dx = \frac{1}{2} e^{2x} \cdot x^2 - \int \frac{1}{2} e^{2x} \cdot 2xdx = \frac{1}{2} x^2 e^{2x} - \int xe^{2x} dx$

For $\int xe^{2x} dx$,

Let $u = x, \ v' = e^{2x} dx$

Then, $u' = dx, \ v = \frac{1}{2} e^{2x}$

$\therefore \ \int xe^{2x} dx = \frac{1}{2} e^{2x} \cdot x - \int \frac{1}{2} e^{2x} \cdot dx = \frac{1}{2} xe^{2x} - \frac{1}{2} \int e^{2x} dx = \frac{1}{2} xe^{2x} - \frac{1}{2} \cdot \frac{1}{2} e^{2x} + C_1$

$= \frac{1}{2} xe^{2x} - \frac{1}{4} e^{2x} + C_1$

Therefore, $\int x^2 e^{2x} dx = \frac{1}{2} x^2 e^{2x} - \left(\frac{1}{2} xe^{2x} - \frac{1}{4} e^{2x} + C_1 \right) = \frac{1}{4} (2x^2 - 2x + 1)e^{2x} + C$

(3) $\int x \cos 2x \, dx$

Let $u = x, \ v' = \cos 2x \, dx$

Then, $u' = dx, \ v = \dfrac{1}{2} \sin 2x$

$\therefore \ \int x \cos 2x \, dx = \dfrac{1}{2} \sin 2x \cdot x - \int \dfrac{1}{2} \sin 2x \cdot dx = \dfrac{1}{2} x \sin 2x - \dfrac{1}{2} \int \sin 2x \, dx$

$\qquad = \dfrac{1}{2} x \sin 2x - \dfrac{1}{2} \left(-\dfrac{1}{2} \cos 2x \right) + C = \dfrac{1}{2} x \sin 2x + \dfrac{1}{4} \cos 2x + C$

(4) $\int x^2 \sin x \, dx$

Let $u = x^2, \ v' = \sin x \, dx$

Then, $u' = 2x \, dx, \ v = -\cos x$

$\therefore \ \int x^2 \sin x \, dx = x^2 \cdot (-\cos x) - \int (-\cos x) \cdot 2x \, dx = -x^2 \cos x + 2 \int x \cos x \, dx$

For $\int x \cos x \, dx$,

Let $u = x, \ v' = \cos x \, dx$

Then, $u' = dx, \ v = \sin x$

$\therefore \ \int x \cos x \, dx = x \cdot (\sin x) - \int \sin x \cdot dx = x \sin x + \cos x + C_1$

Therefore, $\int x^2 \sin x \, dx = -x^2 \cos x + 2(x \sin x + \cos x + C_1)$

$\qquad\qquad = -x^2 \cos x + 2x \sin x + 2\cos x + C = (2 - x^2) \cos x + 2x \sin x + C$

(5) $\int \sin^2 x \, dx$

Let $u = \sin x, \ v' = \sin x \, dx$

Then, $u' = \cos x \, dx, \ v = -\cos x$

$\therefore \ \int \sin^2 x \, dx = \sin x \cdot (-\cos x) - \int (-\cos x) \cdot \cos x \, dx = -\sin x \cos x + \int \cos^2 x \, dx$

$\qquad = -\sin x \cos x + \int (1 - \sin^2 x) dx = -\sin x \cos x + x + C_1 - \int \sin^2 x \, dx$

By transposing the integral on the right, we have

$2 \int \sin^2 x \, dx = -\sin x \cos x + x + C_1$

$\therefore \ \int \sin^2 x \, dx = -\dfrac{1}{2} \sin x \cos x + \dfrac{1}{2} x + \dfrac{1}{2} C_1 = -\dfrac{1}{4} \sin 2x + \dfrac{1}{2} x + C$

(6) $\int x^2 \log x \, dx$

Let $u = \log x, \ v' = x^2 dx$

Then, $u' = \dfrac{1}{x} dx, \ v = \dfrac{1}{3} x^3$

$\therefore \ \int x^2 \log x \, dx = \dfrac{1}{3} x^3 \cdot \log x - \int \dfrac{1}{3} x^3 \cdot \dfrac{1}{x} dx = \dfrac{1}{3} x^3 \log x - \dfrac{1}{3} \int x^2 dx$

$\qquad = \dfrac{1}{3} x^3 \log x - \dfrac{1}{3} \cdot \dfrac{1}{3} x^3 + C$

$\qquad = \dfrac{1}{3} x^3 \log x - \dfrac{1}{9} x^3 + C$

(7) $\int \log(x+2)\,dx$

Let $u = \log(x+2),\ v' = dx$

Then, $u' = \frac{1}{x+2}\,dx,\ v = x$

$\therefore\ \int \log(x+2)\,dx = \log(x+2)\cdot x - \int x\cdot\frac{1}{x+2}\,dx = x\log(x+2) - \int \frac{x}{x+2}\,dx$

$\qquad\qquad = x\log(x+2) - \int \frac{x+2-2}{x+2}\,dx = x\log(x+2) - \int\left(1 - \frac{2}{x+2}\right)dx$

$\qquad\qquad = x\log(x+2) - \{x - 2\log(x+2) + C_1\} = (x+2)\log(x+2) - x + C$

(8) $\int x\sqrt{1+x}\,dx$

Let $u = x,\ v' = \sqrt{1+x}\,dx$

Then, $u' = dx,\ v = \frac{2}{3}(1+x)^{\frac{3}{2}}$

$\int x\sqrt{1+x}\,dx = x\cdot\frac{2}{3}(1+x)^{\frac{3}{2}} - \int \frac{2}{3}(1+x)^{\frac{3}{2}}\,dx = \frac{2}{3}x(1+x)^{\frac{3}{2}} - \frac{2}{3}\cdot\frac{2}{5}(1+x)^{\frac{5}{2}} + C$

$\qquad\qquad = \frac{2}{3}x(1+x)^{\frac{3}{2}} - \frac{4}{15}(1+x)^{\frac{5}{2}} + C$

(9) $\int \frac{\log(x-1)}{\sqrt{x-1}}\,dx$

Let $u = \log(x-1),\ v' = \frac{1}{\sqrt{x-1}}\,dx$

Then, $u' = \frac{1}{x-1}\,dx,\ v = \int \frac{1}{\sqrt{x-1}}\,dx = \int(x-1)^{-\frac{1}{2}}\,dx = 2(x-1)^{\frac{1}{2}} = 2\sqrt{x-1}$

$\int \frac{\log(x-1)}{\sqrt{x-1}}\,dx = \log(x-1)\cdot 2\sqrt{x-1} - \int 2\sqrt{x-1}\cdot\frac{1}{x-1}\,dx = 2\sqrt{x-1}\log(x-1) - 2\int \frac{1}{\sqrt{x-1}}\,dx$

$\qquad\qquad = 2\sqrt{x-1}\log(x-1) - 2(2\sqrt{x-1}) + C = 2\sqrt{x-1}\log(x-1) - 4\sqrt{x-1} + C$

#11 Find the value.

(1) For a function $f(x) = \int e^x \cos x\,dx$ satisfying $f(0) = \frac{1}{2}$,

1) Find $f(x)$.

Let $u = \cos x,\ v' = e^x dx$

Then, $u' = -\sin x\,dx,\ v = e^x$

$\therefore\ \int e^x \cos x\,dx = \cos x\cdot e^x - \int e^x\cdot(-\sin x\,dx) = e^x \cos x + \int e^x \sin x\,dx$

For $\int e^x \sin x\,dx$,

Let $u = \sin x,\ v' = e^x dx$

Then, $u' = \cos x\,dx,\ v = e^x$

$\therefore\ \int e^x \sin x\,dx = \sin x\cdot e^x - \int e^x\cdot\cos x\,dx = e^x \sin x - \int e^x \cos x\,dx$

$\therefore\ \int e^x \cos x\,dx = e^x \cos x + e^x \sin x - \int e^x \cos x\,dx$

By transposing the integral on the right, we have

$$2 \int e^x \cos x \, dx = e^x \cos x + e^x \sin x = e^x(\cos x + \sin x)$$

$$\therefore f(x) = \int e^x \cos x \, dx = \frac{1}{2} e^x(\cos x + \sin x) + C$$

Since $f(0) = \frac{1}{2}$, $\frac{1}{2} e^0(\cos 0 + \sin 0) + C = \frac{1}{2}$ $\quad \therefore \; C = 0$

Therefore, $f(x) = \frac{1}{2} e^x(\cos x + \sin x)$

2) Find the value of x such that $f(x) = \frac{1}{2} e^x$ $(0 < x < 2\pi)$.

Since $f(x) = \frac{1}{2} e^x$, $\cos x + \sin x = 1$

$$\therefore \; \sqrt{2} \sin\left(x + \frac{\pi}{4}\right) = 1 \; ; \; \sin\left(x + \frac{\pi}{4}\right) = \frac{1}{\sqrt{2}}$$

$$\therefore \; x + \frac{\pi}{4} = \frac{\pi}{4} \; \text{ or } \; x + \frac{\pi}{4} = \frac{3\pi}{4}$$

$$\therefore \; x = 0 \; \text{ or } \; x = \frac{\pi}{2}$$

Since $x > 0$, $x = \frac{\pi}{2}$

(2) When the slope of the tangent line at (x, y) on the graph of $y = f(x)$ is $x \log x$, find $f(x)$ passing through a point $(1, 3)$.

Since $f'(x) = x \log x$, $f(x) = \int x \log x \, dx$

Let $u = \log x$, $v' = x \, dx$

Then, $u' = \frac{1}{x} dx$, $v = \frac{1}{2} x^2$

$$\therefore \; f(x) = \int x \log x \, dx = \log x \cdot \frac{1}{2} x^2 - \int \frac{1}{2} x^2 \cdot \frac{1}{x} \, dx = \frac{1}{2} x^2 \log x - \frac{1}{2} \int x \, dx$$

$$= \frac{1}{2} x^2 \log x - \frac{1}{2} \cdot \frac{1}{2} x^2 + C = \frac{1}{2} x^2 \log x - \frac{1}{4} x^2 + C$$

Since $(1, 3)$ lies on the graph of $f(x)$, $f(1) = 3$

$$\therefore \; \frac{1}{2} \log 1 - \frac{1}{4} + C = 3 \; ; \; C = \frac{13}{4}$$

Therefore, $f(x) = \frac{1}{2} x^2 \log x - \frac{1}{4} x^2 + \frac{13}{4}$

(3) For two differentiable functions $f(x)$ and $g(x)$, $f(x) > 0$ and $f(0) = e^2$.

Find $f(x)$ such that $f(x)\{g(x) + g'(x)\} = \big(f(x)g(x)\big)'$. $(g(x) \neq 0)$

Since $f(x)\{g(x) + g'(x)\} = f(x)g(x) + f(x)g'(x)$ and

$$\big(f(x)g(x)\big)' = f'(x)g(x) + f(x)g'(x), \; f(x)g(x) = f'(x)g(x)$$

Since $g(x) \neq 0$, $f(x) = f'(x)$

$$\therefore \; \frac{f'(x)}{f(x)} = 1$$

$\therefore \int \frac{f'(x)}{f(x)} dx = \int 1 dx$

$\therefore \log|f(x)| + C_1 = x + C_2$

Since $f(x) > 0$, $\quad \log f(x) = x + C$

$\therefore f(x) = e^{x+C}$

Since $f(0) = e^2$, $e^{0+C} = e^C = e^2$; $C = 2$

Therefore, $f(x) = e^{x+2}$

(4) For a function $f(x)$ defined on real numbers, $f'(x) = \frac{3x}{x^2+1}$ and $f(0) = 1$.

Find the value of $f(\sqrt{e-1})$.

$f(x) = \int f'(x) dx = \int \frac{3x}{x^2+1} dx = \frac{3}{2} \int \frac{2x}{x^2+1} dx = \frac{3}{2} \int \frac{(x^2+1)'}{x^2+1} dx = \frac{3}{2} \log(x^2+1) + C$

Since $f(0) = 1$, $\frac{3}{2} \log(0+1) + C = 1$; $C = 1$

$\therefore f(x) = \frac{3}{2} \log(x^2+1) + 1$

$\therefore f(\sqrt{e-1}) = \frac{3}{2} \log((e-1)+1) + 1 = \frac{3}{2} \log e + 1 = \frac{3}{2} + 1 = \frac{5}{2}$

(5) For a function $f(x) = \int \frac{\log x}{x} dx$, $f(1) = -1$.

Find the value of x such that $f(x) = \log x - 1$.

$f(x) = \int \frac{\log x}{x} dx$

Let $\log x = t$ Then, $\frac{1}{x} dx = dt$

$\therefore f(x) = \int \frac{\log x}{x} dx = \int t \, dt = \frac{1}{2} t^2 + C = \frac{1}{2} (\log x)^2 + C$

Since $f(1) = -1$, $\frac{1}{2} (\log 1)^2 + C = -1$; $C = -1$

$\therefore f(x) = \frac{1}{2} (\log x)^2 - 1$

Since $f(x) = \log x - 1$, $\frac{1}{2} (\log x)^2 = \log x$

$\therefore \frac{1}{2} \log x (\log x - 2) = 0$; $\log x = 0$ or $\log x = 2$

Since $f(x) = \int \frac{\log x}{x} dx$, $\log x \neq 0$

$\therefore \log x = 2$

Therefore, $x = e^2$

(6) When a function $f(x)$ defined on an open interval $\left(0, \dfrac{3\pi}{4}\right)$ satisfies:

i) $f'(x) = \sin 2x - \cos x$ and

ii) $f(x)$ has a local minimum value 1,

find the difference between the local extreme values.

$f(x) = \int f'(x)dx = \int(\sin 2x - \cos x)dx = \int \sin 2x\, dx - \int \cos x\, dx$

$\qquad = -\dfrac{1}{2}\cos 2x - \sin x + C$

$f'(x) = \sin 2x - \cos x = 2 \sin x \cos x - \cos x = \cos x\,(2\sin x - 1)$

$f''(x) = 2 \cos 2x + \sin x$

$f'(x) = 0 \;\Rightarrow\; \cos x = 0 \text{ or } \sin x = \dfrac{1}{2} \qquad \therefore\; x = \dfrac{\pi}{2} \text{ or } x = \dfrac{\pi}{6} \;\left(\because 0 < x < \dfrac{3\pi}{4}\right)$

Since $f''\left(\dfrac{\pi}{2}\right) = 2 \cos \pi + \sin\dfrac{\pi}{2} = -2 + 1 = -1 < 0$ and $\qquad$ local Max.

$\qquad f''\left(\dfrac{\pi}{6}\right) = 2 \cos\dfrac{\pi}{3} + \sin\dfrac{\pi}{6} = 1 + \dfrac{1}{2} = \dfrac{3}{2} > 0,$

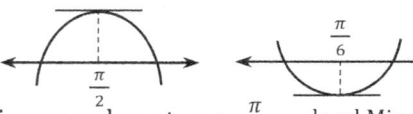

$f(x)$ has a local maximum value at $x = \dfrac{\pi}{2}$ and a local minimum value at $x = \dfrac{\pi}{6}.$ $\quad$ local Min.

Since $f(x)$ has a local minimum vale 1, $f\left(\dfrac{\pi}{6}\right) = 1$

$\therefore\; -\dfrac{1}{2}\cos\dfrac{\pi}{3} - \sin\dfrac{\pi}{6} + C = 1\;;\quad -\dfrac{1}{4} - \dfrac{1}{2} + C = 1\;;\quad C = \dfrac{7}{4}$

$\therefore\; f(x) = -\dfrac{1}{2}\cos 2x - \sin x + \dfrac{7}{4}$

$\therefore$ The maximum value is $f\left(\dfrac{\pi}{2}\right) = -\dfrac{1}{2}\cos\pi - \sin\dfrac{\pi}{2} + \dfrac{7}{4} = \dfrac{1}{2} - 1 + \dfrac{7}{4} = \dfrac{5}{4}$

Therefore, the difference between the local extreme values is $\dfrac{5}{4} - 1 = \dfrac{1}{4}$

(7) Let $F(x) = \int f(x)dx \;\;(x > 0).$

When $F(x) = xf(x) - x^2 \sin x$ and $F\left(\dfrac{\pi}{2}\right) = \dfrac{\pi}{4},$ find the value of $f(\pi).$

$F(x) = xf(x) - x^2 \sin x \;\Rightarrow\; F'(x) = f(x) + xf'(x) - 2x \sin x - x^2 \cos x \;\cdots\cdots\; ①$

Since $F(x) = \int f(x)dx,\; F'(x) = f(x) \;\cdots\cdots\; ②$

By ① and ②, $f(x) = f(x) + xf'(x) - 2x \sin x - x^2 \cos x$

$\therefore\; xf'(x) = 2x \sin x + x^2 \cos x\;;\; f'(x) = 2 \sin x + x \cos x$

$\therefore\; f(x) = \int f'(x)dx = \int(2 \sin x + x \cos x)dx = -2 \cos x + \int x \cos x\, dx$

For $\int x \cos x\, dx,$

Let $u = x,\; v' = \cos x\, dx$

Then, $u' = dx,\; v = \sin x$

$\therefore\; \int x \cos x\, dx = x \sin x - \int \sin x\, dx = x \sin x + \cos x + C$

$\therefore \ f(x) = -2\cos x + (x\sin x + \cos x + C) = x\sin x - \cos x + C$

Since $F\left(\frac{\pi}{2}\right) = \frac{\pi}{4}, \quad \frac{\pi}{2}f\left(\frac{\pi}{2}\right) - \left(\frac{\pi}{2}\right)^2 \sin\frac{\pi}{2} = \frac{\pi}{4}$

Since $f\left(\frac{\pi}{2}\right) = \frac{\pi}{2}\sin\frac{\pi}{2} - \cos\frac{\pi}{2} + C = \frac{\pi}{2} - 0 + C = \frac{\pi}{2} + C, \quad \frac{\pi}{2}\left(\frac{\pi}{2} + C\right) - \left(\frac{\pi}{2}\right)^2 = \frac{\pi}{4}$

$\therefore \ \frac{\pi}{2}C = \frac{\pi}{4} \ ; \ C = \frac{1}{2}$

$\therefore \ f(x) = x\sin x - \cos x + \frac{1}{2}$

Therefore, $f(\pi) = \pi\sin\pi - \cos\pi + \frac{1}{2} = 0 - (-1) + \frac{1}{2} = \frac{3}{2}$

(8) The rate of change of y with respect to x is $2x^3$, and $y = 12$ when $x = 2$.
   Find the value of y when $x = -2$.

Since $\frac{dy}{dx} = 2x^3, \ dy = 2x^3 dx$

$\therefore \ y = \int dy = \int 2x^3 dx = \frac{1}{2}x^4 + C$

When $x = 2, \ y = \frac{1}{2}\cdot 2^4 + C = 12 \ ; \ C = 4$

$\therefore \ y = \frac{1}{2}x^4 + 4$

When $x = -2, \ y = \frac{1}{2}\cdot(-2)^4 + 4 = 12$

(9) A certain quantity a increases at a rate proportional to itself. If $a = 8$ when $t = 0$ and
   $a = 32$ when $t = 2$, find the value of a when $t = 4$.

$\frac{da}{dt} = ka \ \ (k; \text{constant}) \qquad \therefore \ \frac{da}{a} = k\cdot dt \qquad \therefore \ \int\frac{da}{a} = \int k\cdot dt$

$\therefore \ \log a = kt + C$

Letting $C = \log c, \ \log a = kt + \log c$

$\therefore \ a = e^{kt+\log c} = e^{kt}e^{\log c} = ce^{kt}$

When $t = 0, \ a = ce^0 = c \qquad \therefore \ c = 8 \qquad \therefore \ a = 8e^{kt}$

When $t = 2, \ a = 8e^{2k} = 32 \ ; \ e^{2k} = 4 = e^{\log 4} \qquad \therefore \ 2k = \log 4 \qquad \therefore \ k = \frac{\log 4}{2}$

When $t = 4, \ a = 8e^{\frac{\log 4}{2}\cdot 4} = 8e^{2\log 4} = 8e^{\log 4^2} = 8\cdot 4^2 = 128$

(10) A ball is rolled over a level lawn with initial velocity 30 feet/sec. Due to friction, the
   velocity decreases at the rate of 4 feet/sec^2. How far will the ball roll?

Since $\frac{dV}{dt} = -4, \ V = \int dV = \int -4dt = -4t + C_1$

When $t = 0, \ V = 30 \qquad \therefore \ 30 = -4\cdot 0 + C_1 \ ; \ \ C_1 = 30$

$\therefore \ V = -4t + 30 \cdots\cdots \ ①$

Since $V = \frac{dS}{dt}$, $\quad S = \int(-4t + 30)\,dt = -2t^2 + 30t + C$

When $t = 0$, $S = 0$ $\qquad \therefore \quad 0 = -2 \cdot 0 + 30 \cdot 0 + C$; $\quad C = 0$

$\therefore \quad S = -2t^2 + 30t$

When $V = 0$, $\quad 0 = -4t + 30$ (by ①) $\qquad \therefore \quad t = \frac{30}{4} = \frac{15}{2}$

That is, the ball rolls for 7.5 seconds before stopping.

When $t = \frac{15}{2}$, $\quad S = -2\left(\frac{15}{2}\right)^2 + 30\left(\frac{15}{2}\right) = -\frac{15^2}{2} + 15^2 = 15^2\left(1 - \frac{1}{2}\right) = \frac{225}{2}$ feet.

(11) A ball dropped from a balloon 648 feet above the ground.

When the balloon was rising at the rate of 36 feet/sec, find the following:

1) The greatest distance above the ground attained by the ball.

2) The time the ball was in the air.

3) The speed of the ball when it struck the ground.

Assume positive distanced velocity to be directed upward.

$a = \frac{dV}{dt} = -24$ **feet/sec^2, and** $V = -24t + C$

When $t = 0$, $V = 36$ $\qquad \therefore \quad C = 36$ $\qquad \therefore \quad V = -24t + 36$

Since $V = \frac{dS}{dt}$, $\quad S = \int(-24t + 36)\,dt = -12t^2 + 36t + C_1$

When $t = 0$, $S = 648$ $\qquad \therefore \quad 648 = -12 \cdot 0 + 36 \cdot 0 + C_1$; $\quad C_1 = 648$

$\therefore \quad S = -12t^2 + 36t + 648$

1) When $V = 0$, $\quad t = \frac{36}{24} = \frac{3}{2}$

$\qquad \therefore \quad S = -12\left(\frac{3}{2}\right)^2 + 36\left(\frac{3}{2}\right) + 648 = -27 + 54 + 648 = 675$

$\qquad \therefore$ The greatest height attained by the ball was 675 feet.

2) When $S = 0$, $\quad 0 = -12t^2 + 36t + 648$

$\qquad \therefore \quad t^2 - 3t - 54 = 0$ $\qquad \therefore \quad (t - 9)(t + 6) = 0$; $\quad t = 9$ or $t = -6$

$\qquad \therefore$ The ball was in the air for 9 seconds.

3) When $t = 9$, $V = -24 \cdot 9 + 36 = -180$

$\qquad \therefore$ The speed of the ball was 180 feet/sec.

Chapter 6. The Definite Integral

#1 Use the fundamental theorem of calculus to evaluate each definite integral.

(1) $\int_1^3 3x^2\, dx = [x^3]_1^3 = (3^3 - 1^3) = 26$

(2) $\int_{-1}^2 (2x - 6x^2)\, dx = [x^2 - 2x^3]_{-1}^2 = (2^2 - 2\cdot 2^3) - ((-1)^2 - 2(-1)^3) = -12 - 3 = -15$

(3) $\int_0^1 \sqrt{x}\, dx = \int_0^1 x^{\frac{1}{2}}\, dx = \left[\frac{2}{3}x^{\frac{3}{2}}\right]_0^1 = \frac{2}{3} - 0 = \frac{2}{3}$

(4) $\int_0^\pi \cos x\, dx = [\sin x]_0^\pi = \sin \pi - \sin 0 = 0$

(5) $\int_0^{\frac{\pi}{2}} \sin 2x\, dx = \left[-\frac{1}{2}\cos 2x\right]_0^{\frac{\pi}{2}} = -\frac{1}{2}(\cos \pi - \cos 0) = -\frac{1}{2}(-1 - 1) = 1$

(6) $\int_0^2 e^{-x}\, dx = [-e^{-x}]_0^2 = -(e^{-2} - e^0) = -\frac{1}{e^2} + 1$

(7) $\int_{-1}^1 (x - 1)(x^2 + x + 1)\, dx = \int_{-1}^1 (x^3 - 1)\, dx = \left[\frac{1}{4}x^4 - x\right]_{-1}^1 = \left(\frac{1}{4} - 1\right) - \left(\frac{1}{4} + 1\right) = -2$

(8) $\int_0^2 (y + 1)(y^2 - 1)\, dy = \int_0^2 (y^3 + y^2 - y - 1)\, dy = \left[\frac{1}{4}y^4 + \frac{1}{3}y^3 - \frac{1}{2}y^2 - y\right]_0^2$

$$= \frac{1}{4}\cdot 2^4 + \frac{1}{3}\cdot 2^3 - \frac{1}{2}\cdot 2^2 - 2 = 4 + \frac{8}{3} - 2 - 2 = \frac{8}{3}$$

(9) $\int_1^2 \frac{x^2 + 2}{x}\, dx = \int_1^2 \left(x + \frac{2}{x}\right) dx = \left[\frac{1}{2}x^2 + 2\log|x|\right]_1^2 = (2 + 2\log 2) - \left(\frac{1}{2} + 0\right) = \frac{3}{2} + 2\log 2$

(10) $\int_1^2 \frac{1}{x^2 + x}\, dx = \int_1^2 \frac{1}{x(x+1)}\, dx = \int_1^2 \left(\frac{1}{x} - \frac{1}{x+1}\right) dx = [\log|x| - \log|x + 1|]_1^2$

$$= (\log 2 - \log 3) - (0 - \log 2) = 2\log 2 - \log 3$$

(11) $\int_1^2 \frac{x}{x^2 + 1}\, dx = \int_1^2 \frac{2x}{x^2 + 1}\cdot \frac{1}{2}\, dx = \frac{1}{2}\int_1^2 \frac{2x}{x^2 + 1}\, dx = \frac{1}{2}\int_1^2 \frac{(x^2 + 1)'}{x^2 + 1}\, dx = \frac{1}{2}[\log|x^2 + 1|]_1^2$

$$= \frac{1}{2}(\log 5 - \log 2) = \frac{1}{2}\log\frac{5}{2}$$

(12) $\int_0^{\frac{\pi}{3}} \tan^2 x\, dx = \int_0^{\frac{\pi}{3}} (\sec^2 x - 1)\, dx = [\tan x - x]_0^{\frac{\pi}{3}} = \tan\frac{\pi}{3} - \frac{\pi}{3} = \sqrt{3} - \frac{\pi}{3}$

(13) $\int_0^{\frac{\pi}{2}} (e^{4x} - \sin^2 x)\, dx = \int_0^{\frac{\pi}{2}} \left(e^{4x} - \frac{1 - \cos 2x}{2}\right) dx = \left[\frac{1}{4}e^{4x} - \frac{1}{2}x + \frac{1}{2}\cdot\frac{1}{2}\sin 2x\right]_0^{\frac{\pi}{2}}$

$$= \left(\frac{1}{4}e^{2\pi} - \frac{\pi}{4}\right) - \frac{1}{4} = \frac{1}{4}(e^{2\pi} - \pi - 1)$$

(14) $\int_0^1 xe^{-x^2}\, dx = -\frac{1}{2}\int_0^1 -2xe^{-x^2}\, dx = -\frac{1}{2}\int_0^1 \left(e^{-x^2}\right)' dx = -\frac{1}{2}\left[e^{-x^2}\right]_0^1$

$$= -\frac{1}{2}(e^{-1} - e^0) = -\frac{1}{2}\left(\frac{1}{e} - 1\right) = \frac{1}{2}\left(1 - \frac{1}{e}\right)$$

(15) $\int_0^2 (2x+1)\sqrt{x^2+x}\,dx$

Let $x^2 + x = t$ Then, $(2x+1)dx = dt$

$\int (2x+1)\sqrt{x^2+x}\,dx = \int \sqrt{t}\,dt = \int t^{\frac{1}{2}}\,dt = \frac{2}{3}t^{\frac{3}{2}} + C = \frac{2}{3}(x^2+x)^{\frac{3}{2}} + C$

$\int_0^2 (2x+1)\sqrt{x^2+x}\,dx = \left[\frac{2}{3}(x^2+x)^{\frac{3}{2}}\right]_0^2 = \frac{2}{3}(2^2+2)^{\frac{3}{2}} = \frac{2}{3}6^{\frac{3}{2}} = 4\sqrt{6}$

(16) $\int_1^0 5\,dx = -\int_0^1 5\,dx = -5\int_0^1 dx = -5[x]_0^1 = -5$

(17) $\int_4^1 \left(\sqrt{x} + \frac{1}{\sqrt{x}} + 1\right)dx = -\int_1^4 \left(\sqrt{x} + \frac{1}{\sqrt{x}} + 1\right)dx = -\left[\frac{2}{3}x^{\frac{3}{2}} + 2x^{\frac{1}{2}} + x\right]_1^4$

$$= -\left(\frac{16}{3} + 4 + 4\right) + \left(\frac{2}{3} + 2 + 1\right) = -\frac{14}{3} - 5 = -\frac{29}{3}$$

(18) $\int_0^1 \left(\frac{1}{\sqrt{x+1}-\sqrt{x}}\right)dx = \int_0^1 \left(\frac{\sqrt{x+1}+\sqrt{x}}{x+1-x}\right)dx = \int_0^1 (\sqrt{x+1}+\sqrt{x})dx = \int_0^1 \left\{(x+1)^{\frac{1}{2}} + x^{\frac{1}{2}}\right\}dx$

$$= \left[\frac{2}{3}(x+1)^{\frac{3}{2}} + \frac{2}{3}x^{\frac{3}{2}}\right]_0^1 = \frac{2}{3}2^{\frac{3}{2}} + \frac{2}{3} - \frac{2}{3} = \frac{4}{3}\sqrt{2}$$

(19) $\int_{-1}^{-2} \left(\frac{2}{x^2} + \frac{1}{x}\right)dx = 2\int_{-1}^{-2}\frac{1}{x^2}\,dx + \int_{-1}^{-2}\frac{1}{x}\,dx = 2\int_{-1}^{-2}x^{-2}\,dx + \int_{-1}^{-2}\frac{1}{x}\,dx$

$$= -2\int_{-2}^{-1}x^{-2}\,dx - \int_{-2}^{-1}\frac{1}{x}\,dx = -2[-x^{-1}]_{-2}^{-1} - [\log|x|]_{-2}^{-1}$$

$$= 2\left(-1 + \frac{1}{2}\right) - (0 - \log 2) = -1 + \log 2$$

(20) $\int_0^1 \frac{(x+2)^2}{x+1}\,dx$

$(x+2)^2 \div (x+1) = x + 1 + \frac{3}{x+1}$

$\int_0^1 \frac{(x+2)^2}{x+1}\,dx = \int_0^1 \left(x + 1 + \frac{3}{x+1}\right)dx = \left[\frac{1}{2}x^2 + x + 3\log|x+1|\right]_0^1 = \frac{1}{2} + 1 + 3\log 2$

$$= \frac{3}{2} + 3\log 2$$

(21) $\int_2^4 \frac{1}{x^2-3x+2}\,dx$

$\frac{1}{x^2-3x+2} = \frac{1}{(x-1)(x-2)} = \frac{a}{x-1} + \frac{b}{x-2} = \frac{a(x-2)+b(x-1)}{(x-1)(x-2)} = \frac{(a+b)x-2a-b}{(x-1)(x-2)}$

$\therefore\ a + b = 0,\ -2a - b = 1$ $\therefore\ a = -1,\ b = 1$

$\int_2^4 \frac{1}{x^2-3x+2}\,dx = \int_2^4 \left(\frac{-1}{x-1} + \frac{1}{x-2}\right)dx = -\int_2^4 \frac{1}{x-1}\,dx + \int_2^4 \frac{1}{x-2}\,dx$

$= -[\log|x-1|]_2^4 + [\log|x-2|]_2^4 = -(\log 3 - \log 1) + (\log 2 - \log 0) = \log 2 - \log 3$

(22) $\int_1^3 \frac{x^2(x^2+2x+4)}{x+2}\,dx + \int_3^1 \frac{4(y^2+2y+4)}{y+2}\,dy$

$= \int_1^3 \frac{x^2(x^2+2x+4)}{x+2}\,dx - \int_1^3 \frac{4(x^2+2x+4)}{x+2}\,dx = \int_1^3 \left\{\frac{x^2(x^2+2x+4)}{x+2} - \frac{4(x^2+2x+4)}{x+2}\right\}dx$

$$= \int_1^3 \left\{ \frac{(x^2-4)(x^2+2x+4)}{x+2} \right\} dx = \int_1^3 (x-2)(x^2+2x+4)dx = \int_1^3 (x^3-8)dx$$

$$= \left[\frac{1}{4}x^4 - 8x \right]_1^3 = \left(\frac{1}{4}3^4 - 24 \right) - \left(\frac{1}{4} - 8 \right) = \frac{80}{4} - 16 = 4$$

(23) $\int_{-\pi}^{\pi} (\sin x + \cos x)^2 \, dx = \int_{-\pi}^{\pi} (\sin^2 x + \cos^2 x + 2\sin x \cos x) \, dx$

$$= \int_{-\pi}^{\pi} (1 + \sin 2x) \, dx = \left[x - \frac{1}{2}\cos 2x \right]_{-\pi}^{\pi}$$

$$= \left(\pi - \frac{1}{2}\cos 2\pi \right) - \left(-\pi - \frac{1}{2}\cos 2(-\pi) \right)$$

$$= \pi - \frac{1}{2} + \pi + \frac{1}{2} = 2\pi$$

(24) $\int_0^{\frac{\pi}{4}} (\sin^3 2x \cos 2x) \, dx$

Let $\sin 2x = t$ Then, $2\cos 2x \, dx = dt$; $\cos 2x \, dx = \frac{1}{2} dt$

$$\int (\sin^3 2x \cos 2x)dx = \int t^3 \cdot \frac{1}{2} dt = \frac{1}{2} \int t^3 dt = \frac{1}{2} \cdot \frac{1}{4} t^4 + C = \frac{1}{8} t^4 + C = \frac{1}{8} \sin^4 2x + C$$

$$\int_0^{\frac{\pi}{4}} (\sin^3 2x \cos 2x) \, dx = \left[\frac{1}{8} \sin^4 2x \right]_0^{\frac{\pi}{4}} = \frac{1}{8} \sin^4 \frac{\pi}{2} = \frac{1}{8}$$

(25) $\int_0^{2\pi} \cos 3x \cos 2x \, dx = \int_0^{2\pi} \frac{1}{2} \{\cos(3x+2x) + \cos(3x-2x)\} dx$

$$= \frac{1}{2} \int_0^{2\pi} \{\cos 5x + \cos x\} dx = \frac{1}{2} \left[\frac{1}{5} \sin 5x + \sin x \right]_0^{2\pi} = \frac{1}{2} \left(\frac{1}{5} \sin 10\pi + \sin 2\pi \right) = 0$$

(26) $\frac{1}{\pi} \int_{-\pi}^{\pi} \sin 3x \, (\sin 3x + \sin 5x) \, dx = \frac{1}{\pi} \int_{-\pi}^{\pi} (\sin^2 3x + \sin 3x \sin 5x) \, dx$

$$= \frac{1}{\pi} \int_{-\pi}^{\pi} \left\{ \frac{1}{2}(1 - \cos 6x) - \frac{1}{2}(\cos 8x - \cos 2x) \right\} dx$$

$$= \frac{1}{2\pi} \left[x - \frac{1}{6} \sin 6x \right]_{-\pi}^{\pi} - \frac{1}{2\pi} \left[\frac{1}{8} \sin 8x - \frac{1}{2} \sin 2x \right]_{-\pi}^{\pi}$$

$$= \frac{1}{2\pi} (2\pi) - 0 = 1$$

(27) $\int_0^1 \frac{x^2}{x+1} dx - \int_0^1 \frac{1}{x+1} dx$

$$= \int_0^1 \left(\frac{x^2}{x+1} - \frac{1}{x+1} \right) dx = \int_0^1 \frac{x^2-1}{x+1} dx = \int_0^1 \frac{(x+1)(x-1)}{x+1} dx = \int_0^1 (x-1)dx = \left[\frac{1}{2}x^2 - x \right]_0^1 = -\frac{1}{2}$$

(28) $\int_0^{\log 2} \frac{e^{3x}}{e^x+1} dx - \int_{\log 2}^0 \frac{1}{e^t+1} dt = \int_0^{\log 2} \frac{e^{3x}}{e^x+1} dx + \int_0^{\log 2} \frac{1}{e^t+1} dt$

$$= \int_0^{\log 2} \left(\frac{e^{3x}}{e^x+1} + \frac{1}{e^x+1} \right) dx = \int_0^{\log 2} \left(\frac{e^{3x}+1}{e^x+1} \right) dx = \int_0^{\log 2} \frac{(e^x+1)(e^{2x}-e^x+1)}{e^x+1} dx$$

$$= \int_0^{\log 2} (e^{2x} - e^x + 1)dx = \left[\frac{1}{2}e^{2x} - e^x + x \right]_0^{\log 2}$$

$$= \left(\frac{1}{2}e^{2\log 2} - e^{\log 2} + \log 2 \right) - \left(\frac{1}{2}e^0 - e^0 \right) = \frac{1}{2} \cdot 4 - 2 + \log 2 - \frac{1}{2} + 1 = \log 2 + \frac{1}{2}$$

#2 Evaluate each limit.

(1) $\displaystyle\lim_{n\to\infty}\frac{1}{n\sqrt{n}}\sum_{k=1}^{n}(\sqrt{n}+\sqrt{k})$

$\displaystyle=\lim_{n\to\infty}\sum_{k=1}^{n}\left(\frac{1}{n}+\frac{\sqrt{k}}{n\sqrt{n}}\right)=\lim_{n\to\infty}\sum_{k=1}^{n}\left(1+\sqrt{\frac{k}{n}}\right)\frac{1}{n}=\int_0^1(1+\sqrt{x})dx=\left[x+\frac{2}{3}x^{\frac{3}{2}}\right]_0^1=1+\frac{2}{3}=\frac{5}{3}$

(2) $\displaystyle\lim_{n\to\infty}\sum_{i=1}^{n}\frac{3}{n}\sin^2\frac{\pi i}{n}$

$\displaystyle=\lim_{n\to\infty}\sum_{i=1}^{n}3\sin^2\frac{\pi i}{n}\cdot\frac{1}{n}$

$\displaystyle=\int_0^1 3\sin^2\pi x\,dx=3\int_0^1\left(\frac{1-\cos 2\pi x}{2}\right)dx=\frac{3}{2}\int_0^1(1-\cos 2\pi x)dx=\frac{3}{2}\left[x-\frac{1}{2\pi}\sin 2\pi x\right]_0^1$

$\displaystyle=\frac{3}{2}\left(1-\frac{1}{2\pi}\sin 2\pi\right)=\frac{3}{2}$

(3) $\displaystyle\lim_{n\to\infty}\sum_{j=1}^{n}\frac{j}{n^2}\cos\frac{\pi j^2}{3n^2}$

$\displaystyle=\lim_{n\to\infty}\sum_{j=1}^{n}\left\{\frac{j}{n}\cos\frac{\pi}{3}\left(\frac{j}{n}\right)^2\right\}\frac{1}{n}=\int_0^1 x\cos\left(\frac{\pi}{3}x^2\right)dx$

For $\displaystyle\int x\cos\left(\frac{\pi}{3}x^2\right)dx$, let $\dfrac{\pi}{3}x^2=t$ Then, $\dfrac{\pi}{3}2xdx=dt$; $xdx=\dfrac{3}{2\pi}dt$

$\displaystyle\therefore\int x\cos\left(\frac{\pi}{3}x^2\right)dx=\int\cos t\frac{3}{2\pi}dt=\frac{3}{2\pi}\int\cos t\,dt=\frac{3}{2\pi}\sin t+C=\frac{3}{2\pi}\sin\left(\frac{\pi}{3}x^2\right)+C$

$\displaystyle\lim_{n\to\infty}\sum_{j=1}^{n}\frac{j}{n^2}\cos\frac{\pi j^2}{3n^2}=\int_0^1 x\cos\left(\frac{\pi}{3}x^2\right)dx=\left[\frac{3}{2\pi}\sin\left(\frac{\pi}{3}x^2\right)\right]_0^1=\frac{3}{2\pi}\sin\frac{\pi}{3}=\frac{3}{2\pi}\cdot\frac{\sqrt{3}}{2}=\frac{3\sqrt{3}}{4\pi}$

(4) $\displaystyle\lim_{n\to\infty}\frac{1}{n^2}\sum_{l=1}^{n}le^{\frac{l}{n}}=\lim_{n\to\infty}\sum_{l=1}^{n}\frac{l}{n}e^{\frac{l}{n}}\frac{1}{n}=\int_0^1 xe^x dx$

For $\displaystyle\int xe^x dx,$

Let $u=x,$ $v'=e^x dx$

Then, $u'=dx,$ $v=e^x$

$\displaystyle\therefore\int xe^x dx=xe^x-\int e^x dx=xe^x-e^x+C$

$\displaystyle\therefore\lim_{n\to\infty}\frac{1}{n^2}\sum_{l=1}^{n}le^{\frac{l}{n}}=\int_0^1 xe^x dx=[xe^x-e^x]_0^1=(e^1-e^1)-(0-e^0)=1$

(5) $\displaystyle\lim_{n\to\infty}\sum_{i=1}^{2n}\frac{1}{2n+i} = \lim_{n\to\infty}\sum_{i=1}^{2n}\frac{\frac{1}{2n}}{\frac{2n+i}{2n}} = \lim_{n\to\infty}\sum_{i=1}^{2n}\frac{\frac{1}{2n}}{1+\frac{i}{2n}} = \lim_{n\to\infty}\sum_{i=1}^{2n}\frac{1}{1+\frac{i}{2n}}\cdot\frac{1}{2n} = \int_0^1\left(\frac{1}{1+x}\right)dx$

$\qquad\qquad = [\log|1+x|]_0^1 = \log 2 - \log 1 = \log 2$

(6) $\displaystyle\lim_{n\to\infty}\frac{1}{n^{k+1}}\sum_{i=1}^{n}i^k = \lim_{n\to\infty}\sum_{i=1}^{n}\frac{1}{n^k}i^k\frac{1}{n} = \lim_{n\to\infty}\sum_{i=1}^{n}\left(\frac{i}{n}\right)^k\frac{1}{n} = \int_0^1 x^k dx = \left[\frac{1}{k+1}x^{k+1}\right]_0^1 = \frac{1}{k+1}$

(7) $\displaystyle\lim_{n\to\infty}\log\left\{\frac{(2n)!}{n^n n!}\right\}^{\frac{1}{n}}$

$\left\{\frac{(2n)!}{n^n n!}\right\}^{\frac{1}{n}} = \left\{\frac{2n(2n-1)\cdots\cdots(n+2)(n+1)n!}{n^n n!}\right\}^{\frac{1}{n}} = \left\{\frac{2n(2n-1)\cdots\cdots(n+2)(n+1)}{n^n}\right\}^{\frac{1}{n}}$

$\therefore\ \log\left\{\frac{(2n)!}{n^n n!}\right\}^{\frac{1}{n}} = \log\left\{\frac{(n+1)(n+2)\cdots\cdots(n+n)}{n\cdot n\cdot n\cdot\ \cdots\cdots\ \cdot n}\right\}^{\frac{1}{n}}$

That is, $\quad\frac{1}{n}\log\left\{\frac{(2n)!}{n^n n!}\right\} = \frac{1}{n}\log\left\{\frac{(n+1)(n+2)\cdots\cdots(n+n)}{n\cdot n\cdot n\cdot\ \cdots\cdots\ \cdot n}\right\}$

$\qquad\qquad\qquad = \frac{1}{n}\left\{\log\left(\frac{n+1}{n}\right) + \log\left(\frac{n+2}{n}\right) + \cdots\cdots + \log\left(\frac{n+n}{n}\right)\right\}$

$\therefore\ \lim_{n\to\infty}\frac{1}{n}\log\left\{\frac{(2n)!}{n^n n!}\right\} = \lim_{n\to\infty}\frac{1}{n}\left\{\log\left(\frac{n+1}{n}\right) + \log\left(\frac{n+2}{n}\right) + \cdots\cdots + \log\left(\frac{n+n}{n}\right)\right\}$

$\therefore\ \lim_{n\to\infty}\frac{1}{n}\sum_{i=1}^{n}\left\{\log\left(\frac{i+1}{i}\right)\right\} = \lim_{n\to\infty}\sum_{i=1}^{n}\left\{\log\left(\frac{i+1}{i}\right)\right\}\frac{1}{n} = \int_0^1\log(1+x)\,dx$

For $\displaystyle\int\log x\,dx$

Let $\qquad u = \log x,\qquad v' = dx$

Then, $\qquad u' = \frac{1}{x}dx,\qquad v = x$

$\therefore\ \displaystyle\int\log x\,dx = x\log x - \int x\frac{1}{x}dx = x\log x - x + C$

$\therefore\ \displaystyle\lim_{n\to\infty}\frac{1}{n}\sum_{i=1}^{n}\left\{\log\left(\frac{i+1}{i}\right)\right\} = \int_0^1\log(1+x)\,dx = [(1+x)\log(1+x) - (1+x)]_0^1$

$\qquad\qquad\qquad = 2\log 2 - 2 + 1 = 2\log 2 - 1$

(8) $\displaystyle\lim_{n\to\infty}\int_0^1\left(\sum_{i=1}^{n}\frac{1}{i}x^i\right)dx$

$\displaystyle\sum_{i=1}^{n}\frac{1}{i}x^i = x + \frac{1}{2}x^2 + \frac{1}{3}x^3 + \cdots\cdots + \frac{1}{n}x^n$

$\therefore\ \displaystyle\lim_{n\to\infty}\int_0^1\left(\sum_{i=1}^{n}\frac{1}{i}x^i\right)dx = \left[\frac{1}{2}x^2 + \frac{1}{2}\cdot\frac{1}{3}x^3 + \frac{1}{3}\cdot\frac{1}{4}x^4 + \cdots\cdots + \frac{1}{n}\cdot\frac{1}{n+1}x^{n+1}\right]_0^1$

$= \frac{1}{1\cdot 2} + \frac{1}{2\cdot 3} + \frac{1}{3\cdot 4} + \cdots\cdots + \frac{1}{n\cdot n+1} = \left(1 - \frac{1}{2}\right) + \left(\frac{1}{2} - \frac{1}{3}\right) + \left(\frac{1}{3} - \frac{1}{4}\right) + \cdots\cdots + \left(\frac{1}{n} - \frac{1}{n+1}\right) = 1 - \frac{1}{n+1}$

$$\therefore \ \lim_{n\to\infty} \int_0^1 \left(\sum_{i=1}^{n} \frac{1}{i} x^i \right) dx = \lim_{n\to\infty} \left(1 - \frac{1}{n+1} \right) = 1$$

Another Approach:

$$\int_0^1 \left(\sum_{i=1}^{n} \frac{1}{i} x^i \right) dx = \sum_{i=1}^{n} \int_0^1 \left(\frac{1}{i} x^i \right) dx = \sum_{i=1}^{n} \frac{1}{i} \int_0^1 x^i dx = \sum_{i=1}^{n} \frac{1}{i} \left[\frac{1}{i+1} x^{i+1} \right]_0^1 = \sum_{i=1}^{n} \frac{1}{i} \left(\frac{1}{i+1} \right)$$

$$= \frac{1}{1\cdot 2} + \frac{1}{2\cdot 3} + \frac{1}{3\cdot 4} + \cdots\cdots + \frac{1}{n\cdot n+1} = 1 - \frac{1}{n+1}$$

$$\therefore \ \lim_{n\to\infty} \int_0^1 \left(\sum_{i=1}^{n} \frac{1}{i} x^i \right) dx = \lim_{n\to\infty} \left(1 - \frac{1}{n+1} \right) = 1$$

(9) $\displaystyle \lim_{n\to\infty} \frac{1}{n} \left\{ \log\left(1 + \frac{1}{n} \right) + \log\left(1 + \frac{2}{n} \right) + \cdots\cdots + \log\left(1 + \frac{n}{n} \right) \right\}$

$$= \lim_{n\to\infty} \sum_{i=1}^{n} \frac{1}{n} \log\left(1 + \frac{i}{n} \right) = \int_0^1 \log(1 + x)\, dx$$

For $\displaystyle \int \log(1 + x)\, dx$,

Let $\qquad u = \log(1 + x), \quad v' = dx$

Then, $\qquad u' = \dfrac{1}{1+x}\, dx, \qquad v = x$

$$\therefore \ \int \log(1 + x)\, dx = x\log(1 + x) - \int \frac{x}{1+x}\, dx = x\log(1 + x) - \int \frac{1 + x - 1}{1+x}\, dx$$

$$= x\log(1 + x) - \int \left(1 - \frac{1}{1+x} \right) dx = x\log(1 + x) - (x - \log|1 + x|) + C$$

$$= (x + 1)\log(1 + x) - x + C$$

$$\therefore \ \lim_{n\to\infty} \frac{1}{n} \left\{ \log\left(1 + \frac{1}{n} \right) + \log\left(1 + \frac{2}{n} \right) + \cdots\cdots + \log\left(1 + \frac{n}{n} \right) \right\} = \int_0^1 \log(1 + x)\, dx$$

$$= \left[(x + 1)\log(1 + x) - x \right]_0^1 = 2\log 2 - 1$$

(10) $\displaystyle \lim_{n\to\infty} \frac{\pi}{n^2} \left(\sin\frac{\pi}{n} + 2\sin\frac{2\pi}{n} + 3\sin\frac{3\pi}{n} + \cdots\cdots + n\sin\frac{n\pi}{n} \right)$

$$= \lim_{n\to\infty} \sum_{i=1}^{n} \frac{\pi}{n^2} i \sin\frac{i\pi}{n} = \frac{1}{\pi} \lim_{n\to\infty} \sum_{i=1}^{n} \frac{\pi i}{n} \sin\frac{\pi i}{n} \cdot \frac{\pi}{n} = \frac{1}{\pi} \int_0^{\pi} x\sin x\, dx$$

For $\displaystyle \int x\sin x\, dx$,

Let $\qquad u = x, \quad v' = \sin x\, dx$

Then, $\qquad u' = dx, \qquad v = -\cos x$

$$\therefore \ \int x\sin x\, dx = -x\cos x + \int \cos x\, dx = -x\cos x + \sin x + C$$

$$\therefore \lim_{n\to\infty} \frac{\pi}{n^2}\left(\sin\frac{\pi}{n} + 2\sin\frac{2\pi}{n} + 3\sin\frac{3\pi}{n} + \cdots\cdots + n\sin\frac{n\pi}{n}\right) = \frac{1}{\pi}\int_0^\pi x\sin x\, dx$$

$$= \frac{1}{\pi}[-x\cos x + \sin x]_0^\pi = \frac{1}{\pi}\{(-\pi\cos\pi + \sin\pi) - (0 + \sin 0)\} = \frac{1}{\pi}\cdot\pi = 1$$

(11) $\displaystyle\lim_{n\to\infty}\frac{1}{n^3}\left\{\sqrt{n^2-1^2} + 2\sqrt{n^2-2^2} + \cdots\cdots + (n-1)\sqrt{n^2-(n-1)^2}\right\}$

$$= \lim_{n\to\infty}\sum_{k=1}^{n-1}\frac{1}{n^3}k\sqrt{n^2-k^2} = \lim_{n\to\infty}\sum_{k=1}^{n-1}\frac{k}{n}\sqrt{\frac{n^2-k^2}{n^2}}\cdot\frac{1}{n} = \lim_{n\to\infty}\sum_{k=1}^{n}\frac{k}{n}\sqrt{1-\left(\frac{k}{n}\right)^2}\cdot\frac{1}{n} = \int_0^1 x\sqrt{1-x^2}\,dx$$

For $\displaystyle\int_0^1 x\sqrt{1-x^2}\,dx$,

Let $1 - x^2 = t$ Then, $-2x\,dx = dt$; $x\,dx = -\frac{1}{2}dt$

$t = 1$ when $x = 0$

$t = 0$ when $x = 1$

$$\therefore \int_0^1 x\sqrt{1-x^2}\,dx = \int_1^0\sqrt{t}\left(-\frac{1}{2}dt\right) = -\frac{1}{2}\int_1^0 t^{\frac{1}{2}}dt = \frac{1}{2}\int_0^1 t^{\frac{1}{2}}dt = \frac{1}{2}\left[\frac{2}{3}t^{\frac{3}{2}}\right]_0^1$$

$$= \frac{1}{3}(1-0) = \frac{1}{3}$$

(12) $\displaystyle\lim_{n\to\infty}\left(\frac{n}{1^2+3n^2} + \frac{n}{2^2+3n^2} + \frac{n}{3^2+3n^2} + \cdots\cdots + \frac{n}{n^2+3n^2}\right)$

$$= \lim_{n\to\infty}\sum_{k=1}^{n}\frac{n}{k^2+3n^2} = \lim_{n\to\infty}\sum_{k=1}^{n}\frac{\frac{n}{n^2}}{\frac{k^2+3n^2}{n^2}} = \lim_{n\to\infty}\sum_{k=1}^{n}\frac{1}{\left(\frac{k}{n}\right)^2+3}\cdot\frac{1}{n} = \int_0^1\frac{1}{x^2+3}\,dx$$

Let $x = \sqrt{3}\tan\theta$

Then, $dx = \sqrt{3}\sec^2\theta\,d\theta$

$\theta = 0$ when $x = 0$

$\theta = \frac{\pi}{6}$ when $x = 1$

$$\int_0^1\frac{1}{x^2+3}\,dx = \int_0^{\frac{\pi}{6}}\frac{\sqrt{3}\sec^2\theta}{3\tan^2\theta+3}\,d\theta = \int_0^{\frac{\pi}{6}}\frac{1}{\sqrt{3}}\cdot\frac{\sec^2\theta}{\tan^2\theta+1}\,d\theta = \int_0^{\frac{\pi}{6}}\frac{1}{\sqrt{3}}\,d\theta = \frac{1}{\sqrt{3}}[\theta]_0^{\frac{\pi}{6}}$$

$$= \frac{1}{\sqrt{3}}\cdot\frac{\pi}{6} = \frac{\sqrt{3}\pi}{18}$$

(13) $\displaystyle\lim_{n\to\infty} \frac{1}{n}\left\{\left(\frac{n}{n}\right)^3 + \left(\frac{n+1}{n}\right)^3 + \left(\frac{n+2}{n}\right)^3 + \cdots\cdots + \left(\frac{2n-1}{n}\right)^3\right\}$

$= \displaystyle\lim_{n\to\infty} \frac{1}{n}\left\{1^3 + \left(1+\frac{1}{n}\right)^3 + \left(1+\frac{2}{n}\right)^3 + \cdots\cdots + \left(1+\frac{n-1}{n}\right)^3\right\}$

$= \displaystyle\lim_{n\to\infty} \frac{1}{n}\sum_{i=0}^{n-1}\left(1+\frac{i}{n}\right)^3 = \int_0^1 (1+x)^3 dx = \int_0^1 (x^3 + 3x^2 + 3x + 1)dx$

$= \left[\frac{1}{4}x^4 + x^3 + \frac{3}{2}x^2 + x\right]_0^1 = \frac{1}{4} + 1 + \frac{3}{2} + 1 = \frac{15}{4}$

#3 Suppose $\int_a^b \sin x\, dx = A$, $\int_b^c \sin x\, dx = B$, **and** $\int_a^c \sin\left(x+\frac{\pi}{4}\right)dx = C$.

Use properties of definite integrals to calculate each of the following integrals:

(1) $\int_a^a \sin x\, dx = 0$ $\left(\because \int_a^a f(x)dx = 0\right)$

(2) $\int_b^a \sin x\, dx = -\int_a^b \sin x\, dx = -A$

(3) $\int_a^c \sin x\, dx = \int_a^b \sin x\, dx + \int_b^c \sin x\, dx = A + B$

(4) $\int_a^c \left\{\sin x + 2\sin\left(x+\frac{\pi}{4}\right)\right\}dx = \int_a^c \sin x\, dx + 2\int_a^c \sin\left(x+\frac{\pi}{4}\right)dx = A + B + 2C$

(5) $\int_a^c \cos x\, dx$

$C = \int_a^c \sin\left(x+\frac{\pi}{4}\right)dx = \int_a^c \frac{1}{\sqrt{2}}(\sin x + \cos x)dx = \frac{1}{\sqrt{2}}\int_a^c (\sin x + \cos x)dx$

$= \frac{1}{\sqrt{2}}\left(\int_a^c \sin x\, dx + \int_a^c \cos x\, dx\right)$

$\therefore \int_a^c \sin x\, dx + \int_a^c \cos x\, dx = \sqrt{2}C$

$\therefore \int_a^c \cos x\, dx = \sqrt{2}C - \int_a^c \sin x\, dx = \sqrt{2}C - (A+B)$

#4 Find the value.

(1) When $f(a) = \int_a^{a+1}(x^2 + a^2)dx$, **find the value of** $\int_0^1 f(x)dx$.

$f(a) = \int_a^{a+1}(x^2 + a^2)dx = \left[\frac{1}{3}x^3 + a^2 x\right]_a^{a+1} = \left\{\frac{1}{3}(a+1)^3 + a^2(a+1)\right\} - \left(\frac{1}{3}a^3 + a^2 \cdot a\right)$

$= \frac{1}{3}(a^3 + 3a^2 + 3a + 1) + a^3 + a^2 - \frac{1}{3}a^3 - a^3 = 2a^2 + a + \frac{1}{3}$

$\therefore \int_0^1 f(x)dx = \int_0^1 \left(2x^2 + x + \frac{1}{3}\right)dx = \left[\frac{2}{3}x^3 + \frac{1}{2}x^2 + \frac{1}{3}x\right]_0^1 = \frac{2}{3} + \frac{1}{2} + \frac{1}{3} = \frac{3}{2}$

(2) For a sequence $\{a_n\}$, $\displaystyle\sum_{i=1}^{n} a_i = \int_0^n (2x+3)dx$. Find the value of a_{10}.

$$S_n = \sum_{i=1}^{n} a_i = \int_0^n (2x+3)dx = [x^2 + 3x]_0^n = n^2 + 3n$$

$$a_{10} = S_{10} - S_9 = (10^2 + 30) - (9^2 + 27) = 130 - 108 = 22$$

(3) For a function $f(x) = ax^3 + bx + c$ such that i) $\displaystyle\lim_{x \to 1} \frac{f(x)}{x-1} = 2$ and ii) $\int_0^1 f(x)dx = 2$,

Find the value of $a - b - c$. (a, b, c; constants)

By i), $f(1) = 0$

$\therefore$ $a + b + c = 0$; $c = -(a+b)$

$$\lim_{x \to 1} \frac{f(x)}{x-1} = \lim_{x \to 1} \frac{ax^3 + bx - (a+b)}{x-1} = \lim_{x \to 1} \frac{a(x^3-1) + b(x-1)}{x-1} = \lim_{x \to 1} \frac{a(x-1)(x^2+x+1) + b(x-1)}{x-1}$$

$$= \lim_{x \to 1}\{a(x^2 + x + 1) + b\} = 2$$

$\therefore$ $3a + b = 2$ $\cdots\cdots$ ①

By ii), $\int_0^1 f(x)dx = \int_0^1 (ax^3 + bx + c)dx = \left[\frac{a}{4}x^4 + \frac{b}{2}x^2 + cx\right]_0^1 = \frac{a}{4} + \frac{b}{2} + c = 2$

$a + 2b + 4c = 8$; $a + 2b - 4(a+b) = 8$; $-3a - 2b = 8$ $\cdots\cdots$ ②

By ① and ②, $-b = 10$; $b = -10$, $a = 4$, $c = 6$

$\therefore$ $a - b - c = 4 + 10 - 6 = 8$

(4) When $f(a) = \int_0^a |x - 1|dx$, find the value of $f'(2)$.

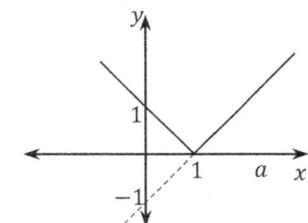

When $a \geq 1$, $f(a) = \int_0^a |x - 1|dx = \int_0^1 |x - 1|dx + \int_1^a |x - 1|dx$

$$= \int_0^1 (-x + 1)dx + \int_1^a (x - 1)dx = \left[-\frac{1}{2}x^2 + x\right]_0^1 + \left[\frac{1}{2}x^2 - x\right]_1^a$$

$$= \left(-\frac{1}{2} + 1\right) + \left(\frac{1}{2}a^2 - a\right) - \left(\frac{1}{2} - 1\right) = \frac{1}{2}a^2 - a + 1$$

$\therefore$ $f'(a) = a - 1$ $\qquad$ $\therefore$ $f'(2) = 2 - 1 = 1$

(5) When a function $f(x) = |x + 1| + |x| + |x - 1|$ has the minimum value m,

find the value of $\int_{-2}^{m} f(x)dx$.

From the graph, $m = 2$

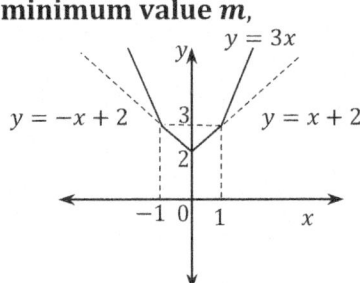

$\therefore$ $\int_{-2}^{m} f(x)dx = \int_{-2}^{2} f(x)dx = 2\int_0^2 f(x)dx$

$$= 2\left[\int_0^1 f(x)dx + \int_1^2 f(x)dx\right]$$

$$= 2\left[\int_0^1 (x + 2)dx + \int_1^2 3xdx\right] = 2\left\{\left[\frac{1}{2}x^2 + 2x\right]_0^1 + \left[\frac{3}{2}x^2\right]_1^2\right\}$$

$$= 2\left\{\left(\frac{1}{2} + 2\right) + \left(6 - \frac{3}{2}\right)\right\} = 14$$

(6) When $a_n = (\log 2)\int_0^n 2^x dx$ $(n = 0, 1, 2, \cdots\cdots)$, **find the value of** $\displaystyle\sum_{n=0}^{\infty} \frac{1}{1+a_n}$.

$a_n = (\log 2)\int_0^n 2^x dx = (\log 2)\left[\frac{2^x}{\log 2}\right]_0^n = \log 2\left(\frac{2^n}{\log 2} - \frac{2^0}{\log 2}\right) = \log 2\left(\frac{2^n}{\log 2} - \frac{1}{\log 2}\right) = 2^n - 1$

$\displaystyle\sum_{n=0}^{\infty} \frac{1}{1+a_n} = \sum_{n=0}^{\infty} \frac{1}{2^n} = \frac{1}{1 - \frac{1}{2}} = 2$

(7) When $f(x) = e^{-3x}$ **and** $g(x) = \frac{1}{1+x}$, **find the value of** $\int_0^{\log 2} g(f(x))dx$.

$g(f(x)) = \frac{1}{1+f(x)} = \frac{1}{1+e^{-3x}} = \frac{e^{3x}}{e^{3x}+1}$

$\therefore \int_0^{\log 2} g(f(x))dx = \int_0^{\log 2} \frac{e^{3x}}{e^{3x}+1}dx = \int_0^{\log 2} \frac{3e^{3x}}{e^{3x}+1}\cdot\frac{1}{3}dx = \frac{1}{3}\int_0^{\log 2} \frac{(e^{3x}+1)'}{e^{3x}+1}dx$

$= \frac{1}{3}\left[\log(e^{3x}+1)\right]_0^{\log 2} = \frac{1}{3}\{\log(e^{3\log 2}+1) - \log(e^0 + 1)\}$

$= \frac{1}{3}\{\log 9 - \log 2\} = \frac{1}{3}\log\frac{9}{2} = \log\sqrt[3]{\frac{9}{2}}$

(8) For a function $f(x)$, $F(x) = \int f(x)dx$.

When $F(x) = xf(x) - 4x^2(x + 1)$ **and** $f(1) = 10$, **find the value of** $\int_0^1 f(x)dx$.

Since $F(x) = \int f(x)dx$, $F'(x) = f(x)$

Since $F(x) = xf(x) - 4x^2(x + 1)$, $F'(x) = f(x) + xf'(x) - 8x(x + 1) - 4x^2$

$\therefore f(x) = f(x) + xf'(x) - 8x(x+1) - 4x^2$; $xf'(x) = 8x(x+1) + 4x^2 = 12x^2 + 8x$

$\therefore f'(x) = 12x + 8$

$\therefore f(x) = \int f'(x)dx = \int(12x + 8)dx = 6x^2 + 8x + C$

Since $f(1) = 10$, $6 + 8 + C = 10$; $C = -4$ $\qquad \therefore f(x) = 6x^2 + 8x - 4$

Therefore, $\int_0^1 (6x^2 + 8x - 4)dx = [2x^3 + 4x^2 - 4x]_0^1 = 2 + 4 - 4 = 2$

#5 Show that $\int_m^n a(x - m)(x - n)dx = -\frac{a}{6}(n - m)^3$.

$(x - m)(x - n) = x^2 - (m + n)x + mn$

$\therefore \int_m^n a(x - m)(x - n)dx = a\int_m^n \{x^2 - (m + n)x + mn\}dx = a\left[\frac{1}{3}x^3 - \frac{m+n}{2}x^2 + mnx\right]_m^n$

$= a\left\{\frac{1}{3}(n^3 - m^3) - \frac{m+n}{2}(n^2 - m^2) + mn(n - m)\right\}$

$= \frac{a}{6}(n - m)\{2(n^2 + mn + m^2) - 3(m + n)^2 + 6mn\}$

$= \frac{a}{6}(n - m)(-m^2 - n^2 + 2mn)$

$= -\frac{a}{6}(n - m)(m^2 + n^2 - 2mn) = -\frac{a}{6}(n - m)(n - m)^2$

$= -\frac{a}{6}(n - m)^3$

#6 Given $f(x) = |x - 1|$, evaluate the following integrals:

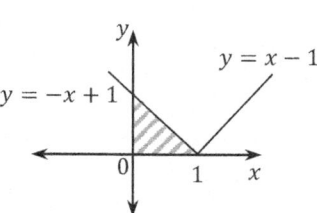

(1) $\int_0^1 f(x)dx = \int_0^1 (-x + 1)dx = \left[-\frac{1}{2}x^2 + x\right]_0^1 = -\frac{1}{2} + 1 = \frac{1}{2}$

(2) $\int_0^3 f(x)dx = \int_0^1 f(x)dx + \int_1^3 f(x)dx$

$= \int_0^1 (-x + 1)dx + \int_1^3 (x - 1)dx$

$= \left[-\frac{1}{2}x^2 + x\right]_0^1 + \left[\frac{1}{2}x^2 - x\right]_1^3$

$= \frac{1}{2} + \left(\frac{9}{2} - 3\right) - \left(\frac{1}{2} - 1\right) = \frac{5}{2}$

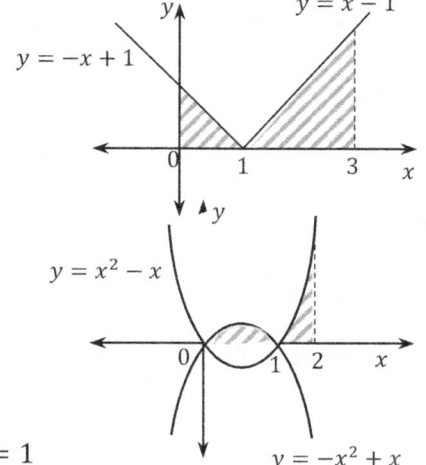

(3) $\int_0^2 |x|f(x)dx = \int_0^1 x(-x + 1)dx + \int_1^2 x(x - 1)dx$

$= \int_0^1 (-x^2 + x)dx + \int_1^2 (x^2 - x)dx$

$= \left[-\frac{1}{3}x^3 + \frac{1}{2}x^2\right]_0^1 + \left[\frac{1}{3}x^3 - \frac{1}{2}x^2\right]_1^2$

$= -\frac{1}{3} + \frac{1}{2} + \left(\frac{1}{3} \cdot 2^3 - \frac{1}{2} \cdot 2^2\right) - \left(\frac{1}{3} - \frac{1}{2}\right) = 1$

#7 Suppose $f(x) = \begin{cases} x^2 + 2, & x \leq 1 \\ -x + 4, & x \geq 1 \end{cases}$.

Use the interval additive property to evaluate the integral.

(1) $\int_{-1}^1 f(x)dx$

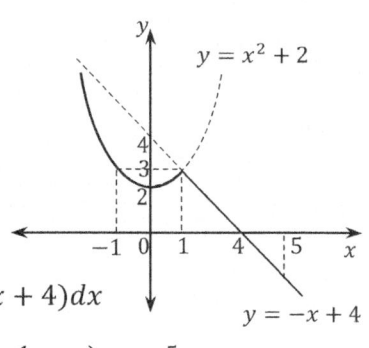

$= \int_{-1}^1 (x^2 + 2)dx = 2\int_0^1 (x^2 + 2)dx = 2\left[\frac{1}{3}x^3 + 2x\right]_0^1$

$= 2\left(\frac{1}{3} + 2\right) = \frac{14}{3}$

(2) $\int_0^2 f(x)dx = \int_0^1 f(x)dx + \int_1^2 f(x)dx = \int_0^1 (x^2 + 2)dx + \int_1^2 (-x + 4)dx$

$= \left[\frac{1}{3}x^3 + 2x\right]_0^1 + \left[-\frac{1}{2}x^2 + 4x\right]_1^2 = \left(\frac{1}{3} + 2\right) + \left(-\frac{1}{2} \cdot 4 + 8\right) - \left(-\frac{1}{2} + 4\right) = 4\frac{5}{6}$

(3) $\int_0^5 f(x)dx = \int_0^1 f(x)dx + \int_1^5 f(x)dx = \int_0^1 (x^2 + 2)dx + \int_1^5 (-x + 4)dx$

$= \left[\frac{1}{3}x^3 + 2x\right]_0^1 + \left[-\frac{1}{2}x^2 + 4x\right]_1^5 = \left(\frac{1}{3} + 2\right) + \left(-\frac{1}{2} \cdot 5^2 + 20\right) - \left(-\frac{1}{2} + 4\right) = 6\frac{1}{3}$

#8 Find the limit.

(1) $\lim\limits_{x \to 1} \frac{1}{x-1}\int_1^x |t - 2|dt = |1 - 2| = |-1| = 1$ $\quad \left(\because \lim\limits_{x \to a} \frac{1}{x-a}\int_a^x f(t)dt = f(a)\right)$

(2) $\lim\limits_{x \to 3} \frac{1}{x-3}\int_3^x t^3 e^t dt = 3^3 e^3 = 27e^3$

(3) $\lim\limits_{x \to 1} \frac{1}{x-1} \int_1^x \sqrt{3 + 2t^3}\, dt = \sqrt{3 + 2 \cdot 1^3} = \sqrt{5}$

(4) $\lim\limits_{x \to 2} \frac{1}{x-2} \int_2^x (t-1)(2t+3)\, dt$

Let $\int (t-1)(2t+3)\, dt = F(t)$ ······ ①

Then, $\int_2^x (t-1)(2t+3)\, dt = [F(t)]_2^x = F(x) - F(2)$

∴ $\lim\limits_{x \to 2} \frac{1}{x-2} \int_2^x (t-1)(2t+3)\, dt = \lim\limits_{x \to 2} \frac{F(x)-F(2)}{x-2} = F'(2)$

By ①, $F'(t) = (t-1)(2t+3)$ ∴ $F'(2) = 1 \cdot (4+3) = 7$

∴ $\lim\limits_{x \to 2} \frac{1}{x-2} \int_2^x (t-1)(2t+3)\, dt = 7$

(5) $\lim\limits_{x \to 2} \frac{1}{x^2-4} \int_2^x (3t^2 + 3t - 2)\, dt$

Let $\int (3t^2 + 3t - 2)\, dt = F(t)$ ······ ①

Then, $\int_2^x (3t^2 + 3t - 2)\, dt = [F(t)]_2^x = F(x) - F(2)$

∴ $\lim\limits_{x \to 2} \frac{1}{x^2-4} \int_2^x (3t^2 + 3t - 2)\, dt = \lim\limits_{x \to 2} \frac{F(x)-F(2)}{x^2-4} = \lim\limits_{x \to 2} \frac{F(x)-F(2)}{x-2} \cdot \frac{1}{x+2} = \frac{1}{4} F'(2)$

By ①, $F'(t) = 3t^2 + 3t - 2$ ∴ $F'(2) = 12 + 6 - 2 = 16$ ∴ $\frac{1}{4} F'(2) = 4$

∴ $\lim\limits_{x \to 2} \frac{1}{x^2-4} \int_2^x (3t^2 + 3t - 2)\, dt = 4$

(6) $\lim\limits_{x \to 0} \frac{1}{x} \int_0^x \frac{\cos t}{1+\sin t}\, dt$

Let $\int \frac{\cos t}{1+\sin t}\, dt = F(t)$ ······ ①

Then, $\int_0^x \frac{\cos t}{1+\sin t}\, dt = [F(t)]_0^x = F(x) - F(0)$

∴ $\lim\limits_{x \to 0} \frac{1}{x} \int_0^x \frac{\cos t}{1+\sin t}\, dt = \lim\limits_{x \to 0} \frac{F(x)-F(0)}{x} = F'(0)$

By ①, $F'(t) = \frac{\cos t}{1+\sin t}$ ∴ $F'(0) = 1$

∴ $\lim\limits_{x \to 0} \frac{1}{x} \int_0^x \frac{\cos t}{1+\sin t}\, dt = 0$

(7) $\lim\limits_{h \to 0} \frac{1}{h} \int_1^{1+3h} (x \log x + x^2 e^x)\, dx$

Let $\int (x \log x + x^2 e^x)\, dx = F(x)$ ······ ①

Then, $\int_1^{1+3h} (x \log x + x^2 e^x)\, dx = [F(x)]_1^{1+3h} = F(1 + 3h) - F(1)$

∴ $\lim\limits_{h \to 0} \frac{1}{h} \int_1^{1+3h} (x \log x + x^2 e^x)\, dx = \lim\limits_{h \to 0} \frac{F(1+3h)-F(1)}{h} = \lim\limits_{h \to 0} \frac{F(1+3h)-F(1)}{3h} \cdot 3 = 3F'(1)$

By ①, $F'(x) = x \log x + x^2 e^x$ ∴ $F'(1) = e$

∴ $\lim\limits_{h \to 0} \frac{1}{h} \int_1^{1+3h} (x \log x + x^2 e^x)\, dx = 3e$

(8) $\lim\limits_{h \to 0} \dfrac{1}{h} \int_{2-h}^{2+h} \dfrac{2x-1}{x^2+1} dx$

Let $\int \dfrac{2x-1}{x^2+1} dx = F(x)$ ······ ①

Then, $\int_{2-h}^{2+h} \dfrac{2x-1}{x^2+1} dx = [F(x)]_{2-h}^{2+h} = F(2+h) - F(2-h)$

$\therefore \ \lim\limits_{h \to 0} \dfrac{1}{h} \int_{2-h}^{2+h} \dfrac{2x-1}{x^2+1} dx = \lim\limits_{h \to 0} \dfrac{F(2+h)-F(2-h)}{h} = \lim\limits_{h \to 0} \dfrac{F(2+h)-F(2)+F(2)-F(2-h)}{h}$

$\qquad\qquad\qquad\qquad\qquad = \lim\limits_{h \to 0} \dfrac{F(2+h)-F(2)}{h} + \lim\limits_{h \to 0} \dfrac{F(2-h)-F(2)}{-h} = F'(2) + F'(2) = 2F'(2)$

By ①, $F'(x) = \dfrac{2x-1}{x^2+1}$ $\qquad \therefore F'(2) = \dfrac{3}{5}$

$\therefore \ \lim\limits_{h \to 0} \dfrac{1}{h} \int_{2-h}^{2+h} \dfrac{2x-1}{x^2+1} dx = \dfrac{6}{5}$

(9) $\lim\limits_{n \to \infty} \int_0^n (x^k e^{-x}) dx$ $\quad \left(\lim\limits_{x \to \infty} \dfrac{x^n}{e^x} = 0 \right)$

For $\int_0^n (x^k e^{-x}) dx$,

Let $\qquad u = x^k, \qquad\qquad v' = e^{-x} dx$

Then, $\qquad u' = kx^{k-1} dx, \qquad v = -e^{-x}$

$\therefore \ \int (x^k e^{-x}) dx = -x^k e^{-x} + \int (kx^{k-1} e^{-x}) dx$

$\therefore \ \int_0^n (x^k e^{-x}) dx = [-x^k e^{-x}]_0^n + \int_0^n (kx^{k-1} e^{-x}) dx = -n^k e^{-n} + k \int_0^n (x^{k-1} e^{-x}) dx$

$\therefore \ \lim\limits_{n \to \infty} \int_0^n (x^k e^{-x}) dx = \lim\limits_{n \to \infty} \left(-\dfrac{n^k}{e^n} \right) + k \lim\limits_{n \to \infty} \int_0^n (x^{k-1} e^{-x}) dx = k \lim\limits_{n \to \infty} \int_0^n (x^{k-1} e^{-x}) dx$

For $\int_0^n (x^{k-1} e^{-x}) dx$,

Let $\qquad u = x^{k-1}, \qquad\qquad v' = e^{-x} dx$

Then, $\qquad u' = (k-1)x^{k-2} dx, \qquad v = -e^{-x}$

$\therefore \ \int (x^{k-1} e^{-x}) dx = -x^{k-1} e^{-x} + \int \{(k-1)x^{k-2} e^{-x}\} dx$

$\therefore \ \int_0^n (x^{k-1} e^{-x}) dx = [-x^{k-1} e^{-x}]_0^n + \int_0^n \{(k-1)x^{k-2} e^{-x}\} dx$

$\qquad\qquad\qquad\qquad = -n^{k-1} e^{-n} + (k-1) \int_0^n (x^{k-2} e^{-x}) dx$

$\therefore \ \lim\limits_{n \to \infty} \int_0^n (x^{k-1} e^{-x}) dx = \lim\limits_{n \to \infty} \left(-\dfrac{n^{k-1}}{e^n} \right) + (k-1) \lim\limits_{n \to \infty} \int_0^n (x^{k-2} e^{-x}) dx$

$\qquad\qquad\qquad\qquad = (k-1) \lim\limits_{n \to \infty} \int_0^n (x^{k-2} e^{-x}) dx$

$\therefore \ \lim\limits_{n \to \infty} \int_0^n (x^k e^{-x}) dx = k(k-1) \lim\limits_{n \to \infty} \int_0^n (x^{k-2} e^{-x}) dx$

$\qquad\qquad\qquad\qquad \vdots$

$\qquad\qquad\qquad\qquad = k(k-1)(k-2) \cdots\cdots 3 \cdot 2 \cdot 1 \cdot \lim\limits_{n \to \infty} \int_0^n e^{-x} dx$

$\qquad\qquad\qquad\qquad = k! \lim\limits_{n \to \infty} [-e^{-x}]_0^n = k! \lim\limits_{n \to \infty} \{(-e^{-n}) - (-e^0)\} = k! \lim\limits_{n \to \infty} \left(-\dfrac{1}{e^n} + 1 \right) = k!$

#9 Find the indicated integrals.

(1) $\int_0^\pi |\sin 2x| dx$

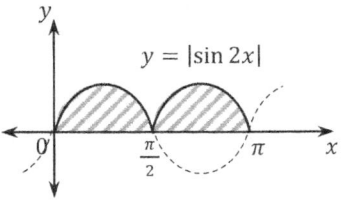

$y = |\sin 2x|$

i) When $0 \le x \le \frac{\pi}{2}$, $\sin 2x \ge 0$ $\therefore |\sin 2x| = \sin 2x$

ii) When $\frac{\pi}{2} \le x \le \pi$, $\sin 2x \le 0$ $\therefore |\sin 2x| = -\sin 2x$

$\therefore \int_0^\pi |\sin 2x| dx = \int_0^{\frac{\pi}{2}} \sin 2x\, dx + \int_{\frac{\pi}{2}}^\pi (-\sin 2x) dx = \int_0^{\frac{\pi}{2}} \sin 2x\, dx - \int_{\frac{\pi}{2}}^\pi \sin 2x\, dx$

$= \left[-\frac{1}{2} \cos 2x \right]_0^{\frac{\pi}{2}} - \left[-\frac{1}{2} \cos 2x \right]_{\frac{\pi}{2}}^\pi = -\frac{1}{2} \left(\cos \left(2 \cdot \frac{\pi}{2} \right) - \cos 0 \right) + \frac{1}{2} \left(\cos 2\pi - \cos \left(2 \cdot \frac{\pi}{2} \right) \right)$

$= -\frac{1}{2}(-1-1) + \frac{1}{2}(1-(-1)) = 1+1 = 2$

(2) $\int_0^1 |e^x - 3| dx$

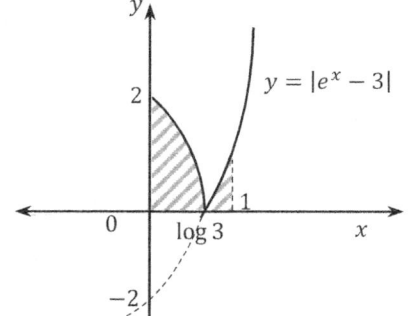

$y = |e^x - 3|$

$e^x - 3 \ge 0 \Rightarrow e^x \ge 3$; $e^x \ge e^{\log 3}$ $\therefore x \ge \log 3$

i) When $0 \le x \le \log 3$,

    $e^x - 3 \le 0$ $\therefore |e^x - 3| = -(e^x - 3)$

ii) When $\log 3 \le x \le 1$,

    $e^x - 3 \ge 0$ $\therefore |e^x - 3| = e^x - 3$

$\therefore \int_0^1 |e^x - 3| dx = \int_0^{\log 3} -(e^x - 3) dx + \int_{\log 3}^1 (e^x - 3) dx$

$= -[e^x - 3x]_0^{\log 3} + [e^x - 3x]_{\log 3}^1$

$= -\left(e^{\log 3} - 3 \log 3 \right) + (e^0 - 0) + (e^1 - 3) - \left(e^{\log 3} - 3 \log 3 \right)$

$= -(3 - 3 \log 3) + 1 + (e - 3) - (3 - 3 \log 3) = 6 \log 3 + e - 8$

(3) $\int_{-3}^3 |(x-1)(x^2 + x - 1)| dx$

$= \int_{-3}^3 |x^3 - 1| dx = \int_{-3}^1 (-x^3 + 1) dx + \int_1^3 (x^3 - 1) dx = \left[-\frac{1}{4}x^4 + x \right]_{-3}^1 + \left[\frac{1}{4}x^4 - x \right]_1^3$

$= \left(-\frac{1}{4} + 1 \right) - \left(-\frac{1}{4} \cdot (-3)^4 - 3 \right) + \left(\frac{1}{4} \cdot 3^4 - 3 \right) - \left(\frac{1}{4} - 1 \right) = \frac{81}{2} + 2 - \frac{1}{2} = 42$

(4) $\int_{-3}^7 ||x - 2| - 3| dx$

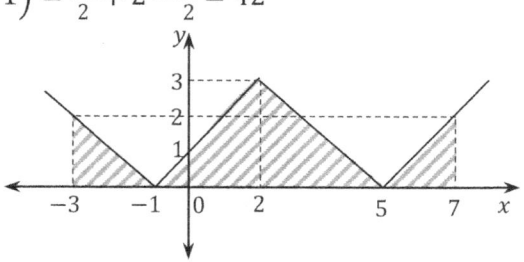

$x \ge 2 \Rightarrow ||x - 2| - 3| = |x - 2 - 3| = |x - 5|$

$x < 2 \Rightarrow ||x - 2| - 3| = |-x + 2 - 3| = |-x - 1|$

$\therefore \frac{1}{2} \cdot 2 \cdot 2 + \frac{1}{2} \cdot 6 \cdot 3 + \frac{1}{2} \cdot 2 \cdot 2 = 13$

(5) $\int_0^2 |x - a|dx \quad (a \geq 0)$

i) $0 \leq a \leq 2 \Rightarrow \int_0^2 |x - a|dx = \int_0^a (-x + a)dx + \int_a^2 (x - a)dx = \left[-\frac{1}{2}x^2 + ax\right]_0^a + \left[\frac{1}{2}x^2 - ax\right]_a^2$

$= \left(-\frac{1}{2}a^2 + a^2\right) + \left(\frac{1}{2} \cdot 2^2 - 2a\right) - \left(\frac{1}{2}a^2 - a^2\right) = a^2 - 2a + 2$

ii) $a \geq 2 \Rightarrow \int_0^2 |x - a|dx = \int_0^2 (-x + a)dx = \left[-\frac{1}{2}x^2 + ax\right]_0^2 = -\frac{1}{2} \cdot 2^2 + 2a = 2a - 2$

(6) $\int_0^2 |x - \sqrt{x}|dx = \int_0^1 (-x + \sqrt{x})dx + \int_1^2 (x - \sqrt{x})dx = \left[-\frac{1}{2}x^2 + \frac{2}{3}x^{\frac{3}{2}}\right]_0^1 + \left[\frac{1}{2}x^2 - \frac{2}{3}x^{\frac{3}{2}}\right]_1^2$

$= -\frac{1}{2} + \frac{2}{3} + \left(\frac{1}{2} \cdot 2^2 - \frac{2}{3} \cdot 2^{\frac{3}{2}}\right) - \left(\frac{1}{2} - \frac{2}{3}\right) = -1 + \frac{4}{3} + 2 - \frac{4}{3}\sqrt{2} = \frac{7}{3} - \frac{4}{3}\sqrt{2} = \frac{1}{3}(7 - 4\sqrt{2})$

(7) $\int_{-1}^4 \left|\frac{x-2}{x+2}\right| dx = \int_{-1}^2 -\left(\frac{x-2}{x+2}\right) dx + \int_2^4 \left(\frac{x-2}{x+2}\right) dx = \int_{-1}^2 -\left(1 - \frac{4}{x+2}\right) dx + \int_2^4 \left(1 - \frac{4}{x+2}\right) dx$

$= -[x - 4\log|x + 2|]_{-1}^2 + [x - 4\log|x + 2|]_2^4$

$= -(2 - 4\log 4) + (-1) + (4 - 4\log 6) - (2 - 4\log 4) = 8\log 4 - 4\log 6 - 1$

$= 4\log 4^2 - 4\log 6 - 1 = 4\log\frac{16}{6} - 1 = 4\log\frac{8}{3} - 1$

(8) $\int_0^\pi |\sin x + \cos x|dx = \int_0^\pi \left|\sqrt{2}\sin\left(x + \frac{\pi}{4}\right)\right| dx$

$= \sqrt{2}\left[\int_0^{\frac{3\pi}{4}} \left|\sin\left(x + \frac{\pi}{4}\right)\right| dx + \int_{\frac{3\pi}{4}}^\pi \left|\sin\left(x + \frac{\pi}{4}\right)\right| dx\right]$

$= \sqrt{2}\left[\int_0^{\frac{3\pi}{4}} \sin\left(x + \frac{\pi}{4}\right) dx - \int_{\frac{3\pi}{4}}^\pi \sin\left(x + \frac{\pi}{4}\right) dx\right] = \sqrt{2}\left\{-\left[\cos\left(x + \frac{\pi}{4}\right)\right]_0^{\frac{3\pi}{4}} + \left[\cos\left(x + \frac{\pi}{4}\right)\right]_{\frac{3\pi}{4}}^\pi\right\}$

$= \sqrt{2}\left\{-\left(\cos\pi - \cos\frac{\pi}{4}\right) + \left(\cos\frac{5\pi}{4} - \cos\pi\right)\right\} = \sqrt{2}\left\{\left(1 + \frac{\sqrt{2}}{2}\right) + \left(-\frac{\sqrt{2}}{2} + 1\right)\right\} = 2\sqrt{2}$

(9) $\int_0^\pi |\sin x \cos x|dx = \int_0^\pi \left|\frac{1}{2}\sin 2x\right| dx = \frac{1}{2}\left[\int_0^{\frac{\pi}{2}} |\sin 2x|dx + \int_{\frac{\pi}{2}}^\pi |\sin 2x|dx\right]$

$= \frac{1}{2}\left[\int_0^{\frac{\pi}{2}} \sin 2x\, dx - \int_{\frac{\pi}{2}}^\pi \sin 2x\, dx\right] = \frac{1}{2}\left\{\left[-\frac{1}{2}\cos 2x\right]_0^{\frac{\pi}{2}} + \left[\frac{1}{2}\cos 2x\right]_{\frac{\pi}{2}}^\pi\right\}$

$= \frac{1}{4}\left\{[-\cos 2x]_0^{\frac{\pi}{2}} + [\cos 2x]_{\frac{\pi}{2}}^\pi\right\} = \frac{1}{4}\{(-\cos\pi) - (-\cos 0) + (\cos 2\pi) - (\cos\pi)\}$

$= \frac{1}{4}(1 + 1 + 1 + 1) = 1$

(10) $\int_{-1}^1 |3^x - 2^x|dx = \int_{-1}^0 -(3^x - 2^x)dx + \int_0^1 (3^x - 2^x)dx$

$= -\left[\frac{3^x}{\log 3} - \frac{2^x}{\log 2}\right]_{-1}^0 + \left[\frac{3^x}{\log 3} - \frac{2^x}{\log 2}\right]_0^1$

$= -\left(\frac{3^0}{\log 3} - \frac{2^0}{\log 2}\right) + \left(\frac{3^{-1}}{\log 3} - \frac{2^{-1}}{\log 2}\right) + \left(\frac{3^1}{\log 3} - \frac{2^1}{\log 2}\right) - \left(\frac{3^0}{\log 3} - \frac{2^0}{\log 2}\right)$

$= -2\left(\frac{1}{\log 3} - \frac{1}{\log 2}\right) + \left(\frac{\frac{1}{3}}{\log 3} - \frac{\frac{1}{2}}{\log 2}\right) + \left(\frac{3^1}{\log 3} - \frac{2^1}{\log 2}\right) = \frac{-2 + \frac{1}{3} + 3}{\log 3} + \frac{2 - \frac{1}{2} - 2}{\log 2} = \frac{\frac{4}{3}}{\log 3} + \frac{-\frac{1}{2}}{\log 2}$

$$= \frac{4}{3\log 3} - \frac{1}{2\log 2}$$

#10 Find the derivative of a definite integral.

(1) $\frac{d}{dx}\left[\int_3^x (t^4 + 3t^2 - 5t + 2)dt\right]$

Note that $\frac{d}{dx}\left[\int_a^x f(t)dt\right] = f(x)$

$\therefore \ \frac{d}{dx}\left[\int_3^x (t^4 + 3t^2 - 5t + 2)dt\right] = x^4 + 3x^2 - 5x + 2$

(2) $\frac{d}{dx}\left[\int_x^3 (\cot^2 t \sin t)dt\right], \ \frac{\pi}{2} < x < \frac{3\pi}{2}$

$\int_x^3 (\cot^2 t \sin t)dt = -\int_3^x (\cot^2 t \sin t)dt$

$\therefore \ \frac{d}{dx}\left[\int_x^3 (\cot^2 t \sin t)dt\right] = \frac{d}{dx}\left[-\int_3^x (\cot^2 t \sin t)dt\right] = -\cot^2 x \sin x$

(3) $\frac{d}{dx}\left[\int_0^{x^2} (2t - 1)dt\right]$

$\int_0^{x^2} (2t - 1)dt = \int_0^u (2t - 1)dt, \ u = x^2 \ \left(\frac{du}{dx} = 2x\right)$

$\therefore \ \frac{d}{dx}\left[\int_0^{x^2} (2t - 1)dt\right] = \frac{du}{dx}\left[\frac{d}{du}\left[\int_0^u (2t - 1)dt\right]\right]$ (By the chain rule)

$\qquad\qquad = \frac{du}{dx}(2u - 1) = 2x \cdot (2u - 1) = 2x(2x^2 - 1) = 4x^3 - 2x$

Another Approach:

$\int_0^{x^2} (2t - 1)dt = [t^2 - t]_0^{x^2} = (x^2)^2 - x^2 = x^4 - x^2$

$\therefore \ \frac{d}{dx}\left[\int_0^{x^2} (2t - 1)dt\right] = \frac{d}{dx}[x^4 - x^2] = 4x^3 - 2x$

(4) $\frac{d}{dx}\left[\int_{3x}^2 \sqrt{t^2 + 3}\,dt\right]$

$\frac{d}{dx}\left[\int_{3x}^2 \sqrt{t^2 + 3}\,dt\right] = \frac{d}{dx}\left[-\int_2^{3x} \sqrt{t^2 + 3}\,dt\right] = -\frac{d}{dx}\left[\int_2^{3x} \sqrt{t^2 + 3}\,dt\right] = -\left(\sqrt{(3x)^2 + 3}\right)(3x)'$

$\qquad\qquad = -3\sqrt{9x^2 + 3}$

#11 Find the function $f(x)$.

(1) Find $f(x)$ such that $f(x) = 3x + \int_0^3 f(x)dx$.

Note: If a and b are constants, then $\int_a^b f(x)dx$ is a constant.

Let $\int_0^3 f(x)dx = k$ $(k; \text{constant})$.

Then, $f(x) = 3x + k$

$\therefore \ 3x + k = 3x + \int_0^3 (3x + k)dx$

$\therefore \ k = \int_0^3 (3x+k)dx = \left[\frac{3}{2}x^2 + kx\right]_0^3 = \frac{3}{2}\cdot 3^2 + 3k = \frac{27}{2} + 3k \ \ ; \ 2k = -\frac{27}{2} \ \ ; \ k = -\frac{27}{4}$

$\therefore \ f(x) = 3x - \frac{27}{4}$

(2) Find $f(x)$ such that $f(x) = \sin x + 2\int_0^{\frac{\pi}{2}} f(x)\cos x\, dx$.

Let $\int_0^{\frac{\pi}{2}} f(x)\cos x\, dx = k$ (k; constant)

Then, $f(x) = \sin x + 2k$

$\therefore \ \sin x + 2k = \sin x + 2\int_0^{\frac{\pi}{2}} (\sin x + 2k)\cos x\, dx$

$\therefore \ k = \int_0^{\frac{\pi}{2}} (\sin x + 2k)\cos x\, dx$

$\quad = \int_0^{\frac{\pi}{2}} \sin x\cos x\, dx + \int_0^{\frac{\pi}{2}} 2k\cos x\, dx = \frac{1}{2}\int_0^{\frac{\pi}{2}} \sin 2x\, dx + 2k\int_0^{\frac{\pi}{2}} \cos x\, dx$

$\quad = \frac{1}{2}\left[-\frac{1}{2}\cos 2x\right]_0^{\frac{\pi}{2}} + 2k[\sin x]_0^{\frac{\pi}{2}} = -\frac{1}{4}(\cos \pi - \cos 0) + 2k(\sin\frac{\pi}{2} - \sin 0)$

$\quad = -\frac{1}{4}(-1-1) + 2k(1-0) = \frac{1}{2} + 2k$

$\therefore \ k = -\frac{1}{2}$

$\therefore \ f(x) = \sin x + 2\left(-\frac{1}{2}\right) = \sin x - 1$

(3) The slope of a tangent line at a point (x, y) on the graph of $y = f(x)$ is $e^x + a$.

Find $f(x)$ such that $f(0) = 1$ and $\int_0^1 f(x)dx = e + 1$.

$f(x) = \int (e^x + a)dx = e^x + ax + C$

Since $f(0) = 1$, $e^0 + 0 + C = 1$; $C = 0$ $\quad \therefore \ f(x) = e^x + ax$

$\int_0^1 f(x)dx = \int_0^1 (e^x + ax)dx = \left[e^x + \frac{a}{2}x^2\right]_0^1 = \left(e^1 + \frac{a}{2}\right) - e^0 = e + \frac{a}{2} - 1 = e + 1$

$\therefore \ \frac{a}{2} - 1 = 1$; $a = 4$ $\qquad \therefore \ f(x) = e^x + 4x$

(4) Find $f(x)$ such that i) $\int_0^x e^t\{f(t) + f'(t)\}dt = f(x)f'(x) - 3$,

$\qquad\qquad\qquad\qquad$ **ii) $f(x) > 0$, and**

$\qquad\qquad\qquad\qquad$ **iii) $f(0) = 3$.**

Since $e^t = (e^t)'$, $e^t\{f(t) + f'(t)\} = (e^t)'f(t) + e^t f'(t) = \{e^t f(t)\}'$

$\int e^t\{f(t) + f'(t)\}dt = \int \{e^t f(t)\}'dt = e^t f(t) + C$

$\therefore \ \int_0^x e^t\{f(t) + f'(t)\}dt = [e^t f(t)]_0^x = e^x f(x) - e^0 f(0) = e^x f(x) - f(0)$

$\qquad\qquad\qquad\qquad = f(x)f'(x) - 3$ (by i))

Since $f(x) > 0$, $f'(x) = e^x$ and $f(0) = 3$ $\qquad \therefore \ f(x) = e^x + C$

Since $f(0) = 3$, $1 + C = 3$ $\qquad \therefore \ C = 2$ $\qquad \therefore \ f(x) = e^x + 2$

(5) Find $f(x)$ such that $f(x) = x + 1 + \int_0^2 g(t)dt$ where $g(x) = 3x - 2 + \int_0^1 f(t)dt$.

Let $\int_0^2 g(t)dt = a$ and $\int_0^1 f(t)dt = b$

Then, $f(x) = x + 1 + a$ and $g(x) = 3x - 2 + b$

$\therefore\ a = \int_0^2 g(t)dt = \int_0^2 (3t - 2 + b)dt = \left[\frac{3}{2}t^2 - 2t + bt\right]_0^2 = \frac{3}{2} \cdot 2^2 - 4 + 2b = 2 + 2b \ \cdots\cdots ①$

$b = \int_0^1 f(t)dt = \int_0^1 (t + 1 + a)dt = \left[\frac{1}{2}t^2 + t + at\right]_0^1 = \frac{1}{2} + 1 + a = \frac{3}{2} + a \cdots\cdots ②$

By ① and ②, $a = 2 + 2b = 2 + 2\left(\frac{3}{2} + a\right) = 5 + 2a$; $a = -5, \ b = -\frac{7}{2}$

$\therefore\ f(x) = x + 1 - 5 = x - 4$ and $g(x) = 3x - 2 - \frac{7}{2} = 3x - \frac{11}{2}$

(6) Find $f(x)$ such that $f(x) = 1 + \int_{-1}^1 (x^2 - t)f(t)dt$.

$f(x) = 1 + \int_{-1}^1 (x^2 - t)f(t)dt = 1 + \int_{-1}^1 x^2 f(t)dt + \int_{-1}^1 -tf(t)dt$

$= 1 + x^2 \int_{-1}^1 f(t)dt - \int_{-1}^1 tf(t)dt$

Let $\int_{-1}^1 f(t)dt = a$ and $\int_{-1}^1 tf(t)dt = b$

Then, $f(x) = 1 + x^2 a - b$

$\therefore\ a = \int_{-1}^1 f(t)dt = \int_{-1}^1 (1 + t^2 a - b)dt = \left[t + \frac{a}{3}t^3 - bt\right]_{-1}^1$

$= \left(1 + \frac{a}{3} - b\right) - \left(-1 - \frac{a}{3} + b\right) = 2 + \frac{2a}{3} - 2b$

$\therefore\ \frac{a}{3} = 2 - 2b$; $a = 6 - 6b$

$b = \int_{-1}^1 tf(t)dt = \int_{-1}^1 t(1 + t^2 a - b)dt = \int_{-1}^1 (t + t^3 a - bt)dt = \left[\frac{1}{2}t^2 + \frac{a}{4}t^4 - \frac{b}{2}t^2\right]_{-1}^1$

$= \left(\frac{1}{2} + \frac{a}{4} - \frac{b}{2}\right) - \left(\frac{1}{2} + \frac{a}{4} - \frac{b}{2}\right) = 0$

$\therefore\ a = 6, \ b = 0$

$\therefore\ f(x) = 1 + x^2 a - b = 6x^2 + 1$

(7) For a continuous function $f(x)$ such that $f(x) = \int_0^x f(t)dt + 1$, find $f(x)$. ($f(x) \neq 0$)

$f(x) = \int_0^x f(t)dt + 1 \ \Rightarrow\ \frac{d}{dx}[f(x)] = \frac{d}{dx}\left[\int_0^x f(t)dt + 1\right]$

$\therefore\ f'(x) = f(x) \qquad \therefore\ \frac{f'(x)}{f(x)} = 1 \qquad \therefore\ \int \frac{f'(x)}{f(x)}dx = \int 1 dx \qquad \therefore\ \log|f(x)| = x + C_1$

$\therefore\ |f(x)| = e^{x+C_1}$; $f(x) = \pm e^{x+C_1} = Ce^x \ (C = \pm e^{C_1})$

Since $f(0) = \int_0^0 f(t)dt + 1 = 0 + 1 = 1, \ f(0) = Ce^0 = C = 1$

$\therefore\ f(x) = Ce^x = e^x$

(8) $f(x) = \tan x - x - \int_\pi^x f'(t)\tan^2 t\, dt$

$\Rightarrow \frac{d}{dx}[f(x)] = \frac{d}{dx}\left[\tan x - x - \int_\pi^x f'(t)\tan^2 t\, dt\right]$

$\therefore f'(x) = \sec^2 x - 1 - f'(x)\tan^2 x$

$\qquad = (1 + \tan^2 x) - 1 - f'(x)\tan^2 x$

$\qquad = \tan^2 x - f'(x)\tan^2 x$

$\therefore (1 + \tan^2 x)f'(x) = \tan^2 x \; ; \; f'(x) = \frac{\tan^2 x}{1+\tan^2 x} = \frac{\tan^2 x}{\sec^2 x} = \cos^2 x \cdot \frac{\sin^2 x}{\cos^2 x} = \sin^2 x = \frac{1-\cos 2x}{2}$

$\therefore f(x) = \int \frac{1-\cos 2x}{2}\, dx = \frac{1}{2}\left(x - \frac{1}{2}\sin 2x\right) + C \cdots\cdots ①$

Note that $\int_a^a f(x)dx = 0$, $f(\pi) = \tan\pi - \pi - \int_\pi^\pi f'(t)\tan^2 t\, dt = \tan\pi - \pi = 0 - \pi = -\pi$

By ①, $f(\pi) = \frac{1}{2}\left(\pi - \frac{1}{2}\sin 2\pi\right) + C = \frac{1}{2}\pi + C = -\pi \qquad \therefore C = -\frac{3}{2}\pi$

$\therefore f(x) = \frac{1}{2}\left(x - \frac{1}{2}\sin 2x\right) - \frac{3}{2}\pi$

(9) Find $f(x)$ and $g(x)$, and the constants a and b such that

    i) $\int_1^x \{f(t) - g(t)\}dt = 3x^2 - 2x + a$ and

    ii) $\int_1^x \{f(t) + g(t)\}dt = 2x^3 - x^2 + 2x + b$.

$\int_1^x \{f(t) - g(t)\}dt = 3x^2 - 2x + a \Rightarrow \frac{d}{dx}\left[\int_1^x \{f(t) - g(t)\}dt\right] = \frac{d}{dx}[3x^2 - 2x + a]$

$\therefore f(x) - g(x) = 6x - 2 \cdots\cdots ①$

$\int_1^x \{f(t) + g(t)\}dt = 2x^3 - x^2 + 2x + b \Rightarrow \frac{d}{dx}\left[\int_1^x \{f(t) + g(t)\}dt\right] = \frac{d}{dx}[2x^3 - x^2 + 2x + b]$

$\therefore f(x) + g(x) = 6x^2 - 2x + 2 \cdots\cdots ②$

①+②; $2f(x) = 6x^2 + 4x \qquad \therefore f(x) = 3x^2 + 2x$

②−①; $2g(x) = 6x^2 - 8x + 4 \qquad \therefore g(x) = 3x^2 - 4x + 2$

Letting $x = 1$ in i), $\int_1^1 \{f(t) - g(t)\}dt = 3 - 2 + a \; ; \; 0 = 1 + a \qquad \therefore a = -1$

Letting $x = 1$ in ii), $\int_1^1 \{f(t) + g(t)\}dt = 2 - 1 + 2 + b \; ; \; 0 = 3 + b \qquad \therefore b = -3$

(10) Find a function $f(x)$ and a constant a such that $\int_a^{\log x} f(u)du = x^2 - 2x$. $(x > 0)$

Let $\log x = t$ Then, $x = e^t$

$\int_a^{\log x} f(u)du = x^2 - 2x \Rightarrow \int_a^t f(u)du = e^{2t} - 2e^t \cdots\cdots ①$

$\qquad\qquad\qquad \Rightarrow \frac{d}{dt}\left[\int_a^t f(u)du\right] = \frac{d}{dt}[e^{2t} - 2e^t]$

$\therefore f(t) = 2e^{2t} - 2e^t \qquad \therefore f(x) = 2e^{2x} - 2e^x$

Lettinf $t = a$ in ①, $\int_a^a f(u)du = e^{2a} - 2e^a \; ; \; 0 = e^{2a} - 2e^a \; ; \; e^a(e^a - 2) = 0$

$\therefore e^a = 2 \qquad \therefore a = \log 2$

(11) When a differentiable function $f(x)$ satisfies $\int_0^x (x-t)f(t)dt = e^x + ax + b$ for any real number x, find $f(x)$ and the constants a and b.

$\int_0^x (x-t)f(t)dt = e^x + ax + b$ $\cdots\cdots$ ①

$\Rightarrow \int_0^x xf(t)dt - \int_0^x tf(t)dt = e^x + ax + b$; $x\int_0^x f(t)dt - \int_0^x tf(t)dt = e^x + ax + b$

$\Rightarrow \frac{d}{dx}\left[x\int_0^x f(t)dt - \int_0^x tf(t)dt\right] = \frac{d}{dx}[e^x + ax + b]$

$\therefore\ 1\cdot\int_0^x f(t)dt + x\cdot\frac{d}{dx}\left[\int_0^x f(t)dt\right] - \frac{d}{dx}\left[\int_0^x tf(t)dt\right] = e^x + a$

$\therefore\ \int_0^x f(t)dt + xf(x) - xf(x) = e^x + a$ $\qquad \therefore\ \int_0^x f(t)dt = e^x + a$ $\cdots\cdots$ ②

Letting $x = 0$ in ① and ②, $0 = e^0 + 0 + b$ and $0 = e^0 + a$

$\therefore\ b = -1,\ a = -1$

From ②, $\frac{d}{dx}\left[\int_0^x f(t)dt\right] = \frac{d}{dx}[e^x + a]$ $\qquad \therefore\ f(x) = e^x$

(12) When a differentiable function $f(x)$ satisfies $\int_a^x f(t)dt = e^{2x} - 4e^x + 3$ for any real number x, find $f(x)$ and the constant a.

Note that $\int_a^a f(x)dx = 0$

Letting $x = a$, $\int_a^a f(t)dt = e^{2a} - 4e^a + 3$

$\therefore\ 0 = e^{2a} - 4e^a + 3 = (e^a)^2 - 4e^a + 3 = (e^a - 1)(e^a - 3)$

$\therefore\ e^a = 1$ or $e^a = 3$

$\therefore\ a = \log 1 = 0$ or $a = \log 3$

$\int_a^x f(t)dt = e^{2x} - 4e^x + 3$ $\Rightarrow$ $\frac{d}{dx}\left[\int_a^x f(t)dt\right] = \frac{d}{dx}[e^{2x} - 4e^x + 3]$

$\therefore\ f(x) = 2e^{2x} - 4e^x$

(13) When a differentiable function $f(x)$ satisfies $\int_{\frac{1}{2}}^x f(t)dt = \sin \pi x - a\cos 2\pi x + 1$ for any real number x, find $f(x)$ and the constant a.

Letting $x = \frac{1}{2}$, $\int_{\frac{1}{2}}^{\frac{1}{2}} f(t)dt = \sin\frac{\pi}{2} - a\cos\pi + 1$

$\therefore\ 0 = 1 + a + 1$; $a = -2$

$\int_{\frac{1}{2}}^x f(t)dt = \sin \pi x - a\cos 2\pi x + 1$ $\Rightarrow$ $\frac{d}{dx}\left[\int_{\frac{1}{2}}^x f(t)dt\right] = \frac{d}{dx}[\sin \pi x - a\cos 2\pi x + 1]$

$\therefore\ f(x) = \pi\cos \pi x + 2\pi a\sin 2\pi x = \pi\cos \pi x - 4\pi\sin 2\pi x$

#12 Solve the equation of x.

(1) $\sin\left\{\frac{\pi}{3}\log_x\left[\frac{d}{dx}\left[\int_2^x t^3 dt\right]\right]\right\} = x^2 - 3x$

Since $\frac{d}{dx}\left[\int_2^x t^3 dt\right] = x^3$,

$\sin\left\{\frac{\pi}{3}\log_x\left[\frac{d}{dx}\left[\int_2^x t^3 dt\right]\right]\right\} = \sin\left\{\frac{\pi}{3}\log_x x^3\right\} = \sin\left\{\frac{\pi}{3}\cdot 3\right\} = \sin\pi = 0$

$\therefore\ 0 = x^2 - 3x\ ;\ \ x(x-3) = 0 \qquad \therefore\ x = 0$ or $x = 3$

Since $x \neq 0$ for the logarithmic function, $x = 3$

(2) $\log_{x^3}\left\{\frac{d}{dx}\left[\int_x^{x+1}\frac{1}{2}(t^2 - t)dt\right]\right\} = x^2 - 4x + \frac{10}{3}$

$\frac{d}{dx}\left[\int_x^{x+1}\frac{1}{2}(t^2 - t)dt\right] = \frac{1}{2}\left[\{(x+1)^2 - (x+1)\} - (x^2 - x)\right] = \frac{1}{2}(2x) = x$

$\therefore\ \log_{x^3}\left\{\frac{d}{dx}\left[\int_x^{x+1}\frac{1}{2}(t^2 - t)dt\right]\right\} = \log_{x^3} x = \frac{1}{3}\log_x x = \frac{1}{3}$

$\therefore\ \frac{1}{3} = x^2 - 4x + \frac{10}{3} \qquad \therefore\ x^2 - 4x + 3 = 0\ ;\ \ (x-1)(x-3) = 0$

$\therefore\ x = 1$ or $x = 3$

Since $x \neq 1$ for the logarithmic function, $x = 3$

#13 Find the value.

(1) When $f(x)$ and $f'(x)$ are continuous functions such that

$f(x) = x^2 e^x - x + \int_0^x (x - t)f'(t)dt,$ **find the value of $f'(1) - f(1)$.**

$\int_0^x (x-t)f'(t)dt = \int_0^x xf'(t)dt - \int_0^x tf'(t)dt = x\int_0^x f'(t)dt - \int_0^x tf'(t)dt$

$\therefore\ \frac{d}{dx}\left[\int_0^x (x-t)f'(t)dt\right] = \frac{d}{dx}\left[x\int_0^x f'(t)dt - \int_0^x tf'(t)dt\right]$

$\qquad\qquad = 1\cdot\int_0^x f'(t)dt + x\cdot\frac{d}{dx}\left[\int_0^x f'(t)dt\right] - \frac{d}{dx}\left[\int_0^x tf'(t)dt\right]$

$\qquad\qquad = \int_0^x f'(t)dt + xf'(x) - xf'(x) = \int_0^x f'(t)dt = [f(t)]_0^x$

$\qquad\qquad = f(x) - f(0)$

$f(x) = x^2 e^x - x + \int_0^x (x-t)f'(t)dt\ \cdots\cdots\ ①$

$\Rightarrow\ \frac{d}{dx}[f(x)] = \frac{d}{dx}\left[x^2 e^x - x + \int_0^x (x-t)f'(t)dt\right]$

$\qquad\qquad = 2xe^x + x^2 e^x - 1 + (f(x) - f(0))$

$\therefore\ f'(x) = 2xe^x + x^2 e^x - 1 + f(x) - f(0)$

When $x = 0$, $\int_0^x (x-t)f'(t)dt = 0 \qquad \therefore\ f(0) = 0$ by ①

$\therefore\ f'(x) = 2xe^x + x^2 e^x - 1 + f(x)\ ;\ \ f'(x) - f(x) = 2xe^x + x^2 e^x - 1$

$\therefore\ f'(1) - f(1) = 2e + e - 1 = 3e - 1$

(2) When $f(x) = \int_0^x t \sin t \, dt$ $(0 \leq x \leq 2\pi)$,

1) Find the value of x so that $f(x)$ is a local extreme value.

$f(x) = \int_0^x t \sin t \, dt$ $(0 \leq x \leq 2\pi)$ $\Rightarrow$ $f'(x) = x \sin x$ $(0 \leq x \leq 2\pi)$

$f'(x) = 0$ $\Rightarrow$ $x = 0$ or $\sin x = 0$ $\therefore$ $x = 0$ or $x = \pi$ or $x = 2\pi$

$f''(x) = \sin x + x \cos x$

$f''(\pi) = \sin \pi + \pi \cos \pi = -\pi < 0$

$f''(2\pi) = \sin 2\pi + 2\pi \cos 2\pi = 2\pi > 0$

$\therefore$ $f(x)$ has a local maximum value at $x = \pi$ and local minimum value at $x = 2\pi$.

2) Find the limit: $\lim\limits_{h \to 0} \dfrac{f(\pi + 2h) - f(\pi)}{h}$.

$\lim\limits_{h \to 0} \dfrac{f(\pi + 2h) - f(\pi)}{h} = \lim\limits_{h \to 0} \dfrac{f(\pi + 2h) - f(\pi)}{2h} \cdot 2 = 2f'(\pi) = 2(\pi \sin \pi) = 0$ (by 1))

(3) Find the value of a so that $f(a) = \int_0^1 (e^x + ax)^2 dx$ has the minimum value.

$f(a) = \int_0^1 (e^x + ax)^2 dx = \int_0^1 (e^{2x} + 2axe^x + a^2x^2) dx$

For $\int xe^x dx$,

Let $u = x,$ $v' = e^x dx$

Then, $u' = dx,$ $v = e^x$

$\therefore$ $\int xe^x dx = xe^x - \int e^x dx = xe^x - e^x + C$

$\therefore$ $f(a) = \int_0^1 (e^{2x} + 2axe^x + a^2x^2) dx = \left[\dfrac{1}{2}e^{2x} + 2a(xe^x - e^x) + \dfrac{a^2}{3}x^3 \right]_0^1$

$= \left\{ \dfrac{1}{2}e^2 + 2a(e - e) + \dfrac{a^2}{3} \right\} - \left\{ \dfrac{1}{2}e^0 + 2a(0 - e^0) + 0 \right\}$

$= \left\{ \dfrac{1}{2}e^2 + \dfrac{a^2}{3} \right\} - \left\{ \dfrac{1}{2} - 2a \right\} = \dfrac{1}{3}(a^2 + 6a) + \dfrac{1}{2}e^2 - \dfrac{1}{2} = \dfrac{1}{3}(a + 3)^2 - 3 + \dfrac{1}{2}e^2 - \dfrac{1}{2}$

$= \dfrac{1}{3}(a + 3)^2 + \dfrac{1}{2}e^2 - \dfrac{7}{2}$

$\therefore$ $f(a)$ has minimum value $\dfrac{1}{2}e^2 - \dfrac{7}{2}$ at $a = -3$

#14 Use the method of substitution in definite integrals to evaluate each of the following:

(1) $\int_1^e x^2 e^x dx$

For $\int x^2 e^x dx$,

Let $u = x^2,$ $v' = e^x dx$

Then, $u' = 2x dx,$ $v = e^x$

$\therefore$ $\int x^2 e^x dx = x^2 e^x - \int e^x 2x dx = x^2 e^x - 2 \int xe^x dx$

For $\int xe^x dx$,

Let $\qquad u = x, \qquad v' = e^x dx$

Then, $\qquad u' = dx, \qquad v = e^x$

$\therefore \int xe^x dx = xe^x - \int e^x dx = xe^x - e^x + C$

$\therefore \int x^2 e^x dx = x^2 e^x - 2\int xe^x dx = x^2 e^x - 2(xe^x - e^x + C)$

$\therefore \int_1^e x^2 e^x dx = [x^2 e^x - 2(xe^x - e^x)]_1^e = \{e^2 e^e - 2(ee^e - e^e)\} - \{e - 2(e - e)\}$

$\qquad\qquad = \{e^2 e^e - 2e^e(e-1)\} - e = e^e(e^2 - 2e + 2) - e$

(2) $\int_0^\pi x \sin x \, dx$

For $\int x \sin x \, dx$,

Let $\qquad u = x, \qquad v' = \sin x \, dx$

Then, $\qquad u' = dx, \qquad v = -\cos x$

$\therefore \int x \sin x \, dx = -x \cos x + \int \cos x \, dx = -x \cos x + \sin x + C$

$\therefore \int_0^\pi x \sin x \, dx = [-x \cos x + \sin x]_0^\pi = (-\pi \cos \pi + \sin \pi) - (0 + \sin 0) = \pi$

(3) $\int_0^2 x^2 \sqrt{x^3 + 1} \, dx$

Let $x^3 + 1 = t \qquad$ Then, $3x^2 dx = dt$

$t = 1$ when $x = 0$

$t = 9$ when $x = 2$

$\therefore \int_0^2 x^2 \sqrt{x^3 + 1} \, dx = \int_1^9 \sqrt{t} \cdot \frac{1}{3} dt = \frac{1}{3} \int_1^9 t^{\frac{1}{2}} dt = \frac{1}{3} \left[\frac{2}{3} t^{\frac{3}{2}} \right]_1^9 = \frac{1}{3} \left(\frac{2}{3} 9^{\frac{3}{2}} - \frac{2}{3} \right) = \frac{2}{9} (3^3 - 1)$

$\qquad\qquad\qquad = 6 - \frac{2}{9} = \frac{52}{9}$

(4) $\int_1^3 \frac{2x}{1+x^2} dx$

Let $1 + x^2 = t \qquad$ Then, $2x dx = dt$

$t = 2 \quad$ when $x = 1$

$t = 10$ when $x = 3$

$\therefore \int_1^3 \frac{2x}{1+x^2} dx = \int_2^{10} \frac{1}{t} dt = [\log|t|]_2^{10} = \log 10 - \log 2 = \log \left(\frac{10}{2} \right) = \log 5$

(5) $\int_{\frac{\pi^2}{9}}^{\frac{\pi^2}{4}} \frac{\sin \sqrt{x}}{\sqrt{x}} dx$

Let $\sqrt{x} = t \qquad$ Then, $\frac{1}{2\sqrt{x}} dx = dt$

$t = \frac{\pi}{3}$ when $x = \frac{\pi^2}{9}$

$t = \frac{\pi}{2}$ when $x = \frac{\pi^2}{4}$

$$\therefore \int_{\frac{\pi^2}{9}}^{\frac{\pi^2}{4}} \frac{\sin\sqrt{x}}{\sqrt{x}}dx = \int_{\frac{\pi}{3}}^{\frac{\pi}{2}} \sin t \cdot 2dt = 2\int_{\frac{\pi}{3}}^{\frac{\pi}{2}} \sin t\, dt = -2[\cos t]_{\frac{\pi}{3}}^{\frac{\pi}{2}} = -2\left(\cos\frac{\pi}{2} - \cos\frac{\pi}{3}\right)$$

$$= -2\left(0 - \frac{1}{2}\right) = 1$$

(6) $\int_0^{\frac{\pi}{2}} \cos^3 x \sin x\, dx$

Let $\cos x = t$ Then, $-\sin x\, dx = dt$

$t = 1$ when $x = 0$

$t = 0$ when $x = \frac{\pi}{2}$

$$\therefore \int_0^{\frac{\pi}{2}} \cos^3 x \sin x\, dx = \int_1^0 t^3(-1)dt = -\int_0^1 t^3(-1)dt = \int_0^1 t^3 dt = \left[\frac{1}{4}t^4\right]_0^1 = \frac{1}{4}$$

(7) $\int_0^{\frac{\pi}{2}} \frac{\sin^3 x}{1+\cos x}dx$

$$\int_0^{\frac{\pi}{2}} \frac{\sin^3 x}{1+\cos x}dx = \int_0^{\frac{\pi}{2}} \frac{\sin^2 x \cdot \sin x}{1+\cos x}dx = \int_0^{\frac{\pi}{2}} \frac{(1-\cos^2 x)\sin x}{1+\cos x}dx = \int_0^{\frac{\pi}{2}}(1-\cos x)\sin x\, dx$$

Let $1 - \cos x = t$ Then, $\sin x\, dx = dt$

$t = 0$ when $x = 0$

$t = 1$ when $x = \frac{\pi}{2}$

$$\therefore \int_0^{\frac{\pi}{2}} \frac{\sin^3 x}{1+\cos x}dx = \int_0^1 t\, dt = \left[\frac{1}{2}t^2\right]_0^1 = \frac{1}{2}$$

(8) $\int_0^{\sqrt{\pi}} x^3 \cos x^2\, dx$

Let $x^2 = t$ Then, $2x\, dx = dt$

$t = 0$ when $x = 0$

$t = \pi$ when $x = \sqrt{\pi}$

$$\therefore \int_0^{\sqrt{\pi}} x^3 \cos x^2\, dx = \int_0^{\pi} t\cos t \cdot \frac{1}{2}dt = \frac{1}{2}\int_0^{\pi} t\cos t\, dt$$

For $\int t\cos t\, dt$,

Let $u = t$, $v' = \cos t\, dt$

Then, $u' = dt$, $v = \sin t$

$$\therefore \int t\cos t\, dt = t\sin t - \int \sin t\, dt = t\sin t + \cos t + C$$

$$\therefore \int_0^{\sqrt{\pi}} x^3 \cos x^2\, dx = \frac{1}{2}\int_0^{\pi} t\cos t\, dt = \frac{1}{2}[t\sin t + \cos t]_0^{\pi}$$

$$= \frac{1}{2}\{(\pi\sin\pi + \cos\pi) - (0 + \cos 0)\} = \frac{1}{2}\{(0-1) - (0+1)\} = -1$$

(9) $\int_0^\pi x|\cos x|dx = \int_0^{\frac{\pi}{2}} x\cos x\, dx + \int_{\frac{\pi}{2}}^\pi x(-\cos x)dx = \int_0^{\frac{\pi}{2}} x\cos x\, dx - \int_{\frac{\pi}{2}}^\pi x\cos x\, dx$

For $\int x\cos x\, dx$,

Let $\quad u = x, \qquad v' = \cos x\, dx$

Then, $\quad u' = dx, \qquad v = \sin x$

$\therefore \int x\cos x\, dx = x\sin x - \int \sin x\, dx = x\sin x + \cos x + C$

$\therefore \int_0^{\frac{\pi}{2}} x\cos x\, dx = [x\sin x + \cos x]_0^{\frac{\pi}{2}} = \left(\frac{\pi}{2}\sin\frac{\pi}{2} + \cos\frac{\pi}{2}\right) - (0 + \cos 0)$

$\qquad\qquad = \left(\frac{\pi}{2} + 0\right) - (0 + 1) = \frac{\pi}{2} - 1$

$\int_{\frac{\pi}{2}}^\pi x\cos x\, dx = [x\sin x + \cos x]_{\frac{\pi}{2}}^\pi = (\pi\sin\pi + \cos\pi) - \left(\frac{\pi}{2}\sin\frac{\pi}{2} + \cos\frac{\pi}{2}\right)$

$\qquad\qquad = (0 - 1) - \left(\frac{\pi}{2} + 0\right) = -1 - \frac{\pi}{2}$

$\therefore \int_0^\pi x|\cos x|dx = \left(\frac{\pi}{2} - 1\right) - \left(-1 - \frac{\pi}{2}\right) = \pi$

(10) $\int_{\frac{\pi}{6}}^{\frac{\pi}{2}} \cos x\log\left(\frac{1}{\sin^2 x}\right)dx$

Let $\sin x = t \quad$ Then, $\cos x\, dx = dt$

$t = \frac{1}{2}$ when $x = \frac{\pi}{6}$

$t = 1$ when $x = \frac{\pi}{2}$

$\therefore \int_{\frac{\pi}{6}}^{\frac{\pi}{2}} \cos x\log\left(\frac{1}{\sin^2 x}\right)dx = \int_{\frac{1}{2}}^1 \log\left(\frac{1}{t^2}\right)dt = \int_{\frac{1}{2}}^1 \log(t^{-2})dt = -2\int_{\frac{1}{2}}^1 \log t\, dt$

For $\int \log t\, dt$,

Let $\quad u = \log t, \qquad v' = dt$

Then, $\quad u' = \frac{1}{t}dt, \qquad v = t$

$\therefore \int \log t\, dt = t\log t - \int t\cdot\frac{1}{t}dt = t\log t - t + C$

$\therefore \int_{\frac{1}{2}}^1 \log t\, dt = [t\log t - t]_{\frac{1}{2}}^1 = (\log 1 - 1) - \left(\frac{1}{2}\log\left(\frac{1}{2}\right) - \frac{1}{2}\right) = -1 - \frac{1}{2}\log\left(\frac{1}{2}\right) + \frac{1}{2}$

$\qquad\qquad = -\frac{1}{2} - \frac{1}{2}\log\left(\frac{1}{2}\right)$

$\therefore \int_{\frac{\pi}{6}}^{\frac{\pi}{2}} \cos x\log\left(\frac{1}{\sin^2 x}\right)dx = -2\int_{\frac{1}{2}}^1 \log t\, dt = -2\left(-\frac{1}{2} - \frac{1}{2}\log\left(\frac{1}{2}\right)\right) = 1 + \log\left(\frac{1}{2}\right) = 1 - \log 2$

(11) $\int_1^e \frac{(\log x)^3}{x}dx$

Let $\log x = t \quad$ Then, $\frac{1}{x}dx = dt$

$t = 0$ when $x = 1$

$t = 1$ when $x = e$

$\therefore \int_1^e \frac{(\log x)^3}{x} dx = \int_0^1 t^3 dt = [\frac{1}{4} t^4]_0^1 = \frac{1}{4}$

(12) $\int_1^4 \frac{1}{\sqrt{x}} e^{\sqrt{x}} dx$

Let $\sqrt{x} = t$ Then, $\frac{1}{2\sqrt{x}} dx = dt$

$t = 1$ when $x = 1$

$t = 2$ when $x = 4$

$\therefore \int_1^4 \frac{1}{\sqrt{x}} e^{\sqrt{x}} dx = \int_1^2 e^t \cdot 2 dt = 2 \int_1^2 e^t dt = 2[e^t]_1^2 = 2(e^2 - e)$

(13) $\int_1^e \frac{\log x}{x(\log x + 1)^2} dx$

Let $\log x + 1 = t$ Then, $\frac{1}{x} dx = dt$

$t = 1$ when $x = 1$

$t = 2$ when $x = e$

$\therefore \int_1^e \frac{\log x}{x(\log x + 1)^2} dx = \int_1^2 \frac{t-1}{t^2} dt = \int_1^2 \left(\frac{1}{t} - \frac{1}{t^2}\right) dt = \int_1^2 \frac{1}{t} dt - \int_1^2 \frac{1}{t^2} dt = \int_1^2 \frac{1}{t} dt - \int_1^2 t^{-2} dt$

$= [\log|t|]_1^2 - [-t^{-1}]_1^2 = (\log 2 - \log 1) - (-2^{-1} + 1^{-1}) = \log 2 + \frac{1}{2} - 1$

$= \log 2 - \frac{1}{2}$

(14) $\int_0^1 \frac{x \log(1+x^2)}{1+x^2} dx$

Let $\log(1 + x^2) = t$ Then, $\frac{2x}{1+x^2} dx = dt$

$t = 0$ when $x = 0$

$t = \log 2$ when $x = 1$

$\therefore \int_0^1 \frac{x \log(1+x^2)}{1+x^2} dx = \int_0^{\log 2} t \cdot \frac{1}{2} dt = \frac{1}{2} \int_0^{\log 2} t dt = \frac{1}{2} \left[\frac{1}{2} t^2\right]_0^{\log 2} = \frac{1}{4} (\log 2)^2$

(15) $\int_0^1 \frac{1}{e^x+1} dx$

Let $e^x = t$ Then, $e^x dx = dt$; $dx = \frac{1}{e^x} dt = \frac{1}{t} dt$

$t = 1$ when $x = 0$

$t = e$ when $x = 1$

$\therefore \int_0^1 \frac{1}{e^x+1} dx = \int_1^e \frac{1}{t+1} \cdot \frac{1}{t} dt = \int_1^e \left(\frac{1}{t} - \frac{1}{t+1}\right) dt = \int_1^e \frac{1}{t} dt - \int_1^e \frac{1}{t+1} dt = [\log|t| - \log|t + 1|]_1^e$

$= (\log e - \log(e + 1)) - (\log 1 - \log 2) = \log e - \log(e + 1) + \log 2$

$= \log \frac{e}{e+1} + \log 2 = \log \frac{2e}{e+1}$

(16) $\int_{\log 2}^{1} \frac{1}{e^x - e^{-x}} dx$

Let $e^x = t$　Then, $e^x dx = dt$; $dx = \frac{1}{e^x} dt = \frac{1}{t} dt$

$t = 2$ when $x = \log 2$

$t = e$ when $x = 1$

$\therefore \int_{\log 2}^{1} \frac{1}{e^x - e^{-x}} dx = \int_{2}^{e} \frac{1}{t - t^{-1}} \cdot \frac{1}{t} dt = \int_{2}^{e} \frac{1}{t^2 - 1} dt = \int_{2}^{e} \frac{1}{2} \left(\frac{1}{t-1} - \frac{1}{t+1} \right) dt$

$\qquad = \frac{1}{2} [\log|t-1| - \log|t+1|]_{2}^{e} = \frac{1}{2} \{(\log|e-1| - \log|e+1|) - (\log 1 - \log 3)\}$

$\qquad = \frac{1}{2} \{(\log(e-1) - \log(e+1)) + \log 3\} = \frac{1}{2} \left\{ \log\left(\frac{e-1}{e+1}\right) + \log 3 \right\} = \frac{1}{2} \log \frac{3(e-1)}{e+1}$

(17) $\int_{0}^{a} \sqrt{a^2 - x^2} dx$　$(a > 0)$

Let $x = a \sin\theta$　Then, $dx = a \cos\theta \, d\theta$

$\theta = 0$　when $x = 0$

$\theta = \frac{\pi}{2}$　when $x = a$

$\therefore \int_{0}^{a} \sqrt{a^2 - x^2} dx = \int_{0}^{\frac{\pi}{2}} \sqrt{a^2 - a^2 \sin^2\theta} \cdot a \cos\theta \, d\theta$

$\qquad = a^2 \int_{0}^{\frac{\pi}{2}} \sqrt{1 - \sin^2\theta} \cdot \cos\theta \, d\theta = a^2 \int_{0}^{\frac{\pi}{2}} \cos^2\theta \, d\theta = a^2 \int_{0}^{\frac{\pi}{2}} \frac{1 + \cos 2\theta}{2} d\theta$

$\qquad = \frac{a^2}{2} \int_{0}^{\frac{\pi}{2}} (1 + \cos 2\theta) d\theta = \frac{a^2}{2} \left[\theta + \frac{1}{2} \sin 2\theta \right]_{0}^{\frac{\pi}{2}}$

$\qquad = \frac{a^2}{2} \left\{ \left(\frac{\pi}{2} + \frac{1}{2} \sin\pi \right) - \left(0 + \frac{1}{2} \sin 0 \right) \right\} = \frac{a^2}{2} \cdot \frac{\pi}{2} = \frac{a^2 \pi}{4}$

(18) $\int_{0}^{a} \frac{1}{x^2 + a^2} dx$　$(a \neq 0)$

Let $x = a \tan\theta$　Then, $dx = a \sec^2\theta \, d\theta$

$\theta = 0$　when $x = 0$

$\theta = \frac{\pi}{4}$　when $x = a$

$\therefore \int_{0}^{a} \frac{1}{x^2 + a^2} dx = \int_{0}^{\frac{\pi}{4}} \frac{a \sec^2\theta}{a^2 \tan^2\theta + a^2} d\theta = \frac{1}{a^2} \int_{0}^{\frac{\pi}{4}} \frac{a \sec^2\theta}{\tan^2\theta + 1} d\theta = \frac{1}{a^2} \int_{0}^{\frac{\pi}{4}} a \, d\theta = \frac{1}{a} \int_{0}^{\frac{\pi}{4}} d\theta = \frac{1}{a} [\theta]_{0}^{\frac{\pi}{4}}$

$\qquad = \frac{1}{a} \cdot \frac{\pi}{4} = \frac{\pi}{4a}$

#15 Find the value.

(1) For a differentiable function $f(x)$ such that $\int_{0}^{1} f(x)dx = 6$,

　find the value of $\int_{0}^{1} x^2 f(x^3)dx$.

Let $x^3 = t$

Then $3x^2 dx = dt$

$t = 0$ when $x = 0$

$t = 1$ when $x = 1$

$\therefore \int_0^1 x^2 f(x^3)dx = \int_0^1 f(t) \cdot \frac{1}{3} dt = \frac{1}{3} \int_0^1 f(t)dt = \frac{1}{3} \int_0^1 f(x)dx = \frac{1}{3} \cdot 6 = 2$

(2) For a continuous function $f(x)$ such that $f(x) + f(-x) = x^2 - 1$,

find the value of $\int_{-1}^1 f(x)dx$.

Let $-x = t$

Then $-dx = dt$

$t = 1$ when $x = -1$

$t = -1$ when $x = 1$

Since $f(x) + f(-x) = x^2 - 1$, $f(x) = x^2 - 1 - f(-x)$

$\therefore \int_{-1}^1 f(x)dx = \int_{-1}^1 \{x^2 - 1 - f(-x)\}dx = \int_{-1}^1 (x^2 - 1)dx - \int_{-1}^1 f(-x)dx$

$\qquad = \int_{-1}^1 (x^2 - 1)dx - \int_1^{-1} f(t)(-1)dt = \int_{-1}^1 (x^2 - 1)dx - \int_{-1}^1 f(t)dt$

$\qquad = \int_{-1}^1 (x^2 - 1)dx - \int_{-1}^1 f(x)dx$

$\therefore 2\int_{-1}^1 f(x)dx = \int_{-1}^1 (x^2 - 1)dx = \left[\frac{1}{3}x^3 - x\right]_{-1}^1 = \left(\frac{1}{3} - 1\right) - \left(-\frac{1}{3} + 1\right) = \frac{2}{3} - 2 = -\frac{4}{3}$

$\therefore \int_{-1}^1 f(x)dx = -\frac{2}{3}$

(3) For a function $f(x) = 2x \log x + x^2 - 5x + 4$,

find the value of $\int_2^5 f(x)dx - \int_4^5 f(x)dx + \int_1^2 f(x)dx$.

$\int_2^5 f(x)dx - \int_4^5 f(x)dx + \int_1^2 f(x)dx = \left[\int_2^5 f(x)dx - \int_4^5 f(x)dx\right] + \int_1^2 f(x)dx$

$\qquad = \int_2^4 f(x)dx + \int_1^2 f(x)dx$

$\qquad = \int_1^2 f(x)dx + \int_2^4 f(x)dx = \int_1^4 f(x)dx$

$\qquad = \int_1^4 (2x \log x + x^2 - 5x + 4)dx$

$\qquad = 2\int_1^4 x \log x \, dx + \int_1^4 (x^2 - 5x + 4)dx$

For $\int x \log x \, dx$,

Let $u = \log x,$ $v' = xdx$

Then, $u' = \frac{1}{x}dx,$ $v = \frac{1}{2}x^2$

$\therefore \int x \log x \, dx = \log x \cdot \frac{1}{2}x^2 - \int \frac{1}{2}x^2 \cdot \frac{1}{x}dx = \frac{1}{2}x^2 \log x - \frac{1}{2}\int xdx = \frac{1}{2}x^2 \log x - \frac{1}{4}x^2 + C$

$\therefore \int_1^4 x \log x \, dx = \left[\frac{1}{2}x^2 \log x - \frac{1}{4}x^2\right]_1^4 = \left(\frac{1}{2} \cdot 16 \cdot \log 4 - \frac{1}{4} \cdot 16\right) - \left(0 - \frac{1}{4}\right) = 8 \log 4 - \frac{15}{4}$

$\therefore 2\int_1^4 x \log x \, dx = 16 \log 4 - \frac{15}{2}$

$$\int_1^4 (x^2 - 5x + 4)dx = \left[\frac{1}{3}x^3 - \frac{5}{2}x^2 + 4x\right]_1^4 = \left(\frac{1}{3}4^3 - \frac{5}{2}4^2 + 16\right) - \left(\frac{1}{3} - \frac{5}{2} + 4\right)$$

$$= \frac{63}{3} - 40 + 16 + \frac{5}{2} - 4 = -7 + \frac{5}{2} = -\frac{9}{2}$$

$$\therefore \int_2^5 f(x)dx - \int_4^5 f(x)dx + \int_1^2 f(x)dx = 16\log 4 - \frac{15}{2} - \frac{9}{2} = 16\log 4 - 12$$

(4) Let S_a be the area surrounded by the tangent line at $x = a$ on the graph of $y = e^x$,

x-axis, and a vertical line $x = a$. Find the value of $\displaystyle\lim_{n\to\infty} \sum_{i=1}^{n} \frac{1}{n} S_{\frac{n+3i}{n}}$.

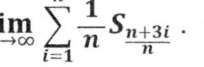

$y = e^x \Rightarrow y' = e^x$

$\therefore$ The slope of the tangent line at $x = a$ is e^a.

$\therefore$ The equation of the tangent line at (a, e^a) is $y - e^a = e^a(x - a)$.

Let $y = 0$ Then, $0 - e^a = e^a(x - a)$; $x - a = -1$; $x = a - 1$

$\therefore$ The intersection point between the tangent line and x-axis is $(a - 1, 0)$.

$\therefore$ The area $S_a = \frac{1}{2} \cdot (a - (a - 1)) \cdot e^a = \frac{1}{2} e^a$

$\therefore S_{\frac{n+3i}{n}} = \frac{1}{2} e^{\frac{n+3i}{n}} = \frac{1}{2} e^{1 + \frac{3}{n}i}$

$$\therefore \lim_{n\to\infty} \sum_{i=1}^{n} \frac{1}{n} S_{\frac{n+3i}{n}} = \lim_{n\to\infty} \sum_{i=1}^{n} \frac{1}{n}\left(\frac{1}{2} e^{1+\frac{3}{n}i}\right) = \int_0^1 \frac{1}{2} e^{1+3x}dx = \frac{1}{2}\int_0^1 e^{1+3x}dx = \frac{1}{2}\left[\frac{1}{3}e^{1+3x}\right]_0^1$$

$$= \frac{1}{6}(e^4 - e)$$

(5) For a differentiable function $f(x)$ such that $f(1) = -1$ and $f'(1) = 2$,

find the value of $\displaystyle\lim_{x\to1} \frac{1}{x-1}\int_1^x \{f(t)\}^2 f'(t)dt$.

Let $\int \{f(t)\}^2 f'(t)dx = F(t)$

Then, $\int_1^x \{f(t)\}^2 f'(t)dt = [F(t)]_1^x = F(x) - F(1)$

$\therefore \displaystyle\lim_{x\to1} \frac{1}{x-1}\int_1^x \{f(t)\}^2 f'(t)dt = \lim_{x\to1} \frac{F(x)-F(1)}{x-1} = F'(1)$

Since $F'(t) = \{f(t)\}^2 f'(t)$, $F'(x) = \{f(x)\}^2 f'(x)$

$\therefore F'(1) = \{f(1)\}^2 f'(1) = (-1)^2 \cdot 2 = 2$

$\therefore \displaystyle\lim_{x\to1} \frac{1}{x-1}\int_1^x \{f(t)\}^2 f'(t)dt = 2$

(6) For a differentiable function $f(x)$ on an open interval $(1, 3)$, $f(x)$ and $f'(x)$ are continuous on the closed interval $[1, 3]$. When $f(1) = 0$, and the minimum and maximum values of $f'(x)$ in the interval $[1, 3]$ are 2 and 5, respectively, find the range of $\int_1^3 f(x)dx$.

By the Mean Value Theorem, there is c $(1 < c < x)$ such that $\frac{f(x)-f(1)}{x-1} = f'(c)$.

Since $f(1) = 0$, $f(x) = (x - 1)f'(c)$

Since $2 \le f'(c) \le 5$, $2(x - 1) \le (x - 1)f'(c) \le 5(x - 1)$; $2(x - 1) \le f(x) \le 5(x - 1)$

$\therefore$ $\int_1^3 2(x - 1)dx \le \int_1^3 f(x)dx \le \int_1^3 5(x - 1)dx$

$\therefore$ $2\left[\frac{1}{2}x^2 - x\right]_1^3 \le \int_1^3 f(x)dx \le 5\left[\frac{1}{2}x^2 - x\right]_1^3$

$\therefore$ $2\left\{\left(\frac{1}{2}\cdot 9 - 3\right) - \left(\frac{1}{2} - 1\right)\right\} \le \int_1^3 f(x)dx \le 5\left\{\left(\frac{1}{2}\cdot 9 - 3\right) - \left(\frac{1}{2} - 1\right)\right\}$

$\therefore$ $2 \cdot 2 \le \int_1^3 f(x)dx \le 5 \cdot 2$ $\qquad$ $\therefore$ $4 \le \int_1^3 f(x)dx \le 10$

(7) The graph of a cubic function $y = f(x)$ is shown as the Figure. When $f(x)$ has a local maximum value 2 at $x = 1$, a local minimum value -2 at $x = 4$, and $f(0) = -2$, find the value of $\int_0^4 |f'(x)|dx$.

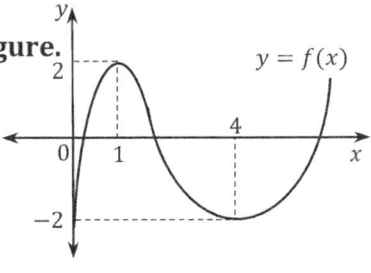

$y = f(x)$ is increasing in $[0, 1]$ and decreasing in $(1, 4)$.

$\therefore$ $f'(x) > 0$ when $0 \le x \le 1$

$f'(x) < 0$ when $1 < x < 4$

$\therefore$ $\int_0^4 |f'(x)|dx = \int_0^1 f'(x)dx - \int_1^4 f'(x)dx = [f(x)]_0^1 - [f(x)]_1^4$

$\qquad = \{f(1) - f(0)\} - \{f(4) - f(1)\} = 2f(1) - f(0) - f(4)$

Note that $f(1) = 2$, $f(4) = -2$, and $f(0) = -2$.

$\therefore$ $\int_0^4 |f'(x)|dx = 2 \cdot 2 - (-2) - (-2) = 8$

(8) For a positive number a, a function $f(x) = -2x(x + a)(x - a)$ has a local maximum value at $x = b$. When $A = \int_{-b}^a f(x)dx$ and $B = \int_b^{a+b} f(x - b)dx$, find the value of $\int_{-b}^a |f(x)|dx$.

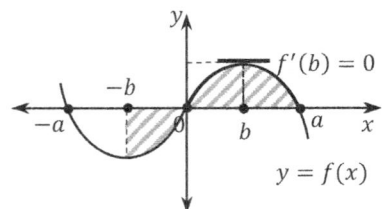

$y = f(x)$

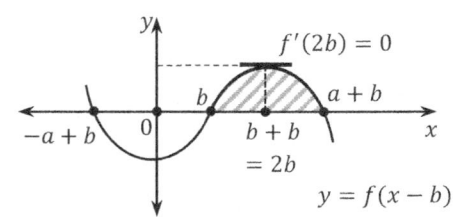

$y = f(x - b)$

$A = \int_{-b}^{a} f(x)dx$

$B = \int_{b}^{a+b} f(x-b)dx = \int_{0}^{a} f(x)dx$

$\therefore \; A - B = \int_{-b}^{0} f(x)dx$

$\therefore \; \int_{-b}^{a} |f(x)|dx = \int_{-b}^{0} -f(x)dx + \int_{0}^{a} f(x)dx = -\int_{-b}^{0} f(x)dx + \int_{0}^{a} f(x)dx = -(A-B) + B$

$= -A + 2B$

#16 Use symmetry to help you evaluate the given integral.

(1) $\int_{-2}^{2} (x^5 - 3x^2 + 4x)dx$

Note that x^5 and $4x$ are odd functions, and x^2 is an even function.

$\int_{-2}^{2}(x^5 - 3x^2 + 4x)dx = \int_{-2}^{2}(x^5 + 4x)dx - 3\int_{-2}^{2}x^2 dx = 0 - 3\left[2\int_{0}^{2}x^2 dx\right] = -6\int_{0}^{2}x^2 dx$

$= -6\left[\frac{1}{3}x^3\right]_{0}^{2} = -6\left(\frac{1}{3}\cdot 2^3\right) = -16$

(2) $\int_{-2}^{0}(x^3 + 2x)dx - \int_{2}^{0}(x^3 + 2x)dx$

$= \int_{-2}^{0}(x^3 + 2x)dx + \int_{0}^{2}(x^3 + 2x)dx = \int_{-2}^{2}(x^3 + 2x)dx = 0$ ($\because x^3 + 2x$ is an odd function.)

(3) $\int_{-\pi}^{\pi}(\sin x + \cos x)dx$

$= \int_{-\pi}^{\pi}\sin x \, dx + \int_{-\pi}^{\pi}\cos x \, dx = 0 + 2\int_{0}^{\pi}\cos x \, dx = 2[\sin x]_{0}^{\pi} = 2(\sin\pi - \sin 0) = 0$

($\because \sin x$ is an odd function and $\cos x$ is an even function.)

(4) $\int_{-\pi}^{\pi}\cos\left(\frac{x}{2}\right)dx$

$= 2\int_{0}^{\pi}\cos\left(\frac{x}{2}\right)dx$ ($\because \cos\frac{x}{2}$ is an even function.)

Let $\frac{x}{2} = t$ Then, $\frac{1}{2}dx = dt$

$t = 0$ when $x = 0$

$t = \frac{\pi}{2}$ when $x = \pi$

$\therefore \; \int_{-\pi}^{\pi}\cos\left(\frac{x}{2}\right)dx = 2\int_{0}^{\pi}\cos\left(\frac{x}{2}\right)dx = 2\int_{0}^{\frac{\pi}{2}}\cos t \cdot 2dt = 4\int_{0}^{\frac{\pi}{2}}\cos t \, dt = 4[\sin t]_{0}^{\frac{\pi}{2}}$

$= 4\left(\sin\frac{\pi}{2} - \sin 0\right) = 4$

(5) $\int_{-3}^{3}\frac{x^5}{x^2+2}dx$

Let $f(x) = \frac{x^5}{x^2+2}$ Then, $f(-x) = -\frac{x^5}{x^2+2} = -f(x)$ $\therefore \; \frac{x^5}{x^2+2}$ is an odd function.

$\therefore \; \int_{-3}^{3}\frac{x^5}{x^2+2}dx = 0$

(6) $\int_{-2}^{2}(x\sin^2 x + x^3 - x^2)dx$

$x\sin^2 x + x^3$ is an odd function and x^2 is an even function.

∴ $\int_{-2}^{2}(x\sin^2 x + x^3 - x^2)dx = \int_{-2}^{2}(x\sin^2 x + x^3)dx - \int_{-2}^{2}x^2 dx = 0 - 2\int_{0}^{2}x^2 dx$

$$= -2\left[\frac{1}{3}x^3\right]_{0}^{2} = -2\left(\frac{1}{3}\cdot 2^3\right) = -\frac{16}{3}$$

#17 Use periodicity to calculate $\int_{0}^{4\pi}|\cos x|dx$.

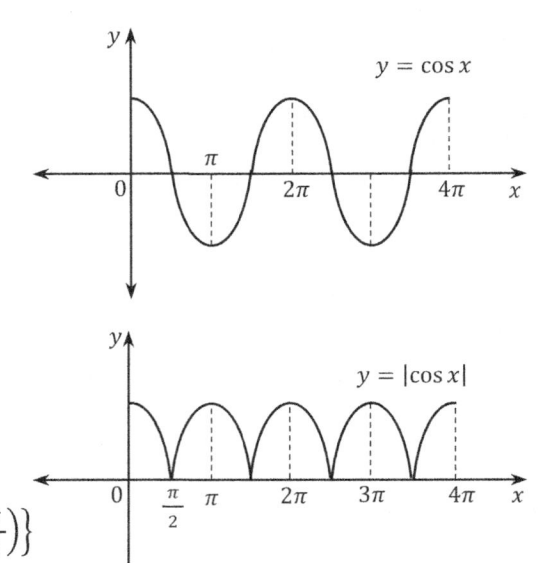

$\int_{0}^{4\pi}|\cos x|dx = \int_{0}^{2\pi}|\cos x|dx + \int_{2\pi}^{4\pi}|\cos x|dx$

$= \int_{0}^{2\pi}|\cos x|dx + \int_{0}^{2\pi}|\cos x|dx$

$= 2\int_{0}^{2\pi}|\cos x|dx = 2\cdot 2\int_{0}^{\pi}|\cos x|dx$

$= 4\left[\int_{0}^{\frac{\pi}{2}}|\cos x|dx + \int_{\frac{\pi}{2}}^{\pi}|\cos x|dx\right]$

$= 4\left[\int_{0}^{\frac{\pi}{2}}\cos x\, dx + \int_{\frac{\pi}{2}}^{\pi}(-\cos x)dx\right]$

$= 4\left\{[\sin x]_{0}^{\frac{\pi}{2}} - [\sin x]_{\frac{\pi}{2}}^{\pi}\right\}$

$= 4\left\{\left(\sin\frac{\pi}{2} - \sin 0\right) - \left(\sin\pi - \sin\frac{\pi}{2}\right)\right\}$

$= 4\{(1-0) - (0-1)\} = 8$

#18 Find the value.

(1) For a function $f(x) = ax^2 + bx + 1$ $(a \neq 0)$ such that $\int_{-\pi}^{\pi}f(x)\sin x\, dx = 0$, the graph
of $y = f(x)$ and a line $y = x$ intersect only at one point. Find the value of $a + b$.

$0 = \int_{-\pi}^{\pi}f(x)\sin x\, dx = \int_{-\pi}^{\pi}(ax^2 + bx + 1)\sin x\, dx$

$= \int_{-\pi}^{\pi}(ax^2\sin x + bx\sin x + \sin x)dx$

Note that $ax^2\sin x$ and $\sin x$ are odd functions.

∴ $\int_{-\pi}^{\pi}ax^2\sin x\, dx = 0$, $\int_{-\pi}^{\pi}\sin x\, dx = 0$

∴ $0 = \int_{-\pi}^{\pi}f(x)\sin x\, dx = \int_{-\pi}^{\pi}bx\sin x\, dx$

Since $bx\sin x$ is an even function, $\int_{-\pi}^{\pi}bx\sin x\, dx = 2\int_{0}^{\pi}bx\sin x\, dx = 2b\int_{0}^{\pi}x\sin x\, dx$

For $\int x\sin x\, dx$,

Let $\qquad u = x, \qquad\quad v' = \sin x\, dx$

Then, $\quad u' = dx, \qquad v = -\cos x$

∴ $\int x\sin x\, dx = -x\cos x - \int(-\cos x)dx = -x\cos x + \int\cos x\, dx = -x\cos x + \sin x + C$

$\therefore \int_0^\pi x \sin x \, dx = [-x \cos x + \sin x]_0^\pi = (-\pi \cos \pi + \sin \pi) - (0 + \sin 0) = \pi$

$\therefore \ 0 = \int_{-\pi}^\pi f(x) \sin x \, dx = 2b \int_0^\pi x \sin x \, dx = 2b\pi \qquad \therefore \ b = 0$

$\therefore \ f(x) = ax^2 + 1$

Since the graph of $y = ax^2 + 1 \ (a \neq 0)$ and a line $y = x$ intersect only at one point,

$ax^2 + 1 = x \ \Rightarrow \ ax^2 - x + 1 = 0 \ \cdots\cdots ①$

Let D be the discriminant of the equation ①.

Then, $D = 0$

$\therefore \ D = (-1)^2 - 4 \cdot a \cdot 1 = 1 - 4a = 0 \ ; \ a = \dfrac{1}{4}$

Therefore, $a + b = \dfrac{1}{4} + 0 = \dfrac{1}{4}$

(2) For any real number x, a function $f(x)$ satisfies $f(x) = f(x + 2)$.

When $f(x) = |x| \ \ (-1 \leq x \leq 1)$, find the value of $\int_0^2 e^{2x} f(x) dx$.

Since $f(x) = f(x + 2)$, $f(x)$ is periodic with period 2.

From the graph, $0 \leq x \leq 1 \ \Rightarrow \ f(x) = x$

$\qquad\qquad\qquad 1 \leq x \leq 2 \ \Rightarrow \ f(x) = -x + 2$

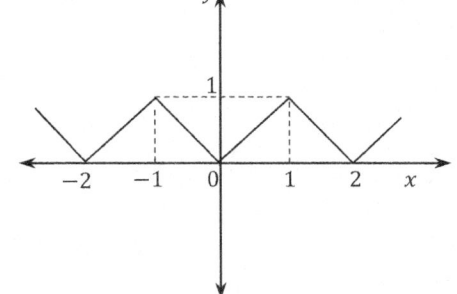

$\therefore \ \int_0^2 e^{2x} f(x) dx = \int_0^1 e^{2x} \cdot x \, dx + \int_1^2 e^{2x}(-x + 2) dx$

$\qquad\qquad\qquad = \int_0^1 xe^{2x} dx - \int_1^2 xe^{2x} dx + 2 \int_1^2 e^{2x} dx$

For $\int xe^{2x} dx$,

Let $\qquad u = x, \qquad v' = e^{2x} dx$

Then, $\qquad u' = dx, \qquad v = \dfrac{1}{2}e^{2x}$

$\therefore \ \int xe^{2x} dx = \dfrac{1}{2}xe^{2x} - \int \dfrac{1}{2}e^{2x} dx = \dfrac{1}{2}xe^{2x} - \dfrac{1}{2}\int e^{2x} dx = \dfrac{1}{2}xe^{2x} - \dfrac{1}{2}\left(\dfrac{1}{2}e^{2x}\right) + C$

$\therefore \ \int_0^1 xe^{2x} dx = \left[\dfrac{1}{2}xe^{2x} - \dfrac{1}{4}e^{2x}\right]_0^1 = \left(\dfrac{1}{2}e^2 - \dfrac{1}{4}e^2\right) - \left(0 - \dfrac{1}{4}e^0\right) = \dfrac{1}{4}e^2 + \dfrac{1}{4}$

$\int_1^2 xe^{2x} dx = \left[\dfrac{1}{2}xe^{2x} - \dfrac{1}{4}e^{2x}\right]_1^2 = \left(\dfrac{1}{2} \cdot 2e^4 - \dfrac{1}{4}e^4\right) - \left(\dfrac{1}{2}e^2 - \dfrac{1}{4}e^2\right) = \dfrac{3}{4}e^4 - \dfrac{1}{4}e^2$

$\int_1^2 e^{2x} dx = \left[\dfrac{1}{2}e^{2x}\right]_1^2 = \dfrac{1}{2}(e^4 - e^2)$

$\therefore \ \int_0^2 e^{2x} f(x) dx = \left(\dfrac{1}{4}e^2 + \dfrac{1}{4}\right) - \left(\dfrac{3}{4}e^4 - \dfrac{1}{4}e^2\right) + 2 \cdot \dfrac{1}{2}(e^4 - e^2) = \dfrac{1}{4}e^4 - \dfrac{1}{2}e^2 + \dfrac{1}{4}$

$\qquad\qquad\qquad = \dfrac{1}{4}(e^4 - 2e^2 + 1) = \dfrac{1}{4}(e^2 - 1)^2$

(3) For a function $f(x) = ax \log x + b$ **such that**

i) $\lim\limits_{x \to e} \dfrac{f(x)-f(e)}{x-e} = 4$ **and**

ii) $\int_1^e f(x)dx = \dfrac{1}{2}e(e+1)$,

find the value of $a + b$.

Note that $\int_a^b uv' = [uv]_a^b - \int_a^b u'v$

$f(x) = ax \log x + b \;\Rightarrow\; f'(x) = a \log x + ax \cdot \dfrac{1}{x} = a \log x + a$

By i), $\lim\limits_{x \to e} \dfrac{f(x)-f(e)}{x-e} = f'(e) = a \log e + a = 2a = 4 \qquad \therefore\; a = 2$

$\therefore\; f(x) = 2x \log x + b$

$\int_1^e f(x)dx = \int_1^e (2x \log x + b)dx = 2 \int_1^e x \log x \, dx + \int_1^e b \, dx$

For $\int x \log x \, dx$,

Let $\qquad u = \log x, \qquad\qquad v' = x dx$

Then, $\qquad u' = \dfrac{1}{x}dx, \qquad\quad v = \dfrac{1}{2}x^2$

$\therefore\; \int x \log x \, dx = \dfrac{1}{2}x^2 \log x - \int \dfrac{1}{2}x^2 \cdot \dfrac{1}{x}dx = \dfrac{1}{2}x^2 \log x - \dfrac{1}{2}\int x dx = \dfrac{1}{2}x^2 \log x - \dfrac{1}{2}\cdot\dfrac{1}{2}x^2 + C$

$\therefore\; \int_1^e x \log x \, dx = \left[\dfrac{1}{2}x^2 \log x - \dfrac{1}{4}x^2\right]_1^e = \left(\dfrac{1}{2}e^2 - \dfrac{1}{4}e^2\right) - \left(-\dfrac{1}{4}\right) = \dfrac{1}{4}e^2 + \dfrac{1}{4}$

$\therefore\; \int_1^e f(x)dx = 2\left(\dfrac{1}{4}e^2 + \dfrac{1}{4}\right) + b(e-1) = \dfrac{1}{2}e^2 + \dfrac{1}{2} + b(e-1) = \dfrac{1}{2}e(e+1) \qquad \therefore\; b = \dfrac{1}{2}$

Therefore, $a + b = 2 + \dfrac{1}{2} = \dfrac{5}{2}$

(4) When $S_n = \sum\limits_{i=1}^{n} \int_i^{i+1} 3x^2 dx$, **find the value of** $\lim\limits_{n \to \infty} \dfrac{S_n}{n^3}$.

$S_n = \int_1^2 3x^2 dx + \int_2^3 3x^2 dx + \int_3^4 3x^2 dx + \cdots \cdots + \int_n^{n+1} 3x^2 dx = \int_1^{n+1} 3x^2 dx = [x^3]_1^{n+1}$

$\qquad = (n+1)^3 - 1$

$\therefore\; \lim\limits_{n \to \infty} \dfrac{S_n}{n^3} = \lim\limits_{n \to \infty} \dfrac{(n+1)^3 - 1}{n^3} = 1$

(5) Find the minimum value of the positive integer n **such that** $S_n = \int_0^{\frac{1}{2}} \left(\sum\limits_{i=1}^{n} ix^{i-1}\right)dx \geq 0.99$

$\int_0^{\frac{1}{2}} \left(\sum\limits_{i=1}^{n} ix^{i-1}\right)dx = \int_0^{\frac{1}{2}}(1 + 2x + 3x^2 + 4x^3 + \cdots \cdots + nx^{n-1})dx$

$= [x + x^2 + x^3 + \cdots \cdots + x^n]_0^{\frac{1}{2}} = \dfrac{1}{2} + \left(\dfrac{1}{2}\right)^2 + \left(\dfrac{1}{2}\right)^3 + \cdots \cdots + \left(\dfrac{1}{2}\right)^n = \dfrac{\left(\frac{1}{2}\right)\left\{1-\left(\frac{1}{2}\right)^n\right\}}{1-\left(\frac{1}{2}\right)} = 1 - \left(\dfrac{1}{2}\right)^n$

$\therefore\; 1 - \left(\dfrac{1}{2}\right)^n \geq 0.99 \qquad \therefore\; \left(\dfrac{1}{2}\right)^n \leq 0.01 = \dfrac{1}{100} \qquad \therefore\; 2^n \geq 100$

Since $2^6 = 64$ and $2^7 = 128$, the minimum value of the positive number n is 7.

(6) For a sequence $\{a_n\}$, the nth term is $a_n = \int_n^{n+1} \left(-\frac{1}{2^n}\sin \pi x\right) dx$.

Find the value of $\sum_{n=1}^{\infty} a_n$.

$a_n = \int_n^{n+1} \left(-\frac{1}{2^n}\sin \pi x\right) dx = \left[\frac{1}{2^n \pi}\cos \pi x\right]_n^{n+1} = \frac{1}{2^n \pi}(\cos(n+1)\pi - \cos n\pi)$

$= \frac{1}{2^n \pi}\{(-1)^{n+1} - (-1)^n\} = \frac{1}{2^n \pi}(-1)^{n+1}\cdot 2 = \frac{1}{2^n \pi}(-1)^n \cdot (-1) \cdot 2$

$= \frac{1}{\pi}\left(-\frac{1}{2}\right)^n \cdot (-2) = \frac{1}{\pi}\left(-\frac{1}{2}\right)^{n-1}$

$\therefore \sum_{n=1}^{\infty} a_n = \frac{1}{\pi}\sum_{n=1}^{\infty}\left(-\frac{1}{2}\right)^{n-1} = \frac{1}{\pi}\frac{1}{1-\left(-\frac{1}{2}\right)} = \frac{2}{3\pi}$

(7) When $S_n = \int_1^{e^n}\left(\frac{\log x}{x}\right) dx$ $(n = 1, 2, 3, \cdots\cdots)$, find the value of $\sum_{n=1}^{\infty}\frac{1}{\sqrt{S_n S_{n+1}}}$.

Let $\log x = t$ Then, $\frac{1}{x}dx = dt$

$t = 0$ when $x = 1$, $t = n$ when $x = e^n$

$\therefore$ $S_n = \int_1^{e^n}\left(\frac{\log x}{x}\right) dx = \int_0^n t\, dt = \left[\frac{1}{2}t^2\right]_0^n = \frac{1}{2}n^2$, $S_{n+1} = \frac{1}{2}(n+1)^2$

$\therefore$ $\sqrt{S_n S_{n+1}} = \sqrt{\frac{1}{2}n^2 \cdot \frac{1}{2}(n+1)^2} = \frac{1}{2}n(n+1)$

$\therefore$ $\sum_{n=1}^{\infty}\frac{1}{\sqrt{S_n S_{n+1}}} = 2\sum_{n=1}^{\infty}\frac{1}{n(n+1)} = 2\lim_{n\to\infty}\left(\frac{1}{1\cdot 2}+\frac{1}{2\cdot 3}+\frac{1}{3\cdot 4}+\cdots\cdots+\frac{1}{n\cdot(n+1)}\right)$

$= 2\lim_{n\to\infty}\left(1-\frac{1}{2}+\frac{1}{2}-\frac{1}{3}+\frac{1}{4}-\cdots\cdots+\frac{1}{n}-\frac{1}{n+1}\right) = 2\lim_{n\to\infty}\left(1-\frac{1}{n+1}\right) = 2\cdot 1 = 2$

(8) For a positive integer n, $f(x) = \int_0^1 \frac{1}{n}x^n dx$. Find the value of $\sum_{n=1}^{100} f(n)$.

$f(x) = \int_0^1 \frac{1}{n}x^n dx = \frac{1}{n}\int_0^1 x^n dx = \frac{1}{n}\left[\frac{1}{n+1}x^{n+1}\right]_0^1 = \frac{1}{n}\cdot\frac{1}{n+1}$

$\therefore$ $\sum_{n=1}^{100} f(n) = \sum_{n=1}^{100}\frac{1}{n(n+1)} = \sum_{n=1}^{100}\left(\frac{1}{n}-\frac{1}{n+1}\right) = \left(1-\frac{1}{2}\right)+\left(\frac{1}{2}-\frac{1}{3}\right)+\cdots\cdots+\left(\frac{1}{100}-\frac{1}{101}\right)$

$= 1 - \frac{1}{101} = \frac{100}{101}$

(9) When $f(x) = x^3 + x$, find the value of $\lim_{n\to\infty}\frac{1}{n}\sum_{i=1}^{n} f\left(1+\frac{2i}{n}\right)$.

$\lim_{n\to\infty}\frac{1}{n}\sum_{i=1}^{n} f\left(1+\frac{2i}{n}\right) = \lim_{n\to\infty}\sum_{i=1}^{n} f\left(1+\frac{2i}{n}\right)\frac{2}{n}\cdot\frac{1}{2} = \frac{1}{2}\int_0^2 f(1+x)dx = \frac{1}{2}\int_0^2\{(1+x)^3 + (1+x)\}dx$

$= \frac{1}{2}\int_0^2\{x^3 + 3x^2 + 3x + 1 + 1 + x\}dx = \frac{1}{2}\int_0^2\{x^3 + 3x^2 + 4x + 2\}dx$

$= \frac{1}{2}\left[\frac{1}{4}x^4 + x^3 + 2x^2 + 2x\right]_0^2 = \frac{1}{2}\left(\frac{1}{4}\cdot 2^4 + 8 + 8 + 4\right) = \frac{1}{2}\cdot 24 = 12$

(10) Find a local extreme value of $f(x) = \int_1^x (1 - \log t)dt.$ $(x > 0)$

$f'(x) = 1 - \log x$

$f'(x) = 0 \Rightarrow \log x = 1 \qquad \therefore x = e$

That is, $f(x)$ has a local extreme value at $x = e$.

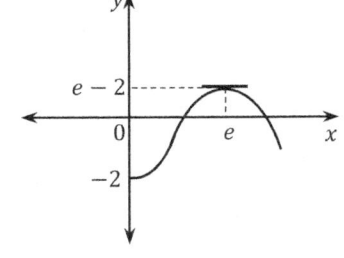

$f(e) = \int_1^e (1 - \log t)dt = [t - (t\log t - t)]_1^e = [2t - t\log t]_1^e$

$\qquad = (2e - e) - (2 - 0) = e - 2$

(11) Let $F(x) = \int_0^x (1 - t)e^t dt.$

Find the maximum value of the function F defined on real numbers.

$F'(x) = (1 - x)e^x$

$F'(x) = 0 \Rightarrow x = 1$

$\therefore$ F has a local extreme value at $x = 1$.

$F(1) = \int_0^1 (1 - t)e^t dt = \int_0^1 e^t dt - \int_0^1 te^t dt = [e^t]_0^1 - \int_0^1 te^t dt$

For $\int te^t dt$

Let $\qquad u = t, \qquad v' = e^t dt$

Then, $\qquad u' = dt, \qquad v = e^t$

$\therefore \int te^t dt = te^t - \int e^t dt = te^t - e^t + C$

$\therefore \int_0^1 te^t dt = [te^t - e^t]_0^1$

$\therefore F(1) = [e^t]_0^1 - [te^t - e^t]_0^1 = (e - 1) - (e - e) + (0 - 1) = e - 2$

Let $x = 0$

Then, $F(0) = \int_0^0 (1 - t)e^t dt = 0$

$\therefore$ $F(x)$ has maximum value $e - 2$ at $x = 1$.

(12) For a function $(x) = x^3 + ax$, $\frac{d}{dx}\left(\int_0^x f(t)dt\right) = \int_1^x \left(\frac{d}{dx}f(x)\right)dx$.

Find the value of the constant a.

Since $\int \left(\frac{d}{dx}f(x)\right)dx = f(x) + C,$ $\int_1^x \left(\frac{d}{dx}f(x)\right)dx = [f(x)]_1^x = f(x) - f(1)$

Note that $\frac{d}{dx}\left(\int_0^x f(t)dt\right) = f(x)$

$\therefore \frac{d}{dx}\left(\int_0^x f(t)dt\right) = \int_1^x \left(\frac{d}{dx}f(x)\right)dx \Rightarrow f(x) = f(x) - f(1) \qquad \therefore f(1) = 0$

$f(x) = x^3 + ax \Rightarrow f(1) = 1 + a = 0 \qquad \therefore a = -1$

(13) When $f(x) = \int_0^x e^{t^2} dt$ **and** $g(x) = \log f'(x),$

find the value of a **such that** $\int_0^2 g(x)dx = 2g(a).$

$f(x) = \int_0^x e^{t^2} dt \Rightarrow f'(x) = e^{x^2}$

$\therefore \ g(x) = \log f'(x) = \log e^{x^2} = x^2$

$\therefore \ \int_0^2 g(x)dx = \int_0^2 x^2 dx = \left[\frac{1}{3}x^3\right]_0^2 = \frac{1}{3} \cdot 2^3 = \frac{8}{3} = 2g(a) = 2a^2$

$\therefore \ a^2 = \frac{8}{3} \cdot \frac{1}{2} = \frac{4}{3} \qquad \therefore \ a = \pm\frac{2\sqrt{3}}{3}$

(14) For a function $f(x)$ **such that** $\int_0^x f(t)dt = xe^{-x},$ **find the value of** $\lim\limits_{h\to 0}\frac{f(1+2h)-f(1)}{h}.$

$\int_0^x f(t)dt = xe^{-x} \Rightarrow \frac{d}{dx}\left(\int_0^x f(t)dt\right) = \frac{d}{dx}(xe^{-x})$

$\therefore \ f(x) = e^{-x} - xe^{-x} = (1-x)e^{-x}$

$\therefore \ f'(x) = -e^{-x} - (1-x)e^{-x} = (-2+x)e^{-x}$

$\therefore \ \lim\limits_{h\to 0}\frac{f(1+2h)-f(1)}{h} = \lim\limits_{h\to 0}\frac{f(1+2h)-f(1)}{2h} \cdot 2 = 2f'(1) = 2(-2+1)e^{-1} = -\frac{2}{e}$

(15) For a function $f(x) = \int_x^{2x} e^{\frac{2t}{\pi}} \sin t \, dt,$ **find the value of** $f'\left(\frac{\pi}{2}\right).$

$f(x) = \int_x^{2x} e^{\frac{2t}{\pi}} \sin t \, dt = \int_0^{2x} e^{\frac{2t}{\pi}} \sin t \, dt - \int_0^x e^{\frac{2t}{\pi}} \sin t \, dt$

Let $g(x) = \int_0^{2x} e^{\frac{2t}{\pi}} \sin t \, dt$ and $2x = u$

Then, $g\left(\frac{u}{2}\right) = \int_0^u e^{\frac{2t}{\pi}} \sin t \, dt$

$\therefore \ \frac{d}{du}\left(g\left(\frac{u}{2}\right)\right) = \frac{d}{du}\left(\int_0^u e^{\frac{2t}{\pi}} \sin t \, dt\right) \ ; \ \frac{1}{2}g'\left(\frac{u}{2}\right) = e^{\frac{2u}{\pi}} \sin u$

$\therefore \ g'(x) = 2e^{\frac{4x}{\pi}} \sin 2x$

$\therefore \ f'(x) = 2e^{\frac{4x}{\pi}} \sin 2x - e^{\frac{2x}{\pi}} \sin x$

$\therefore \ f'\left(\frac{\pi}{2}\right) = 2e^{\frac{4\frac{\pi}{2}}{\pi}} \sin \pi - e^{\frac{2\frac{\pi}{2}}{\pi}} \sin\frac{\pi}{2} = 0 - e = -e$

(16) Find the value of x **such that** $\frac{d}{dx}\int_0^x (x-t)\cos^2 t \, dt = \frac{1}{2}x + \frac{1}{4}. \quad (0 < x < \pi)$

$\int_0^x (x-t)\cos^2 t \, dt = x\int_0^x \cos^2 t \, dt - \int_0^x t\cos^2 t \, dt$

$\therefore \ \frac{d}{dx}\int_0^x (x-t)\cos^2 t \, dt = 1 \cdot \int_0^x \cos^2 t \, dt + x\frac{d}{dx}\left(\int_0^x \cos^2 t \, dt\right) - \frac{d}{dx}\left(\int_0^x t\cos^2 t \, dt\right)$

$= \int_0^x \cos^2 t \, dt + x\cos^2 x - x\cos^2 x = \int_0^x \cos^2 t \, dt = \int_0^x \frac{1+\cos 2t}{2}dt$

$= \frac{1}{2}\int_0^x dt + \frac{1}{2}\int_0^x \cos 2t \, dt = \frac{1}{2}[t]_0^x + \frac{1}{2}\left[\frac{1}{2}\sin 2t\right]_0^x = \frac{1}{2}x + \frac{1}{4}\sin 2x$

$$\therefore \frac{1}{2}x + \frac{1}{4} = \frac{1}{2}x + \frac{1}{4}\sin 2x \quad (0 < x < \pi)$$

$$\therefore \sin 2x = 1 \; ; \; 2x = \frac{\pi}{2} \qquad \therefore x = \frac{\pi}{4}$$

(17) The graph of the function $y = f(x)$ is shown as the Figure.

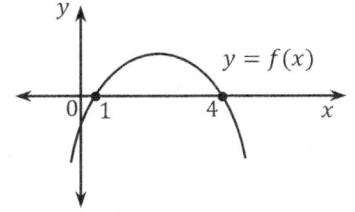

For a function $g(x) = \int_x^{x+2} f(x)dx$,

find the maximum value of $g(x)$.

From the graph, $f(x) = a(x-1)(x-4), \; a < 0$

$g(x) = \int_x^{x+2} f(x)dx$

$\Rightarrow g'(x) = f(x+2) - f(x) = a(x+1)(x-2) - a(x-1)(x-4)$

$\qquad = a(x^2 - x - 2 - x^2 + 5x - 4) = a(4x - 6) = 2a(x - 3)$

$g'(x) = 0 \;\Rightarrow\; x = \frac{3}{2}$

Since $a < 0$, $g(x)$ has maximum value at $x = \frac{3}{2}$

$\therefore g\left(\frac{3}{2}\right)$ is the maximum value.

(18) For a function $f(x) = 3x^2 - x + 3\int_0^1 f(x)dx$, find the product of all solutions of the equation $f(x) = 0$.

Let $\int_0^1 f(x)dx = a$ (a; constant)

Then, $f(x) = 3x^2 - x + 3a$

$\therefore a = \int_0^1 f(x)dx = \int_0^1 (3x^2 - x + 3a)dx = \left[x^3 - \frac{1}{2}x^2 + 3ax\right]_0^1 = 1 - \frac{1}{2} + 3a = \frac{1}{2} + 3a$

$\therefore 2a = -\frac{1}{2} \; ; \; a = -\frac{1}{4}$

$\therefore f(x) = 3x^2 - x + 3\left(-\frac{1}{4}\right) = 3x^2 - x - \frac{3}{4}$

By the relationship between the roots and coefficients, the product of the solutions of the

equation $f(x) = 0$ is $\dfrac{-\frac{3}{4}}{3} = -\frac{1}{4}$

(19) When $f(x) = x^3 - 2x + \int_0^2 g(t)dt$ and $g(x) = 3x^2 - 5 + \int_{-1}^1 f(t)dt$,

find the value of $f(1) + g(-1)$.

Let $\int_0^2 g(t)dt = a$ and $\int_{-1}^1 f(t)dt = b$

Then, $f(x) = x^3 - 2x + a$ and $g(x) = 3x^2 - 5 + b$

$\therefore a = \int_0^2 g(t)dt = \int_0^2 (3t^2 - 5 + b)dt = [t^3 - 5t + bt]_0^2 = 8 - 10 + 2b = 2b - 2 \cdots\cdots ①$

$b = \int_{-1}^{1} f(t)dt = \int_{-1}^{1}(t^3 - 2t + a)dt = \left[\frac{1}{4}t^4 - t^2 + at\right]_{-1}^{1} = \left(\frac{1}{4} - 1 + a\right) - \left(\frac{1}{4} - 1 - a\right)$

$= 2a \cdots\cdots ②$

By ① and ②, $a = 2b - 2 = 4a - 2$; $3a = 2$ $\qquad \therefore a = \frac{2}{3}, b = \frac{4}{3}$

$\therefore f(x) = x^3 - 2x + \frac{2}{3}$ and $g(x) = 3x^2 - 5 + \frac{4}{3}$

$\therefore f(1) + g(-1) = \left(1 - 2 + \frac{2}{3}\right) + \left(3 - 5 + \frac{4}{3}\right) = -1$

(20) For a function $f(x)$ such that $\int_1^x f(t)dt = x^5 - 3x + a$,

find the value of $f'(-1) + a$. (a; constant)

$\int_1^x f(t)dt = x^5 - 3x + a \Rightarrow \int_1^1 f(t)dt = 1^5 - 3\cdot 1 + a$ (Letting $x = 1$)

$\therefore 0 = -2 + a$; $a = 2$

$\int_1^x f(t)dt = x^5 - 3x + a \Rightarrow \frac{d}{dx}\left(\int_1^x f(t)dt\right) = \frac{d}{dx}(x^5 - 3x + a)$

$\therefore f(x) = 5x^4 - 3$

$\therefore f'(x) = 20x^3$; $f'(-1) = -20$

$\therefore f'(-1) + a = -20 + 2 = -18$

(21) Find the relationship between the real numbers a and b such that

$f(x) = \int_{x-1}^{x}(t^2 + at + b)dt > 0$ for any real number x.

$f(x) = \int_{x-1}^{x}(t^2 + at + b)dt = \left[\frac{1}{3}t^3 + \frac{a}{2}t^2 + bt\right]_{x-1}^{x}$

$\qquad = \frac{1}{3}\{x^3 - (x-1)^3\} + \frac{a}{2}\{x^2 - (x-1)^2\} + b\{x - (x-1)\}$

$\qquad = \frac{1}{3}\{x^3 - (x^3 - 3x^2 + 3x - 1)\} + \frac{a}{2}\{x^2 - (x^2 - 2x + 1)\} + b$

$\qquad = \frac{1}{3}(3x^2 - 3x + 1) + \frac{a}{2}(2x - 1) + b$

$\qquad = x^2 - x + \frac{1}{3} + ax - \frac{a}{2} + b$

$\qquad = x^2 + (a-1)x + \frac{1}{3} - \frac{a}{2} + b$

Let D be the discriminant of the equation $x^2 + (a-1)x + \frac{1}{3} - \frac{a}{2} + b = 0$.

Since $f(x) > 0$ for any real number x, $D < 0$

$\therefore (a-1)^2 - 4\left(\frac{1}{3} - \frac{a}{2} + b\right) = a^2 - 2a + 1 - \frac{4}{3} + 2a - 4b = a^2 - \frac{1}{3} - 4b < 0$

$\therefore a^2 - 4b < \frac{1}{3}$

(22) When a differentiable function $f(x)$ satisfies:

i) $f(-x) = -f(x)$ **for any real number** x **and**

ii) $f(2) = 3$,

find the value of $\int_{-2}^{2} f'(x)(3-x)dx$.

Since $f(-x) = -f(x)$, $f(x)$ is an odd function.

Since $-f'(-x) = -f'(x)$; $f'(-x) = f'(x)$, $f'(x)$ is an even function.

$\therefore \int_{-a}^{a} f'(x)dx = 2\int_{0}^{a} f'(x)dx$

Since $xf'(x)$ is an odd function, $\int_{-a}^{a} xf'(x)dx = 0$

$f(-x) = -f(x) \Rightarrow f(0) = -f(0)$; $2f(0) = 0$; $f(0) = 0$

$\therefore \int_{-2}^{2} f'(x)(3-x)dx = 3\int_{-2}^{2} f'(x)dx - \int_{-2}^{2} xf'(x)dx = 3\left(2\int_{0}^{2} f'(x)dx\right) - 0 = 6\int_{0}^{2} f'(x)dx$

$= 6[f(x)]_{0}^{2} = 6\{f(2) - f(0)\} = 6(3-0) = 18$

(23) When a polynomial function $f(x)$ defined on real numbers satisfies:

i) $f(1) = 10$,

ii) $f(x) = \frac{1}{2}\int_{x}^{x+1} f(t)dt - \frac{1}{2}\int_{x}^{x-1} f(t)dt - \int_{0}^{1} f(t)dt$, **and**

iii) $f(x+y) + f(x-y) = 3\{f(x) + f(y)\}$,

find the value of $f'(1)$.

From iii), let $x = y = 0$.

Then, $f(0) + f(0) = 3\{f(0) + f(0)\}$; $2f(0) = 6f(0)$; $f(0) = 0$

Let $x = y = 1$.

Then, $f(1+1) + f(1-1) = 3\{f(1) + f(1)\}$; $f(2) + f(0) = 6f(1)$

$\therefore f(2) = 6 \cdot 10 = 60$ (by i))

From ii), $f'(x) = \frac{1}{2}\{f(x+1) - f(x)\} - \frac{1}{2}\{f(x-1) - f(x)\} - 0$ $\left(\because \int_{0}^{1} f(t)dt \text{ is a constant}\right)$

$= \frac{1}{2}\{f(x+1) - f(x-1)\}$

$\therefore f'(1) = \frac{1}{2}\{f(1+1) - f(1-1)\} = \frac{1}{2}\{f(2) - f(0)\} = \frac{1}{2} \cdot 60 = 30$

(24) For a function $f(x)$ such that $f(x+2) = -f(x)$ and $\int_{0}^{2} f(x)dx = 10$,

find the value of $\int_{-2}^{4} f(x)dx$.

$f(x+2) = -f(x) \Rightarrow f(x+2) + f(x) = 0$

$\therefore \int_{-2}^{4} f(x)dx = \int_{-2}^{0} f(x)dx + \int_{0}^{2} f(x)dx + \int_{2}^{4} f(x)dx$

$= -\int_{-2}^{0} f(x+2)dx + \int_{0}^{2} f(x)dx + \int_{2}^{4} f(x)dx \quad \cdots\cdots ①$

Let $x + 2 = t \Rightarrow dx = dt$

$t = 0$ when $x = -2$

$t = 2$ when $x = 0$

$\therefore \int_{-2}^{0} f(x+2)dx = \int_{0}^{2} f(t)dt = \int_{0}^{2} f(x)dx$

$\therefore ① = -\int_{0}^{2} f(x)dx + \int_{0}^{2} f(x)dx + \int_{2}^{4} f(x)dx = \int_{2}^{4} f(x)dx$

$\quad = \int_{0}^{2} f(x+2)dx = -\int_{0}^{2} f(x)dx = -10$

(25) For a polynomial function $f(x)$ such that $f'(-a) = f'(a)$ and $\int_{-a}^{a} f(x)dx = 4a$, find the value of $f(0)$.

Since $f'(-a) = f'(a)$, $f'(x)$ is an even function.

$\therefore f'(x)$ is the polynomial of even degrees and constant.

$\therefore f(x)$ is the polynomial of odd degrees and constant.

Let $g(x)$ be the polynomial of odd degrees.

Then, $f(x) = g(x) + C$ $(C; \text{constant})$

$\therefore \int_{-a}^{a} f(x)dx = \int_{-a}^{a} (g(x) + C)dx = \int_{-a}^{a} g(x)dx + \int_{-a}^{a} Cdx = 0 + C\int_{-a}^{a} dx = 2C\int_{0}^{a} dx$

$\quad = 2C(a - 0) = 2aC = 4a \qquad \therefore C = 2$

$\therefore f(x) = g(x) + 2$

$\therefore f(0) = g(0) + 2 = 0 + 2 = 2$

(26) For a polynomial function $g(x)$, let $f(x) = x^2 + ax + \int_{2}^{x} g(t)dt$.

When $f(x)$ is divided by $(x-2)^2$, there is no remainder.

Find the remainder when $g(x)$ is divided by $x - 2$.

Letting $x = 2$, $f(2) = 2^2 + 2a + \int_{2}^{2} g(t)dt = 4 + 2a + 0 = 4 + 2a \cdots\cdots ①$

When $f(x)$ is divided by $(x-2)^2$, let $Q(x)$ be the quotient;

i.e., $f(x) = (x-2)^2 Q(x) + 0 \cdots\cdots ②$

Then $f(2) = 0$

$\therefore 0 = 4 + 2a$ (By ①) $\qquad \therefore a = -2$

$\therefore f(x) = x^2 - 2x + \int_{2}^{x} g(t)dt$

$\therefore f'(x) = 2x - 2 + g(x)$; $g(x) = f'(x) - 2x + 2$

$\therefore$ When $g(x)$ is divided by $x - 2$, the remainder is $g(2) = f'(2) - 2\cdot 2 + 2 = f'(2) - 2$

By ②, $f'(x) = 2(x-2)Q(x) + (x-2)^2 Q'(x)$ $\qquad \therefore f'(2) = 0$

Therefore, $g(2) = 0 - 2 = -2$

(27) For a quadratic function $f(x)$, $f(x) = \frac{12}{7}x^2 - 2x \int_1^2 f(t)dt + \left[\int_1^2 f(t)dt\right]^2$.

Find the value of $\int_1^2 f(x)dx$.

Let $\int_1^2 f(t)dt = a$ $(a;\text{constant})$

Then, $f(x) = \frac{12}{7}x^2 - 2xa + a^2$

$\therefore\ a = \int_1^2 f(t)dt = \int_1^2 \left(\frac{12}{7}t^2 - 2at + a^2\right)dt = \left[\frac{12}{7}\cdot\frac{1}{3}t^3 - 2a\cdot\frac{1}{2}t^2 + a^2t\right]_1^2$

$= \left[\frac{4}{7}t^3 - at^2 + a^2t\right]_1^2 = \left(\frac{4}{7}\cdot 2^3 - 4a + 2a^2\right) - \left(\frac{4}{7} - a + a^2\right) = a^2 - 3a + 4$

$\therefore\ a^2 - 4a + 4 = 0$; $(a-2)^2 = 0$; $a = 2$

$\therefore\ \int_1^2 f(x)dx = 2$

Chapter 7. Applications of the Integral

#1 Sketch the region bounded by the graphs of the given equations and calculate the area S of the region.

(1) $y = x^2 - 4x + 3$, $x = 0$, and $y = 0$

$y = x^2 - 4x + 3 = (x - 1)(x - 3)$

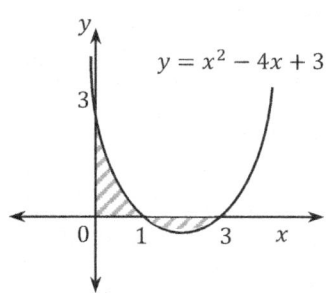

$\therefore \ S = \int_0^1 (x^2 - 4x + 3)dx - \int_1^3 (x^2 - 4x + 3)dx$

$= \left[\frac{1}{3}x^3 - 2x^2 + 3x\right]_0^1 - \left[\frac{1}{3}x^3 - 2x^2 + 3x\right]_1^3$

$= \left(\frac{1}{3} - 2 + 3\right) - 0 - \left(\frac{1}{3}\cdot 3^3 - 2\cdot 3^2 + 3\cdot 3\right) + \left(\frac{1}{3} - 2 + 3\right)$

$= \left(\frac{2}{3} - 4 + 6\right) - (9 - 18 + 9) = \frac{8}{3}$

(2) $y = x^3 - 3x^2 - x + 3$, x-axis between $x = -1$ and $x = 2$

$y = x^3 - 3x^2 - x + 3 = x^2(x - 3) - (x - 3) = (x - 3)(x^2 - 1)$

$= (x - 3)(x + 1)(x - 1)$

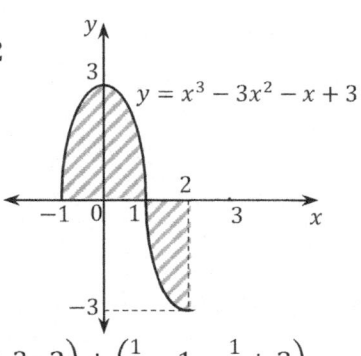

$\therefore \ S = \int_{-1}^1 (x^3 - 3x^2 - x + 3)dx - \int_1^2 (x^3 - 3x^2 - x + 3)dx$

$= \left[\frac{1}{4}x^4 - x^3 - \frac{1}{2}x^2 + 3x\right]_{-1}^1 - \left[\frac{1}{4}x^4 - x^3 - \frac{1}{2}x^2 + 3x\right]_1^2$

$= \left(\frac{1}{4} - 1 - \frac{1}{2} + 3\right) - \left(\frac{1}{4} + 1 - \frac{1}{2} - 3\right) - \left(\frac{1}{4}\cdot 2^4 - 2^3 - \frac{1}{2}\cdot 2^2 + 3\cdot 2\right) + \left(\frac{1}{4} - 1 - \frac{1}{2} + 3\right)$

$= \left(\frac{2}{4} - 2 - 1 + 6\right) - \left(\frac{1}{4} + 1 - \frac{1}{2} - 3\right) - (4 - 8 - 2 + 6) = \frac{23}{4}$

(3) $y = x^3 + 4x^2 + 4x$, x-axis

$y = x^3 + 4x^2 + 4x = x(x^2 + 4x + 4) = x(x + 2)^2$

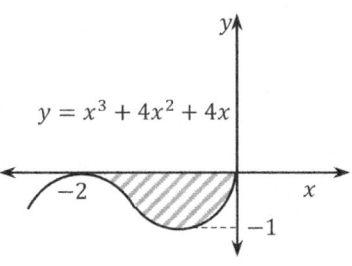

$\therefore \ S = -\int_{-2}^0 (x^3 + 4x^2 + 4x)dx = -\left[\frac{1}{4}x^4 + \frac{4}{3}x^3 + 2x^2\right]_{-2}^0$

$= \frac{1}{4}(-2)^4 + \frac{4}{3}(-2)^3 + 2(-2)^2 = 4 - \frac{32}{3} + 8 = \frac{4}{3}$

(4) $y = x^3 - 3x^2 + 2x$, x-axis

$y = x^3 - 3x^2 + 2x = x(x^2 - 3x + 2) = x(x - 2)(x - 1)$

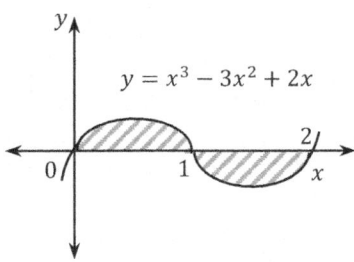

$\therefore \ S = \int_0^1 (x^3 - 3x^2 + 2x)dx - \int_1^2 (x^3 - 3x^2 + 2x)dx$

$= \left[\frac{1}{4}x^4 - x^3 + x^2\right]_0^1 - \left[\frac{1}{4}x^4 - x^3 + x^2\right]_1^2$

$= \left(\frac{1}{4} - 1 + 1\right) - \left(\frac{1}{4}\cdot 2^4 - 2^3 + 2^2\right) + \left(\frac{1}{4} - 1 + 1\right)$

$= \frac{1}{2}$

(5) $y = \frac{1}{x}$ $(1 \le x \le e)$, **x-axis**

$S = \int_1^e \frac{1}{x} dx = [\log x]_1^e = \log e - \log 1 = 1 - 0 = 1$

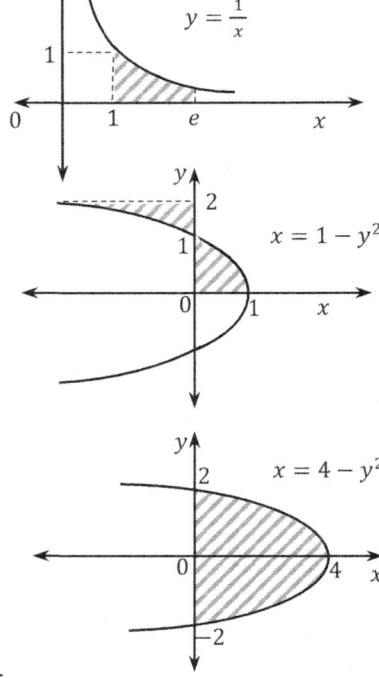

(6) $y = \sqrt{-x+1}$, **x-axis, y-axis, and** $y = 2$

$y = \sqrt{-x+1} \;\Rightarrow\; y^2 = -x+1 \qquad \therefore\; x = 1 - y^2$

$\therefore\; S = \int_0^1 xdy - \int_1^2 xdy = \int_0^1 (1-y^2)dy - \int_1^2 (1-y^2)dy$

$\qquad = \left[1 - \frac{1}{3}y^3\right]_0^1 - \left[1 - \frac{1}{3}y^3\right]_1^2$

$\qquad = \left(1 - \frac{1}{3}\right) - (1) - \left(1 - \frac{1}{3}\cdot 2^3\right) + \left(1 - \frac{1}{3}\right)$

$\qquad = -\frac{2}{3} + \frac{8}{3} = 2$

(7) $y^2 = 4 - x$, $x = 0$

$y^2 = 4 - x \;\Rightarrow\; x = 4 - y^2$

$\therefore\; S = \int_{-2}^2 xdy = \int_{-2}^2 (4-y^2)dy = 2\int_0^2 (4-y^2)dy$

$\qquad = 2\left[4y - \frac{1}{3}y^3\right]_0^2 = 2\left(4\cdot 2 - \frac{1}{3}\cdot 2^3\right) = 2\left(8 - \frac{8}{3}\right) = \frac{32}{3}$

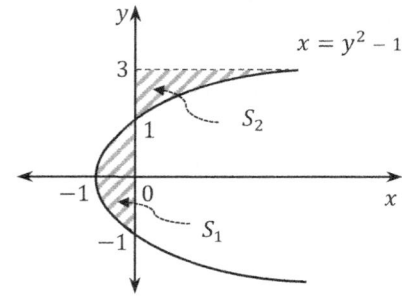

(8) $y^2 = x + 1$, $y = 3$, **and** $x = 0$

$S_1 = -\int_{-1}^1 xdy = -\int_{-1}^1 (y^2 - 1)dy = -\left[\frac{1}{3}y^3 - y\right]_{-1}^1$

$\qquad = -\left(\frac{1}{3} - 1\right) + \left(-\frac{1}{3} + 1\right) = -\frac{2}{3} + 2 = \frac{4}{3}$

$S_2 = \int_1^3 xdy = \int_1^3 (y^2 - 1)dy = \left[\frac{1}{3}y^3 - y\right]_1^3$

$\qquad = \left(\frac{1}{3}3^3 - 3\right) - \left(\frac{1}{3} - 1\right) = 7 - \frac{1}{3} = \frac{20}{3}$

$\therefore\; S = S_1 + S_2 = \frac{4}{3} + \frac{20}{3} = \frac{24}{3} = 8$

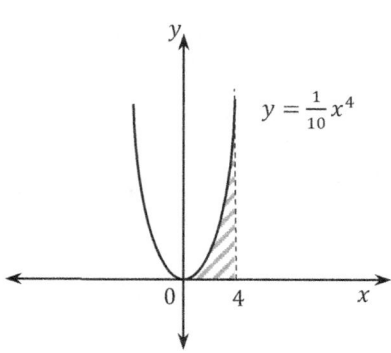

(9) $1 + \log_{10} y = 4\log_{10} x$, $x = 4$, **x-axis**

$1 + \log_{10} y = 4\log_{10} x \;\Rightarrow\; \log_{10} 10 + \log_{10} y = 4\log_{10} x$

$\qquad\qquad\qquad\qquad\qquad \Rightarrow\; \log_{10} 10y = \log_{10} x^4$

$\qquad\qquad\qquad\qquad\qquad \Rightarrow\; 10y = x^4$

$\therefore\; y = \frac{1}{10}x^4$ $(x > 0,\; y > 0)$

$\therefore\; S = \int_0^4 ydx = \int_0^4 \left(\frac{1}{10}x^4\right)dx = \frac{1}{10}\int_0^4 x^4 dx$

$\qquad = \frac{1}{10}\left[\frac{1}{5}x^5\right]_0^4 = \frac{1}{10}\left(\frac{1}{5}\cdot 4^5\right) = \frac{4^5}{50} = \frac{512}{25}$

(10) $y = -\log x \ (1 \leq x \leq e)$, *x*-axis

$S = -\int_1^e (-\log x)dx = \int_1^e \log x \, dx = [x \log x - x]_1^e$

$= (e \log e - e) - (\log 1 - 1) = (e - e) + 1 = 1$

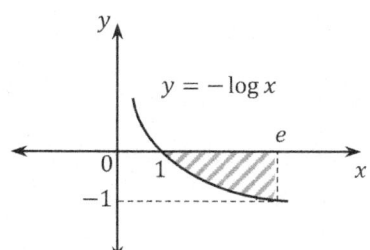

$y = -\log x$

(11) $y = \log(x + 1)$, $x = 0$, $y = \log 3$, $y = -\log 3$

$y = \log(x + 1) \Rightarrow x + 1 = e^y \qquad \therefore \ x = e^y - 1$

$S = \int_{-\log 3}^{\log 3} (e^y - 1)dy = -\int_{-\log 3}^0 (e^y - 1)dy + \int_0^{\log 3} (e^y - 1)dy$

$= -[e^y - y]_{-\log 3}^0 + [e^y - y]_0^{\log 3}$

$= -e^0 + (e^{-\log 3} + \log 3) + (e^{\log 3} - \log 3) - e^0$

$= -2 + \dfrac{1}{3} + 3 = \dfrac{4}{3}$

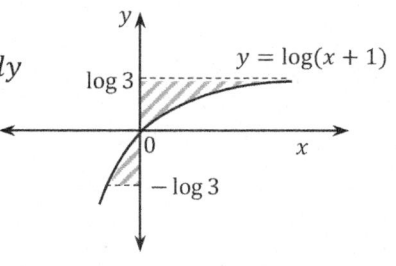

$y = \log(x + 1)$

(12) $y = x \sin x \ (1 \leq x \leq 2\pi)$, *x*-axis

$\begin{cases} y = x \sin x \\ y = 0 \end{cases} \Rightarrow x \sin x = 0 \qquad \therefore \ x = 0, \ x = \pi, \ x = 2\pi$

$0 \leq x \leq \pi \Rightarrow y \geq 0$

$\pi \leq x \leq 2\pi \Rightarrow y \leq 0$

$S = \int_0^\pi x \sin x \, dx - \int_\pi^{2\pi} x \sin x \, dx$

For $\int x \sin x \, dx$,

Let $\qquad u = x, \qquad v' = \sin x \, dx$

Then, $\qquad u' = dx, \qquad v = -\cos x$

$\therefore \ \int x \sin x \, dx = x(-\cos x) - \int(-\cos x)dx = -x \cos x + \int \cos x \, dx$

$\qquad\qquad = -x \cos x + \sin x + C$

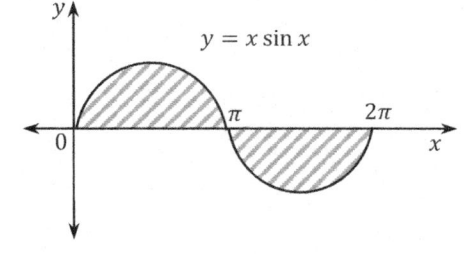

$y = x \sin x$

$\therefore \ S = [-x \cos x + \sin x]_0^\pi - [-x \cos x + \sin x]_\pi^{2\pi}$

$\qquad = (-\pi \cos \pi + \sin \pi) - 0 - (-2\pi \cos 2\pi + \sin 2\pi) + (-\pi \cos \pi + \sin \pi)$

$\qquad = \pi - 0 + 2\pi + \pi = 4\pi$

(13) $y = \tan x$, $x = \dfrac{\pi}{3}$, $y = 0$

$S = \int_0^{\frac{\pi}{3}} \tan x \, dx = \int_0^{\frac{\pi}{3}} \dfrac{\sin x}{\cos x} dx = \int_0^{\frac{\pi}{3}} \dfrac{-(\cos x)'}{\cos x} dx = [-\log|\cos x|]_0^{\frac{\pi}{3}}$

$= \left(-\log\left|\cos\dfrac{\pi}{3}\right|\right) - (-\log|\cos 0|) = -\log\dfrac{1}{2} + \log 1 = -\log 2^{-1} = \log 2$

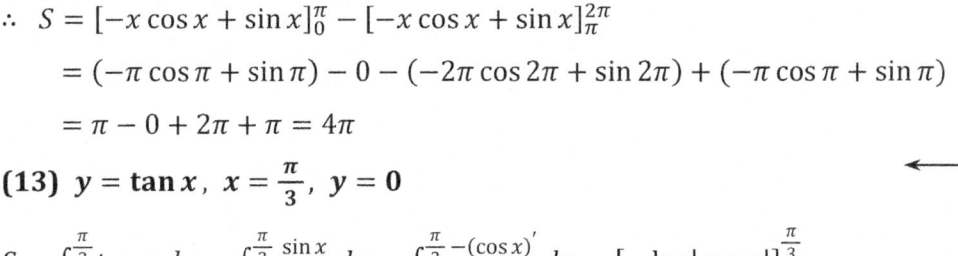

$y = \tan x$

(14) $x = \cos y \ (0 \leq y \leq \pi)$, $y = 0$, $y = \pi$, $x = 0$

$S = \int_0^{\frac{\pi}{2}} \cos y \, dy - \int_{\frac{\pi}{2}}^\pi \cos y \, dy = [\sin y]_0^{\frac{\pi}{2}} - [\sin y]_{\frac{\pi}{2}}^\pi$

$= \sin\dfrac{\pi}{2} - \sin 0 - \sin \pi + \sin\dfrac{\pi}{2} = 2$

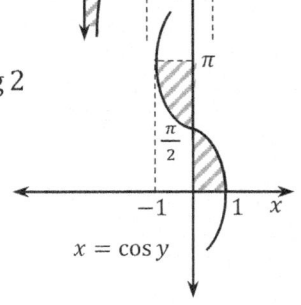

$x = \cos y$

(15) $\sqrt{x} + \sqrt{y} = 1$, $x = 0$, $y = 0$

Since $x \geq 0$ and $y \geq 0$, $\sqrt{y} = 1 - \sqrt{x} \geq 0$ $\quad \therefore \sqrt{x} \leq 1$; $0 \leq x \leq 1$

Since $y = (1 - \sqrt{x})^2$,

$S = \int_0^1 y\,dx = \int_0^1 (1 - \sqrt{x})^2 dx = \int_0^1 (1 - 2\sqrt{x} + x)^2 dx$

$= \left[x - 2 \cdot \frac{2}{3} x^{\frac{3}{2}} + \frac{1}{2} x^2 \right]_0^1 = 1 - 2 \cdot \frac{2}{3} + \frac{1}{2} = \frac{1}{6}$

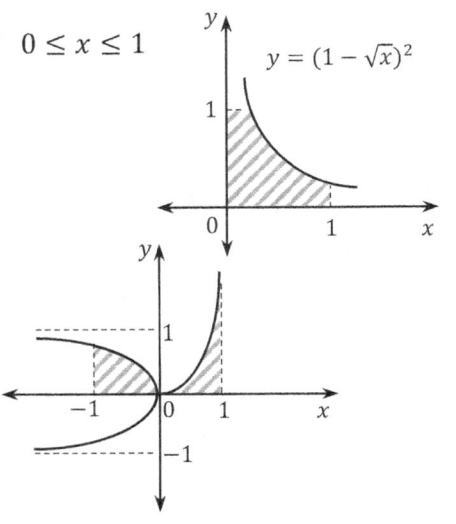

(16) $y = |e^x - 1|$, $y = 0$, $y = -1$, $x = 1$

$S = \int_{-1}^0 -(e^x - 1)dx + \int_0^1 (e^x - 1)dx$

$= -[e^x - x]_{-1}^0 + [e^x - x]_0^1$

$= -e^0 + (e^{-1} + 1) + (e - 1) - e^0 = e + \frac{1}{e} - 2$

#2 Find the value.

(1) For the area S of the region between a curve $y = x(x - 1)(x - a)$, $a > 1$, and the x-axis, find the value of a so that the areas of parts of S are the same.

$y = x(x - 1)(x - a) = 0 \Rightarrow x = 0$, $x = 1$, or $x = a$ $(a > 1)$

Let S_1 and S_2 be the areas of parts of S.

Then $S_1 = S_2$

$\therefore S_2 = -S_1$

$\therefore S = S_1 + S_2 = 0$

$\therefore S = \int_0^a y\,dx = \int_0^a \{x^3 - (1 + a)x^2 + ax\}dx = 0$

$\therefore \left[\frac{1}{4}x^4 - \frac{1}{3}(1 + a)x^3 + \frac{1}{2}ax^2 \right]_0^a = \frac{1}{4}a^4 - \frac{1}{3}(1 + a)a^3 + \frac{1}{2}a^3$

$= a^3 \left(\frac{1}{4}a - \frac{1}{3}(1 + a) + \frac{1}{2} \right) = a^3 \left(-\frac{1}{12}a + \frac{1}{6} \right) = 0$

$\therefore a = 2$

(2) For the area S of the region bounded by $y = (x - a)\sin x$ $(0 \leq x \leq \pi)$ and the x-axis, find the value of a so that the areas of parts of S are the same. $(0 < a < \pi)$

$y = (x - a)\sin x = 0 \Rightarrow x = a$, $x = 0$, or $x = \pi$

From the Figure, $S = \int_0^\pi y\,dx = 0$

$\therefore S = \int_0^\pi y\,dx = \int_0^\pi (x - a)\sin x\,dx = 0$

For $\int (x - a)\sin x\,dx$,

Let $\quad u = x - a$, $\quad v' = \sin x\,dx$

Then, $\quad u' = dx$, $\quad v = -\cos x$

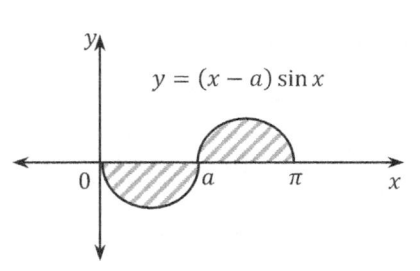

$\therefore \int (x-a)\sin x\,dx = (x-a)(-\cos x) - \int(-\cos x)dx = -(x-a)\cos x + \int \cos x\,dx$

$$= (a-x)\cos x + \sin x + C$$

$\therefore \ S = [(a-x)\cos x + \sin x]_0^\pi = \{(a-\pi)\cos\pi + \sin\pi\} - \{(a)\cos 0 + \sin 0\} = \pi - 2a = 0$

$\therefore \ a = \dfrac{\pi}{2}$

(3) When the area of the region bounded by $y^2 = 4 - ax$ $(a > 0)$ and the y-axis is $\dfrac{1}{3}$, find the value of a.

$y^2 = 4 - ax \ \Rightarrow \ x = \dfrac{1}{a}(4 - y^2)$

$\therefore \ x = 0 \ \Rightarrow \ y = \pm 2$

$S = \int_{-2}^2 x\,dy = 2\int_0^2 x\,dy = 2\int_0^2 \dfrac{1}{a}(4 - y^2)dy$

$= \dfrac{2}{a}\int_0^2 (4 - y^2)dy = \dfrac{2}{a}\left[4y - \dfrac{1}{3}y^3\right]_0^2 = \dfrac{2}{a}\left(4\cdot 2 - \dfrac{1}{3}\cdot 2^3\right)$

$= \dfrac{2}{a}\left(8 - \dfrac{8}{3}\right) = \dfrac{2}{a}\cdot\dfrac{16}{3} = \dfrac{32}{3a}$

$\therefore \ \dfrac{32}{3a} = \dfrac{1}{3} \ ; \ \ 3a = 3\cdot 32 \qquad \therefore \ a = 32$

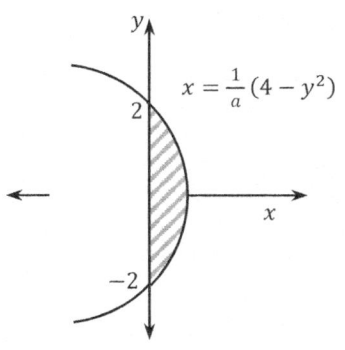

$x = \dfrac{1}{a}(4 - y^2)$

(4) When the area of the region bounded by $y = x(x-a)^2$ $(a < 0)$ and the x-axis is 12, find the value of a.

$S = -\int_a^0 y\,dx = -\int_a^0 x(x-a)^2 dx = -\int_a^0 (x^3 - 2ax^2 + a^2 x)dx$

$= -\left[\dfrac{1}{4}x^4 - \dfrac{2a}{3}x^3 + \dfrac{1}{2}a^2 x^2\right]_a^0 = \dfrac{1}{4}a^4 - \dfrac{2a}{3}a^3 + \dfrac{1}{2}a^2\cdot a^2 = \dfrac{1}{12}a^4$

$\therefore \ \dfrac{1}{12}a^4 = 12 \ ; \ \ a^4 = 12^2 \ ; \ \ a^2 = 12$

$\therefore \ a = \pm 2\sqrt{3}$

Since $a < 0$, $a = -2\sqrt{3}$

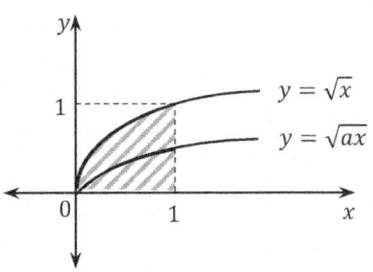

$y = x(x-a)^2$

(5) When a curve $y = \sqrt{ax}$ divides the area of the region bounded by $y = \sqrt{x}$ and the x-axis into two equal parts, find the value of a. $(a > 0)$

Let $S_1 = \int_0^1 \sqrt{x}\,dx$ and $S_2 = \int_0^1 \sqrt{ax}\,dx$

Then, $S_1 = \left[\dfrac{2}{3}x^{\frac{3}{2}}\right]_0^1 = \dfrac{2}{3}$ and

$S_2 = \left[\dfrac{2}{3}(ax)^{\frac{3}{2}}\cdot\dfrac{1}{a}\right]_0^1 = \dfrac{2}{3a}a^{\frac{3}{2}} = \dfrac{2}{3a}a\sqrt{a} = \dfrac{2}{3}\sqrt{a}$

Since $S_1 = 2S_2$, $\dfrac{2}{3} = 2\cdot\dfrac{2}{3}\sqrt{a}$

$\therefore \ \sqrt{a} = \dfrac{1}{2} \qquad \therefore \ a = \dfrac{1}{4}$

$y = \sqrt{x}$

$y = \sqrt{ax}$

(6) Find the value of a so that the area of the region bounded by $y = \sqrt{x + a}$, $x = 0$, and

$y = 0$ **is** $\dfrac{5}{2}$.

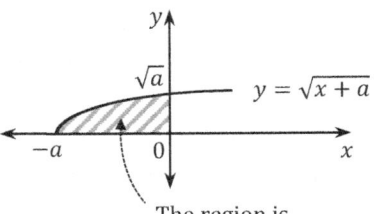

If $x = 0$, then $y = \sqrt{a}$

Since $\sqrt{a} \geq 0$, $a > 0$

$y = \sqrt{x + a} \Rightarrow y^2 = x + a \quad \therefore x = y^2 - a \ (y \geq 0)$

The region is
on the left of the y-axis

The shaded area is $\displaystyle\int_0^{\sqrt{a}} x\,dy = -\int_0^{\sqrt{a}} (y^2 - a)\,dy = -\left[\frac{1}{3}y^3 - ay\right]_0^{\sqrt{a}}$

$$= -\left(\frac{1}{3}a\sqrt{a} - a\sqrt{a}\right) = \frac{2}{3}a\sqrt{a}$$

$\therefore \ \dfrac{2}{3}a\sqrt{a} = \dfrac{5}{2} \ ; \quad a\sqrt{a} = \dfrac{15}{4} \ ; \quad a^3 = \left(\dfrac{15}{4}\right)^2$

$\therefore \ a = \sqrt[3]{\left(\dfrac{15}{4}\right)^2} \quad (\because a > 0)$

(7) For a function $f(x) = e^x - 1$, let g be the inverse of f. When a is a positive constant,

find the value of $\displaystyle\int_0^a f(x)\,dx + \int_0^{f(a)} g(t)\,dt$.

Since g is the inverse of f, f and g are symmetric with respect to the line $y = x$.

Let $S_1 = \int_0^a f(x)\,dx$ and $S_2 = \int_0^{f(a)} g(t)\,dt$

Then, $S_1 + S_2 = $ The area of the rectangle R

$\therefore \ \int_0^a f(x)\,dx + \int_0^{f(a)} g(t)\,dt = a \cdot f(a) = a(e^a - 1)$

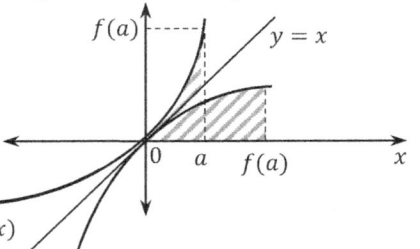

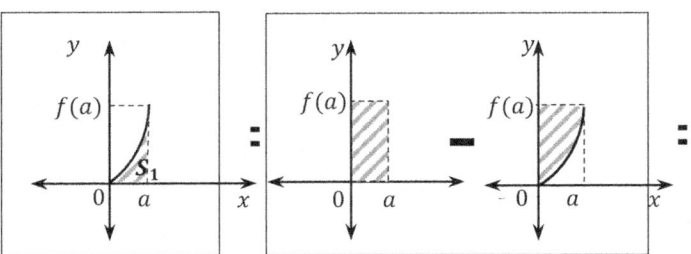

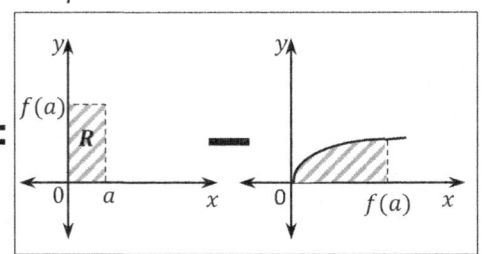

<u>Another Approach:</u>

$y = f(x) = e^x - 1 \Rightarrow e^x = y + 1 \quad \therefore x = \log(y + 1)$

Changing x and y, $y = \log(x + 1) \quad \therefore g(x) = \log(x + 1)$

$\int_0^a f(x)\,dx + \int_0^{f(a)} g(t)\,dt = \int_0^a (e^x - 1)\,dx + \int_0^{f(a)} \log(t + 1)\,dt$

$\qquad = [e^x - x]_0^a + [(t + 1)\log(t + 1) - (t + 1)]_0^{e^a - 1} = (e^a - a - 1) + e^a \log e^a - e^a + 1$

$\qquad = ae^a - a = a(e^a - 1)$

(8) Let S_1 be the area of the region bounded by a curve $y = x^2 - 4x + a$, the x-axis, and the y-axis, and S_2 be the area of the region bounded by the curve y and the x-axis. Find the area of a such that $S_1 : S_2 = 1 : 2$.

Let $f(x) = x^2 - 4x + a$

Then, $f(x) = (x - 2)^2 + a - 4$

$\therefore$ $f(x)$ is symmetric with respect to the line $x = 2$.

Note that the x-coordinates of the intersection points between the curve and the x-axis are the roots of the equation $x^2 - 4x + a = 0$.

Let α and β be the roots.

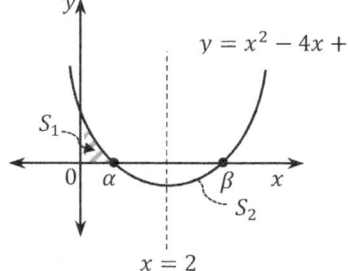

Then, $\int_\alpha^2 f(x)dx = \int_2^\beta f(x)dx$ ($\because$ symmetry)

$\therefore$ $S_2 = \int_\alpha^\beta |f(x)|dx = 2\int_\alpha^2 |f(x)|dx$

Since $S_1 : S_2 = 1 : 2$, $2S_1 = S_2$ $\therefore$ $S_1 = \frac{1}{2}S_2$

$\therefore$ $S_1 = \int_0^\alpha f(x)dx = \int_\alpha^2 |f(x)|dx = -\int_\alpha^2 f(x)dx$ ($\because$ $f(x) < 0$ on $(\alpha, 2)$)

$\therefore$ $\int_0^\alpha f(x)dx + \int_\alpha^2 f(x)dx = 0$

$\therefore$ $\int_0^2 f(x)dx = 0$; that is, $\int_0^2 (x^2 - 4x + a)dx = 0$

$\therefore$ $\left[\frac{1}{3}x^3 - 2x^2 + ax\right]_0^2 = \frac{1}{3} \cdot 2^3 - 2 \cdot 2^2 + 2a = -\frac{16}{3} + 2a = 0$

$\therefore$ $a = \frac{8}{3}$

(9) Let S_1 be the area of the region bounded by a curve $f(x) = x^2 - 7x + 10$, the x-axis, and the y-axis, and S_2 be the area of the region bounded by the curve $f(x)$ and the x-axis, and S_3 be the area of the region bounded by the curve $f(x)$, the x-axis, and $x = a$ $(a > 5)$. When $\{S_1, S_2, S_3\}$ forms an arithmetic sequence, find the value of $\int_0^a f(x)dx$.

$f(x) = x^2 - 7x + 10 = (x - 2)(x - 5)$

$\therefore$ $f(x) = 0$ $\Rightarrow$ $x = 2$ or $x = 5$

$0 \le x \le 2$ $\Rightarrow$ $f(x) \ge 0$

$2 \le x \le 5$ $\Rightarrow$ $f(x) < 0$

$x \ge 5$ $\Rightarrow$ $f(x) \ge 0$

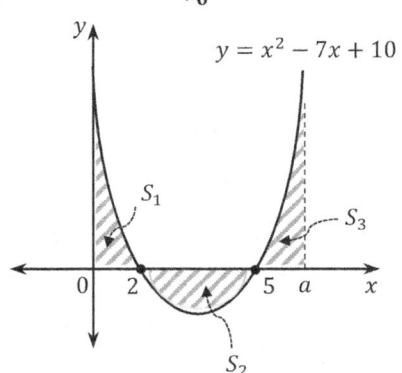

$\therefore$ $S_1 = \int_0^2 |f(x)|dx = \int_0^2 f(x)dx$

$S_2 = \int_2^5 |f(x)|dx = -\int_2^5 f(x)dx$

$S_3 = \int_5^a |f(x)|dx = \int_5^a f(x)dx$ ($\because$ $a > 5$)

Since $\{S_1, S_2, S_3\}$ is an arithmetic sequence, $2S_2 = S_1 + S_3$

$\therefore \int_0^a f(x)dx = \int_0^2 f(x)dx + \int_2^5 f(x)dx + \int_5^a f(x)dx = S_1 - S_2 + S_3 = (S_1 + S_3) - S_2$

$$= 2S_2 - S_2 = S_2 = -\int_2^5 f(x)dx = -\int_2^5 (x^2 - 7x + 10)dx$$

$$= -\left[\frac{1}{3}x^3 - \frac{7}{2}x^2 + 10x\right]_2^5 = -\left\{\frac{1}{3}(5^3 - 2^3) - \frac{7}{2}(5^2 - 2^2) + 10(5 - 2)\right\}$$

$$= -\left(\frac{117}{3} - \frac{7}{2}\cdot 21 + 30\right) = -39 + \frac{147}{2} - 30 = -69 + \frac{147}{2} = \frac{9}{2}$$

(10) For a differentiable function $f(x)$, $f(x) \geq 0$ for all $x \geq 0$.

When the area of the region bounded by $y = f(x)$, the x-axis, the y-axis, and $x = a$ $(a > 0)$ is $F(a) = a \sin a + ab$, find the value of $f(a)$ such that $f(0) = 0$.

Since $F(a) = \int_0^a f(x)dx$, $\int_0^a f(x)dx = a \sin a + ab$

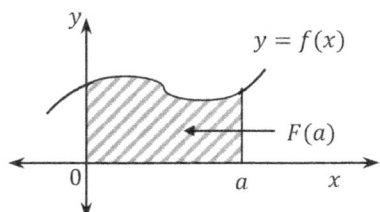

$\therefore \frac{d}{da}\left[\int_0^a f(x)dx\right] = \frac{d}{da}[a \sin a + ab]$

$\therefore f(a) = \sin a + a \cos a + b$

Since $f(0) = 0$, $\sin 0 + 0 \cos 0 + b = 0$; $b = 0$

$\therefore f(a) = \sin a + a \cos a$

(11) For the area S_n of the region bounded by a curve $y = -\frac{1}{n(n+1)}x(x+1)$ and the x-axis, find the value of $\displaystyle\sum_{n=1}^{\infty} S_n$.

$y = -\frac{1}{n(n+1)}x(x+1) \Rightarrow x = 0$ or $x = -1$

$\therefore S_n = \int_{-1}^0 \left\{-\frac{1}{n(n+1)}x(x+1)\right\}dx = -\frac{1}{n(n+1)}\int_{-1}^0 x(x+1)dx$

$$= -\frac{1}{n(n+1)}\left[\frac{1}{3}x^3 + \frac{1}{2}x^2\right]_{-1}^0 = -\frac{1}{n(n+1)}\left(\frac{1}{3} - \frac{1}{2}\right) = \frac{1}{6n(n+1)}$$

$\therefore \displaystyle\sum_{n=1}^{\infty} S_n = \lim_{n\to\infty}\sum_{k=1}^n \frac{1}{6k(k+1)} = \lim_{n\to\infty}\frac{1}{6}\sum_{k=1}^n\left(\frac{1}{k} - \frac{1}{k+1}\right) = \lim_{n\to\infty}\frac{1}{6}\left(1 - \frac{1}{n+1}\right) = \frac{1}{6}$

(12) Find the minimum value of the area of the region bounded by $y = x^3 + (x-a)^2 + 1$, $x = 0$, $x = 1$, and $y = 0$.

The area is $\int_0^1 |x^3 + (x-a)^2 + 1|dx = \int_0^1 (x^3 + (x-a)^2 + 1)dx$

$$= \int_0^1 (x^3 + x^2 - 2ax + a^2 + 1)dx = \left[\frac{1}{4}x^4 + \frac{1}{3}x^3 - ax^2 + a^2x + x\right]_0^1 = \frac{1}{4} + \frac{1}{3} - a + a^2 + 1$$

$$= a^2 - a + \frac{19}{12} = \left(a - \frac{1}{2}\right)^2 - \frac{1}{4} + \frac{19}{12} = \left(a - \frac{1}{2}\right)^2 + \frac{4}{3} \geq \frac{4}{3}$$

$\therefore$ The area of the region has minimum value $\frac{4}{3}$ when $a = \frac{1}{2}$.

#3 Find the area of the region bounded by the given curves.

(1) $y = 3x$, $y = x^2$ when $x > 0$.

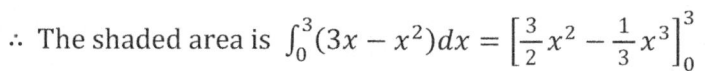

Finding the intersection points, $x^2 = 3x$; $x(x - 3) = 0$

∴ $x = 0$ and $x = 3$ are the intersection points of the graphs.

From the Figure, the graph of $y = 3x$ is above

the graph of $y = x^2$ on $[0, 3]$.

∴ The shaded area is $\int_0^3 (3x - x^2)\,dx = \left[\frac{3}{2}x^2 - \frac{1}{3}x^3\right]_0^3$

$$= \frac{3}{2} \cdot 3^2 - \frac{1}{3} \cdot 3^3 = 3^2\left(\frac{3}{2} - 1\right) = \frac{9}{2}$$

(2) $y = \frac{1}{x}$, $y = x$, $y = \frac{1}{2}x$, $x > 0$

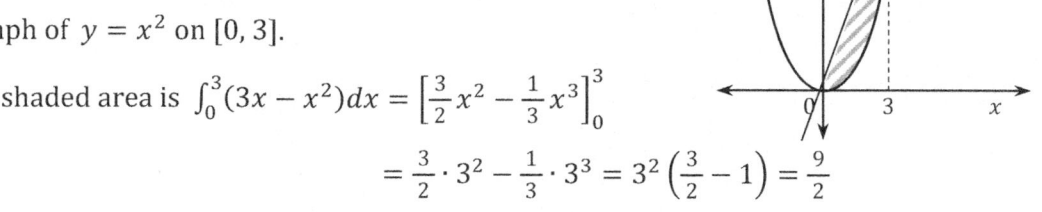

$\frac{1}{x} = \frac{1}{2}x \Rightarrow 1 = \frac{1}{2}x^2$; $x^2 = 2$

∴ $x = \pm\sqrt{2}$

Since $x > 0$, $x = \sqrt{2}$

The shaded area is

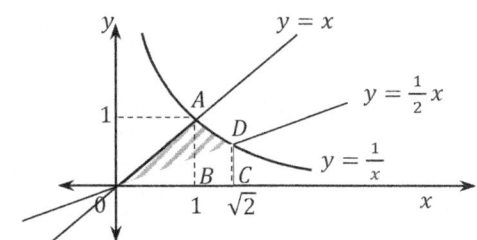

(The area of $_0\triangle{}^A_B$) + (The area of $_B\square{}^D_C$) − (The area of $_0\triangle{}^D_C$)

$$= \left(\frac{1}{2} \cdot 1 \cdot 1\right) + \left(\int_1^{\sqrt{2}} \frac{1}{x}\,dx\right) - \left(\frac{1}{2} \cdot \sqrt{2} \cdot \frac{1}{\sqrt{2}}\right)$$

$$= \frac{1}{2} + \left[\log|x|\right]_1^{\sqrt{2}} - \frac{1}{2} = \log\sqrt{2}$$

(3) $y = -x$, $y = -x^2 + 2x$

$-x = -x^2 + 2x \Rightarrow x^2 - 3x = 0$; $x(x - 3) = 0$

∴ $x = 0$ and $x = 3$ are the intersection points of the curves.

The shaded area is $\int_0^3 \{(-x^2 + 2x) - (-x)\}\,dx$

$$= \int_0^3 (-x^2 + 3x)\,dx$$

$$= \left[-\frac{1}{3}x^3 + \frac{3}{2}x^2\right]_0^3 = -\frac{1}{3} \cdot 3^3 + \frac{3}{2} \cdot 3^2 = 3^2\left(-1 + \frac{3}{2}\right) = \frac{9}{2}$$

(4) $y = x$, $y = x^3 - 6x^2 + 9x$

$x^3 - 6x^2 + 9x = x \Rightarrow x^3 - 6x^2 + 8x = 0 \Rightarrow x(x^2 - 6x + 8) = x(x - 2)(x - 4) = 0$

∴ $x = 0$, $x = 2$, and $x = 4$ are the intersection points of the curves.

On the interval $[0, 2]$, $x^3 - 6x^2 + 9x \geq x$

On the interval $[2, 4]$, $x \geq x^3 - 6x^2 + 9x$

The shaded area is

$$\int_0^2 (x^3 - 6x^2 + 9x - x)\, dx + \int_2^4 \{x - (x^3 - 6x^2 + 9x)\}\, dx$$

$$= \int_0^2 (x^3 - 6x^2 + 8x)\, dx + \int_2^4 (-x^3 + 6x^2 - 8x)\, dx$$

$$= \left[\frac{1}{4}x^4 - 2x^3 + 4x^2\right]_0^2 + \left[-\frac{1}{4}x^4 + 2x^3 - 4x^2\right]_2^4$$

$$= \left(\frac{1}{4}\cdot 2^4 - 2\cdot 2^3 + 4\cdot 2^2\right) + \left(-\frac{1}{4}\cdot 4^4 + 2\cdot 4^3 - 4\cdot 4^2\right) - \left(-\frac{1}{4}\cdot 2^4 + 2\cdot 2^3 - 4\cdot 2^2\right)$$

$$= (4 - 16 + 16) + 0 - (-4) = 4 + 4 = 8$$

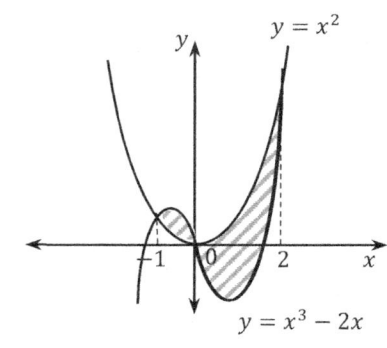

(5) $y = x^2, \ y = x^3 - 2x$

$$x^2 = x^3 - 2x \ \Rightarrow \ x^3 - x^2 - 2x = 0 \ \Rightarrow \ x(x^2 - x - 2) = 0 \ \Rightarrow \ x(x - 2)(x + 1) = 0$$

$\therefore \ x = 0, \ x = 2, \ \text{and} \ x = -1$ are the intersection points of the curves.

On the interval $[-1, 0]$, $x^3 - 2x \geq x^2$

On the interval $[0, 2]$, $x^2 \geq x^3 - 2x$

The shaded area is

$$\int_{-1}^0 (x^3 - 2x - x^2)\, dx + \int_0^2 \{x^2 - (x^3 - 2x)\}\, dx$$

$$= \left[\frac{1}{4}x^4 - x^2 - \frac{1}{3}x^3\right]_{-1}^0 + \left[\frac{1}{3}x^3 - \frac{1}{4}x^4 + x^2\right]_0^2$$

$$= -\left(\frac{1}{4} - 1 + \frac{1}{3}\right) + \left(\frac{1}{3}\cdot 2^3 - \frac{1}{4}\cdot 2^4 + 2^2\right) = \frac{5}{12} + \frac{8}{3} = \frac{37}{12}$$

(6) $y = 2x - x^2, \ y = x^4$

$$2x - x^2 = x^4 \ \Rightarrow \ x^4 + x^2 - 2x = 0 \ \Rightarrow \ x(x^3 + x - 2) = 0 \ \Rightarrow \ x(x - 1)(x^2 + x - 2) = 0$$

$$\Rightarrow \ x(x - 1)(x + 2)(x - 1) = 0 \ \Rightarrow \ x(x + 2)(x - 1)^2 = 0$$

$\therefore \ x = 0, \ x = -2, \ \text{and} \ x = 1$ are the intersection points of the curves.

The shaded area is

$$\int_0^1 (2x - x^2 - x^4)\, dx = \left[x^2 - \frac{1}{3}x^3 - \frac{1}{5}x^5\right]_0^1$$

$$= 1 - \frac{1}{3} - \frac{1}{5}$$

$$= \frac{15 - 5 - 3}{15} = \frac{7}{15}$$

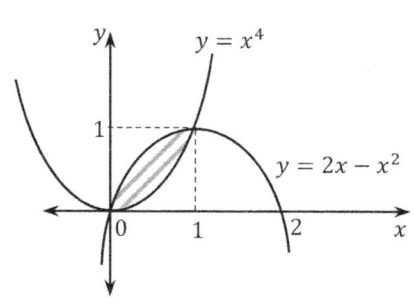

(7) $y = \frac{1}{3}x^2 + 1$, $y = -\frac{1}{3}x^2 + 3$, $x = 0$, $x = 3$

$\frac{1}{3}x^2 + 1 = -\frac{1}{3}x^2 + 3 \Rightarrow \frac{2}{3}x^2 = 2 \Rightarrow x^2 = 3$

$\therefore x = \pm\sqrt{3}$

$\therefore (\sqrt{3}, 2)$ and $(-\sqrt{3}, 2)$ are the intersection points of the curves.

The shaded area is

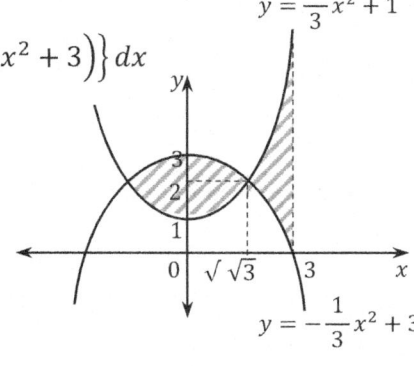

$\int_0^{\sqrt{3}} \left\{-\frac{1}{3}x^2 + 3 - \left(\frac{1}{3}x^2 + 1\right)\right\} dx + \int_{\sqrt{3}}^3 \left\{\frac{1}{3}x^2 + 1 - \left(-\frac{1}{3}x^2 + 3\right)\right\} dx$

$= \int_0^{\sqrt{3}} \left(-\frac{2}{3}x^2 + 2\right) dx + \int_{\sqrt{3}}^3 \left(\frac{2}{3}x^2 - 2\right) dx$

$= \left[-\frac{2}{9}x^3 + 2x\right]_0^{\sqrt{3}} + \left[\frac{2}{9}x^3 - 2x\right]_{\sqrt{3}}^3$

$= \left(-\frac{2}{9} \cdot 3\sqrt{3} + 2\sqrt{3}\right) + \left(\frac{2}{9} \cdot 3^3 - 6\right) - \left(\frac{2}{9} \cdot 3\sqrt{3} - 2\sqrt{3}\right)$

$= -\frac{12}{9}\sqrt{3} + 4\sqrt{3} = \left(4 - \frac{4}{3}\right)\sqrt{3} = \frac{8}{3}\sqrt{3}$

(8) $y = x^3 - 3x^2 + 2x$, $y = 0$, $x = 3$

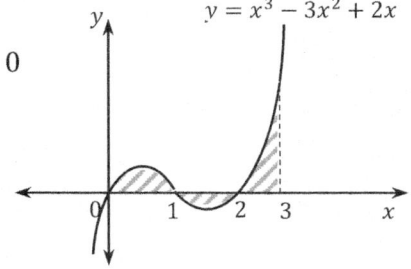

$y = x^3 - 3x^2 + 2x = x(x^2 - 3x + 2) = x(x - 2)(x - 1) = 0$

$\Rightarrow x = 0$, $x = 1$, or $x = 2$

$\therefore x = 0$, $x = 1$, and $x = 2$ are the intersection points

of the curve y and the x-axis.

$\therefore$ The shaded area is

$\int_0^1 y\, dx - \int_1^2 y\, dx + \int_2^3 y\, dx = \left[\frac{1}{4}x^4 - x^3 + x^2\right]_0^1 - \left[\frac{1}{4}x^4 - x^3 + x^2\right]_1^2 + \left[\frac{1}{4}x^4 - x^3 + x^2\right]_2^3$

$= \frac{1}{4} - 1 + 1 - \left(\frac{1}{4} \cdot 2^4 - 2^3 + 2^2\right) + \left(\frac{1}{4} - 1 + 1\right) + \left(\frac{1}{4} \cdot 3^4 - 3^3 + 3^2\right) - \left(\frac{1}{4} \cdot 2^4 - 2^3 + 2^2\right)$

$= \frac{1}{2} + \frac{9}{4} = \frac{11}{4}$

(9) $y = x^2$, $y = x + 2$, $y = x$

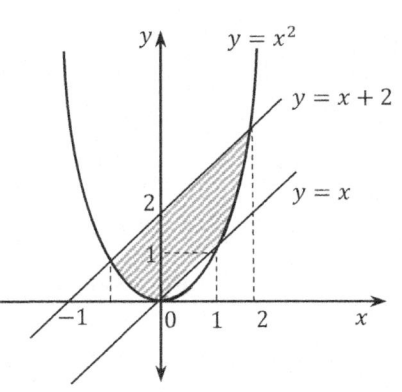

$x^2 = x + 2 \Rightarrow x^2 - x - 2 = (x - 2)(x + 1) = 0$

$\therefore x = 2$ and $x = -1$ are the intersection points of the curves.

From the Figure,

The shaded area is $\int_{-1}^2 (x + 2 - x^2)dx - \int_0^1 (x - x^2)dx$

$= \left[\frac{1}{2}x^2 + 2x - \frac{1}{3}x^3\right]_{-1}^2 - \left[\frac{1}{2}x^2 - \frac{1}{3}x^3\right]_0^1$

$= \left(\frac{1}{2} \cdot 2^2 + 2 \cdot 2 - \frac{1}{3} \cdot 2^3\right) - \left(\frac{1}{2}(-1)^2 + 2(-1) - \frac{1}{3}(-1)^3\right) - \left(\frac{1}{2} \cdot 1^2 - \frac{1}{3} \cdot 1^3\right)$

$= \left(6 - \frac{8}{3}\right) - \left(\frac{1}{2} - 2 + \frac{1}{3}\right) - \left(\frac{1}{2} - \frac{1}{3}\right) = 7 - \frac{8}{3} = \frac{13}{3}$

(10) $y \geq x, \quad x^2 - xy - xy^2 + y^3 \leq 0$

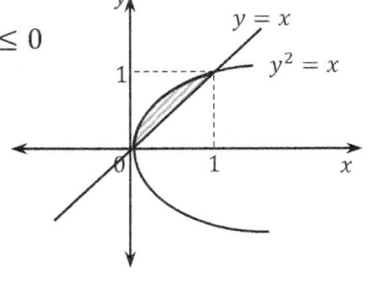

$x^2 - xy - xy^2 + y^3 = x(x - y) - y^2(x - y) = (x - y)(x - y^2) \leq 0$

Since $y \geq x, \quad x - y \leq 0$

$\therefore \ x - y^2 \geq 0 \ ; \quad y^2 \leq x$

The shaded area is $\int_0^1 x dy = \int_0^1 (y - y^2)dy = \left[\frac{1}{2}y^2 - \frac{1}{3}y^3\right]_0^1$

$$= \frac{1}{2} - \frac{1}{3} = \frac{1}{6}$$

Or, $\int_0^1 y dx = \int_0^1 (\sqrt{x} - x)dy = \left[\frac{2}{3}x^{\frac{3}{2}} - \frac{1}{2}x^2\right]_0^1 = \frac{2}{3} - \frac{1}{2} = \frac{1}{6}$

(11) $y = xe^{-x}, \quad y = e^{-2}x$

$xe^{-x} = e^{-2}x \ \Rightarrow \ x(e^{-x} - e^{-2}) = 0 \ \Rightarrow \ x = 0 \text{ or } e^{-x} = e^{-2}$

$\therefore$ The intersection points of the curves are $x = 0$ and $x = 2$.

The shaded area is

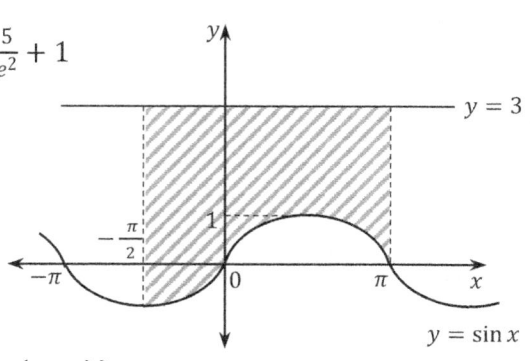

$\int_0^2 (xe^{-x} - e^{-2}x)dy = \int_0^2 xe^{-x}dy - e^{-2}\int_0^2 xdy$

For $\int xe^{-x}dx$,

Let $\quad u = x, \ v' = e^{-x}dx$

Then, $\quad u' = dx, \quad v = -e^{-x}$

$\therefore \ \int xe^{-x}dx = -xe^{-x} + \int e^{-x}dx = -xe^{-x} - e^{-x} + C$

$\therefore \ \int_0^2 xe^{-x}dy = [-xe^{-x} - e^{-x}]_0^2 = (-2e^{-2} - e^{-2}) - (0 - e^{-0}) = -3e^{-2} + 1$

$\int_0^2 xdy = \left[\frac{1}{2}x^2\right]_0^2 = \frac{1}{2} \cdot 2^2 = 2$

Therefore, $\int_0^2 (xe^{-x} - e^{-2}x)dy = -3e^{-2} + 1 - 2e^{-2}$

$$= -5e^{-2} + 1 = -\frac{5}{e^2} + 1$$

(12) $y = \sin x, \quad y = 3, \quad x = -\frac{\pi}{2}, \quad x = \pi$

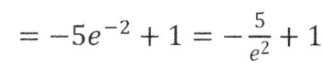
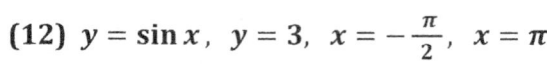

Since $3 > \sin x$ on the interval $\left[-\frac{\pi}{2}, \pi\right]$,

the shaded area is

$\int_{-\frac{\pi}{2}}^{\pi} (3 - \sin x)dx = [3x + \cos x]_{-\frac{\pi}{2}}^{\pi}$

$$= (3\pi + \cos \pi) - \left\{3\left(-\frac{\pi}{2}\right) + \cos\left(-\frac{\pi}{2}\right)\right\}$$

$$= 3\pi - 1 + \frac{3\pi}{2} = \frac{9\pi}{2} - 1$$

(13) $y = \sin x$, $y = \sin 2x$, $x = 0$, $x = \pi$

$\sin x = \sin 2x \implies \sin x = 2 \sin x \cos x \implies \sin x (2 \cos x - 1) = 0$

$\implies \sin x = 0$ or $\cos x = \dfrac{1}{2}$

$\therefore x = 0$, $x = \pi$, or $x = \dfrac{\pi}{3}$

$\therefore$ The intersection points of the curves are

$\quad x = 0$, $x = \pi$, and $x = \dfrac{\pi}{3}$.

The shaded area is

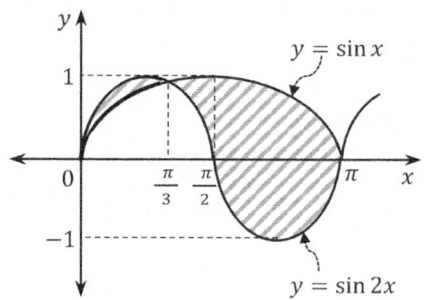

$\int_0^{\frac{\pi}{3}} (\sin 2x - \sin x)dx + \int_{\frac{\pi}{3}}^{\pi} (\sin x - \sin 2x)dx$

$= \left[-\dfrac{1}{2}\cos 2x + \cos x \right]_0^{\frac{\pi}{3}} + \left[-\cos x + \dfrac{1}{2}\cos 2x \right]_{\frac{\pi}{3}}^{\pi}$

$= \left(-\dfrac{1}{2}\cos\dfrac{2\pi}{3} + \cos\dfrac{\pi}{3} \right) - \left(-\dfrac{1}{2}\cos 0 + \cos 0 \right) + \left(-\cos\pi + \dfrac{1}{2}\cos 2\pi \right) - \left(-\cos\dfrac{\pi}{3} + \dfrac{1}{2}\cos\dfrac{2\pi}{3} \right)$

$= \left(-\dfrac{1}{2}\left(-\dfrac{1}{2}\right) + \dfrac{1}{2} \right) - \left(-\dfrac{1}{2} + 1 \right) + \left(1 + \dfrac{1}{2} \right) - \left(-\dfrac{1}{2} + \dfrac{1}{2}\left(-\dfrac{1}{2}\right) \right)$

$= \dfrac{5}{2}$

(14) $y = x^2$, $y^2 = 27x$

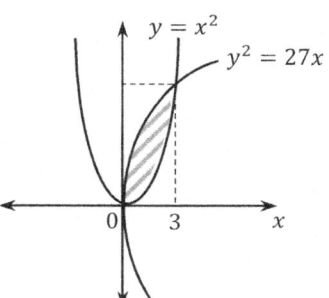

$(x^2)^2 = 27x$; $x^4 = 27x$; $x(x^3 - 27) = 0$; $x(x^3 - 3^3) = 0$

$\therefore x = 0$ or $x = 3$

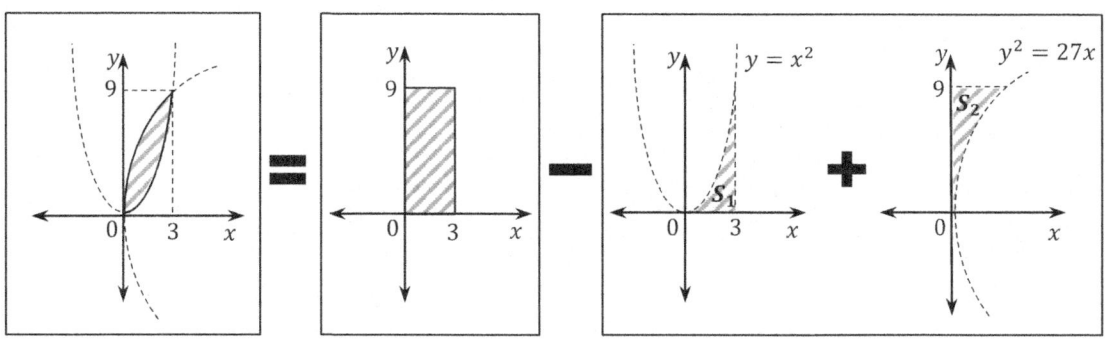

$S_1 = \int_0^3 x^2 dx = \left[\dfrac{1}{3}x^3 \right]_0^3 = \dfrac{1}{3}\cdot 3^3 = 9$

$S_1 = \int_0^9 \dfrac{1}{27}y^2 dy = \dfrac{1}{27}\left[\dfrac{1}{3}y^3 \right]_0^9 = \dfrac{1}{27}\cdot\dfrac{1}{3}\cdot 9^3 = 9$

$\therefore$ The shaded area is $3 \cdot 9 - (9 + 9) = 27 - 18 = 9$

(15) $y = -x + 2, \ y^2 = x$

$y = -x + 2 \ \Rightarrow \ x = -y + 2$

$\therefore \ y^2 = -y + 2 \ ; \ \ y^2 + y - 2 = 0 \ ; \ \ (y + 2)(y - 1) = 0$

$\therefore \ $ The intersection points of the curves are $y = -2$ and $y = 1$.

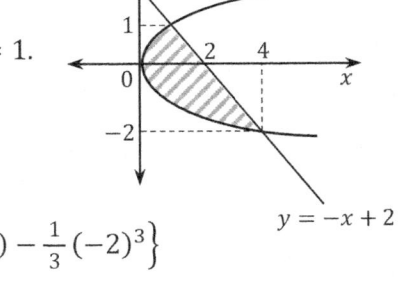

The shaded area is

$\int_{-2}^{1}(-y + 2 - y^2)dy = \left[-\frac{1}{2}y^2 + 2y - \frac{1}{3}y^3\right]_{-2}^{1}$

$\qquad = \left(-\frac{1}{2} + 2 - \frac{1}{3}\right) - \left\{-\frac{1}{2}(-2)^2 + 2(-2) - \frac{1}{3}(-2)^3\right\}$

$\qquad = \left(-\frac{1}{2} + 2 - \frac{1}{3}\right) - \left(-2 - 4 + \frac{8}{3}\right) = \frac{9}{2}$

(16) $y = x^2, \ y = \sqrt{x}$

$x^2 = \sqrt{x} \ \Rightarrow \ x^4 = x \ \Rightarrow \ x(x^3 - 1) = 0$

$\Rightarrow \ x(x - 1)(x^2 + x + 1) = 0$

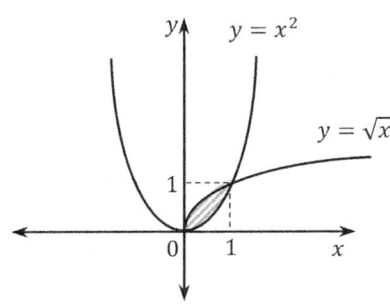

Since $x^2 + x + 1 = \left(x + \frac{1}{2}\right)^2 - \frac{1}{4} + 1 = \left(x + \frac{1}{2}\right)^2 + \frac{3}{4} > 0$,

$x = 0$ or $x = 1$

$\therefore \ $ The intersection points of the curves are

$\qquad x = 0$ and $x = 1$.

The shaded area is

$\int_{0}^{1}(\sqrt{x} - x^2)dx = \left[\frac{2}{3}x^{\frac{3}{2}} - \frac{1}{3}x^3\right]_{0}^{1} = \frac{2}{3} - \frac{1}{3} = \frac{1}{3}$

(17) $x - y = 3, \ x = y^2 - y$

$y + 3 = y^2 - y \ ; \ \ y^2 - 2y - 3 = 0 \ ; \ \ (y - 3)(y + 1) = 0$

$\therefore \ y = 3$ or $y = -1$

$\therefore \ $ The intersection points of the curves are $y = 3$ and $y = -1$.

The shaded area is

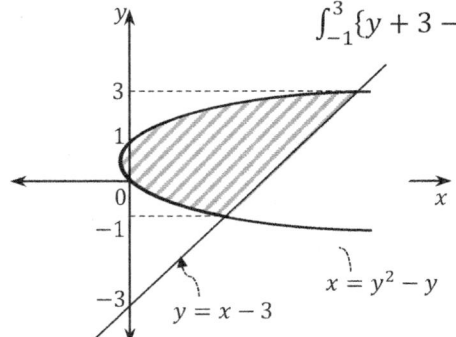

$\int_{-1}^{3}\{y + 3 - (y^2 - y)\}dy = \int_{-1}^{3}(-y^2 + 2y + 3)dy$

$\qquad = \left[-\frac{1}{3}y^3 + y^2 + 3y\right]_{-1}^{3}$

$\qquad = \left(-\frac{1}{3} \cdot 3^3 + 3^2 + 9\right) - \left(\frac{1}{3} + 1 - 3\right)$

$\qquad = \frac{32}{3}$

(18) $y = x^2$, $y = 4\sqrt{x} - 3$, $x = 0$

$x^2 = 4\sqrt{x} - 3 \;\Rightarrow\; x^2 + 3 = 4\sqrt{x}$

Squaring both sides, $(x^2 + 3)^2 = 16x$

$\therefore\;\; x^4 + 6x^2 - 16x + 9 = 0$

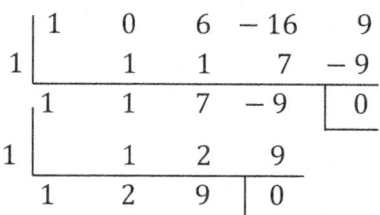

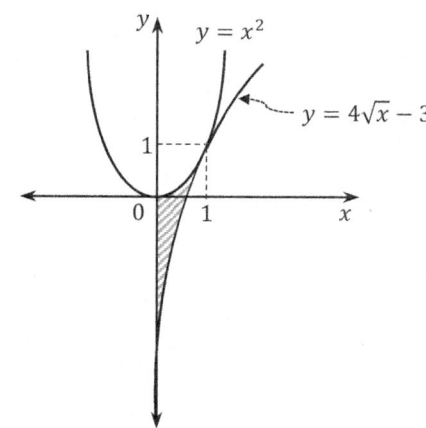

$\therefore\;\; (x-1)^2(x^2 + 2x + 9) = 0$

The shaded area is $\int_0^1 \{x^2 - (4\sqrt{x} - 3)\}dx = \left[\frac{1}{3}x^3 - 4 \cdot \frac{2}{3}x^{\frac{3}{2}} + 3x\right]_0^1 = \frac{1}{3} - \frac{8}{3} + 3 = \frac{2}{3}$

(19) $y = x$, $y = e^x$, $y = 1$, $y = 3$

$y = e^x \;\Rightarrow\; x = \log y$

The shaded area is

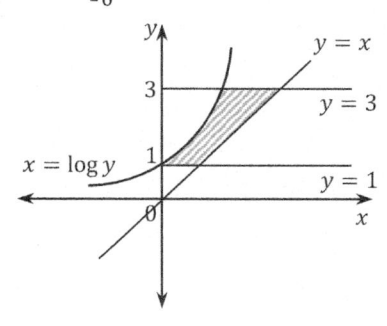

$\int_1^3 (y - \log y)dy = \left[\frac{1}{2}y^2 - (y\log y - y)\right]_1^3$

$\qquad = \left(\frac{1}{2} \cdot 3^2 - (3\log 3 - 3)\right) - \left(\frac{1}{2} - (0 - 1)\right)$

$\qquad = 6 - 3\log 3$

(20) $y = e^x$, $y = ex$, $x = 0$

Since the slope of the line passing through the origin and a point $(1, e)$ on the graph is $\frac{e-0}{1-0} = e$,

the equation of the tangent line at $(0, 0)$ to the graph of $y = e^x$ is $y - 0 = e(x - 0)$; $y = ex$

The shaded area is

$\int_0^1 (e^x - ex)dx = \left[e^x - \frac{1}{2}ex^2\right]_0^1$

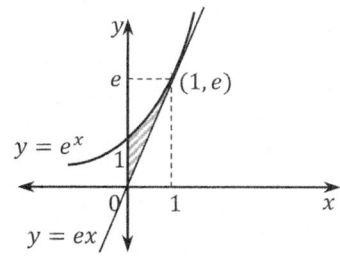

$= \left(e - \frac{1}{2}e\right) - (e^0 - 0) = \frac{1}{2}e - 1$

(21) $y = \log x$, $y = \frac{x}{e}$, $y = 0$

The equation of the tangent line at the origin to the graph of $y = \log x$ is $y = \frac{1}{e}x$ and the point

of tangency is $(e, 1)$.

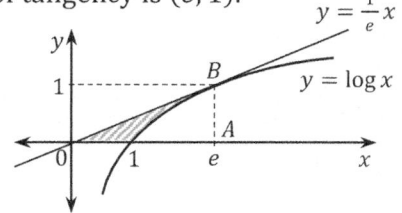

The shaded area is

(The area of the triangle $\triangle OAB$) $- \int_1^e \log x \, dx$

$= \frac{1}{2} \cdot e \cdot 1 - [x\log x - x]_1^e = \frac{e}{2} - (e - e) + (0 - 1) = \frac{e}{2} - 1$

(22) $y = \frac{1}{1+x^2}$, $x = \sqrt{3}$, x-axis, y-axis

The shaded area is $\int_0^{\sqrt{3}} \left(\frac{1}{1+x^2}\right) dx$.

Let $x = \tan\theta$

Then, $dx = \sec^2\theta \, d\theta$

$\theta = 0$ when $x = 0$

$\theta = \frac{\pi}{3}$ when $x = \sqrt{3}$

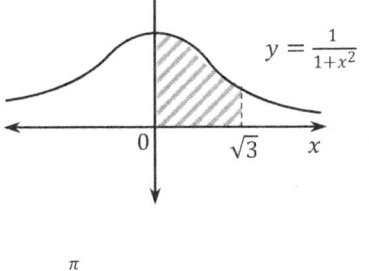

$y = \frac{1}{1+x^2}$

Therefore, $\int_0^{\sqrt{3}} \left(\frac{1}{1+x^2}\right) dx = \int_0^{\frac{\pi}{3}} \left(\frac{1}{1+\tan^2\theta}\right) \sec^2\theta \, d\theta = \int_0^{\frac{\pi}{3}} d\theta = [\theta]_0^{\frac{\pi}{3}} = \frac{\pi}{3}$

(23) $y = x^{2n-1}$, $y = x^{2n+1}$

$x^{2n-1} = x^{2n+1} \Rightarrow x^{2n-1} - x^{2n+1} = 0 \Rightarrow x^{2n-1}(1 - x^2) = 0 \Rightarrow x^{2n-1}(1-x)(1+x) = 0$

$\therefore x = 0$, $x = 1$, or $x = -1$

$\therefore$ The intersection points of the curves are $x = 0$, $x = 1$, and $x = -1$.

i) When $-1 \le x \le 0$, $x^{2n+1} \ge x^{2n-1}$

ii) When $0 \le x \le 1$, $x^{2n+1} \le x^{2n-1}$

Therefore, the area of the region bounded by the curve is

$\int_{-1}^{1} |x^{2n+1} - x^{2n-1}| dx$

$= \int_{-1}^{0} (x^{2n+1} - x^{2n-1}) dx + \int_0^1 (x^{2n-1} - x^{2n+1}) dx$

$= \int_0^1 (x^{2n-1} - x^{2n+1}) dx + \int_0^1 (x^{2n-1} - x^{2n+1}) dx$

$= 2 \int_0^1 (x^{2n-1} - x^{2n+1}) dx$

$= 2 \left[\frac{1}{2n} x^{2n} - \frac{1}{2n+2} x^{2n+2}\right]_0^1 = 2 \left(\frac{1}{2n} - \frac{1}{2n+2}\right) = \frac{1}{n} - \frac{1}{n+1} = \frac{1}{n(n+1)}$

(24) $y = x^n$, $y^n = x$ $(x \ge 0)$

Since $y = x^n$ and $y^n = x$ are inverse functions of each other,

they are symmetric with respect to the line $y = x$.

Thus, the intersection point of two curves is

the intersection point of $y = x^n$ and $y = x$.

$x^n = x \Rightarrow x(x^{n-1} - 1) = 0$ $\therefore x = 0$ or $x = 1$

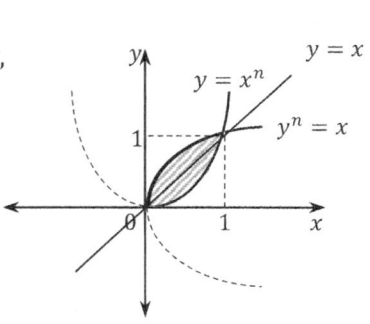

Therefore, the shaded area is

$2 \int_0^1 (x - x^n) dx = 2 \left[\frac{1}{2} x^2 - \frac{1}{n+1} x^{n+1}\right]_0^1 = 2 \left(\frac{1}{2} - \frac{1}{n+1}\right) = 1 - \frac{2}{n+1}$

#4 Find the area.

(1) Find the area of the region bounded by the curves $y = \sqrt{x+1}$, the normal line at

$(3, 2)$ **on the graph of y, and the x-axis.**

$y = \sqrt{x+1} \implies y' = \frac{1}{2\sqrt{x+1}}$

$\therefore$ The slope of the tangent line at $(3, 2)$ is $\frac{1}{2\sqrt{3+1}} = \frac{1}{4}$.

$\therefore$ The slope of the normal line at $(3, 2)$ is -4.

$\therefore$ The equation of the normal line at $(3, 2)$ to the curve is $y - 2 = -4(x - 3)$;

that is, $y = -4x + 14$.

When $y = 0,\ 4x = 14$

$\therefore\ x = \frac{14}{4} = \frac{7}{2}$

Therefore, the shaded area is

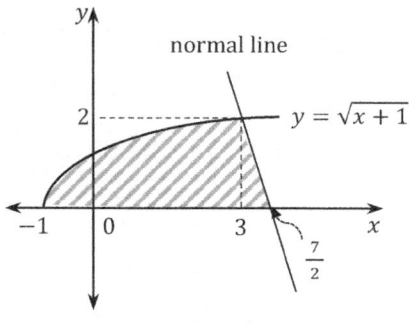

$\int_{-1}^{3} \sqrt{x+1}\, dx + \frac{1}{2} \cdot \left(\frac{7}{2} - 3\right) \cdot 2 = \left[\frac{2}{3}(x+1)^{\frac{3}{2}}\right]_{-1}^{3} + \frac{1}{2}$

$\qquad = \frac{2}{3}(3+1)^{\frac{3}{2}} + \frac{1}{2} = \frac{16}{3} + \frac{1}{2}$

$\qquad = \frac{35}{6}$

(2) Find the area of the region bounded by the curves $y = x^3 - 3x^2 + 5$ and the tangent

line at the local minimum value of the curve.

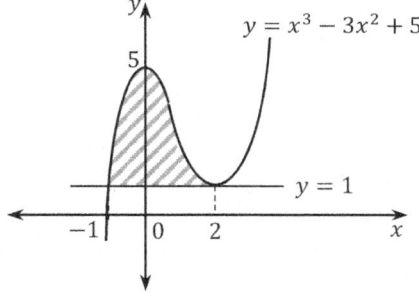

$y = x^3 - 3x^2 + 5 \implies y' = 3x^2 - 6x = 3x(x - 2)$

$\therefore\ y$ has local extreme values at $x = 0$ and $x = 2$.

At $x = 0,\ y = 5$

At $x = 2,\ y = 8 - 12 + 5 = 1$

From the graph, y has local minimum value 1 at $x = 2$.

$\therefore$ The equation of the tangent line is $y = 1$.

$x^3 - 3x^2 + 5 = 1 \implies x^3 - 3x^2 + 4 = 0 \implies (x+1)(x^2 - 4x + 4) = 0$

$\qquad\qquad \implies (x+1)(x-2)^2 = 0$

$\therefore\ x = -1$ or $x = 2$

Therefore, the shaded area is

$\int_{-1}^{2}(x^3 - 3x^2 + 5 - 1)dx = \int_{-1}^{2}(x^3 - 3x^2 + 4)dx = \left[\frac{1}{4}x^4 - x^3 + 4x\right]_{-1}^{2}$

$\qquad\qquad = \left(\frac{1}{4} \cdot 16 - 8 + 8\right) - \left(\frac{1}{4} + 1 - 4\right) = 7 - \frac{1}{4} = \frac{27}{4}$

(3) For a curve $y = x^3 - x$, a line l passing through the origin is perpendicular to the tangent line at the origin to the curve.

Find the area of the region bounded by the curve and the line l.

$y = x^3 - x = x(x^2 - 1) = x(x - 1)(x + 1)$

$y = x^3 - x \Rightarrow y' = 3x^2 - 1$

At $x = 0$, $y' = -1$

$\therefore$ The equation of the tangent line at $(0, 0)$ on the curve is

$\quad y - 0 = -1(x - 0)$; $y = -x$.

$\therefore$ The slope of the normal line l is 1.

Since the line l passes through $(0, 0)$,

the equation of the line l is $y - 0 = 1(x - 0)$; $y = x$.

To find the intersection points between the curve and the line l, set $x^3 - x = x$.

Then, $x^3 - 2x = 0$; $x(x^2 - 2) = 0$; $x(x + \sqrt{2})(x - \sqrt{2}) = 0$

$\therefore$ The intersection points are $x = 0$, $x = -\sqrt{2}$, and $x = \sqrt{2}$.

Therefore, the shaded area is

$2\int_0^{\sqrt{2}}\{x - (x^3 - x)\}dx = 2\int_0^{\sqrt{2}}(-x^3 + 2x)dx = 2\left[-\frac{1}{4}x^4 + x^2\right]_0^{\sqrt{2}} = 2(-1 + 2) = 2$

(4) When the tangent line at $(2, 1)$ on the curve $y = x^3 + ax + b$ is $y = x + c$, find the area of the region bounded by the curve and the line.

Since the curve and the line pass through $(2, 1)$, $1 = 8 + 2a + b$ and $1 = 2 + c$.

$\therefore$ $2a + b = -7$ and $c = -1$

$y = x^3 + ax + b \Rightarrow y' = 3x^2 + a$

When $x = 2$, $y' = 12 + a$

Since the slope of the tangent line at $(2, 1)$ is 1, $12 + a = 1$

$\therefore$ $a = -11$, $b = -2a - 7 = -2(-11) - 7 = 15$

$\therefore$ $y = x^3 - 11x + 15$ and the tangent line is $y = x - 1$

Let $x^3 - 11x + 15 = x - 1$

Then, $x^3 - 12x + 16 = 0$

$\therefore$ $(x - 2)(x^2 + 2x - 8) = 0$; $(x - 2)(x + 4)(x - 2) = 0$

$\therefore$ The intersection points are $x = -4$ and $x = 2$.

The shaded area is $\int_{-4}^{2}\{(x^3 - 11x + 15) - (x - 1)\}dx = \int_{-4}^{2}(x^3 - 12x + 16)dx$

$= \left[\frac{1}{4}x^4 - 6x^2 + 16x\right]_{-4}^{2} = (4 - 24 + 32) - (4^3 - 96 - 64) = 12 + 96 = 108$

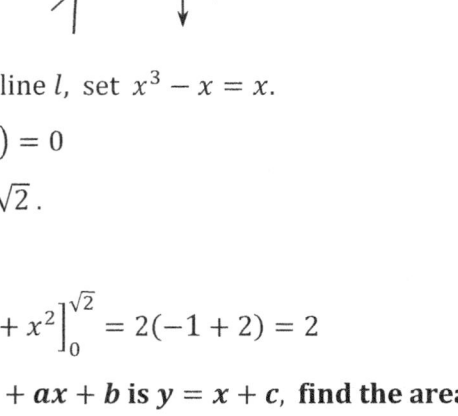

(5) When $x = 2$ is the point of tangency for the curves $f(x) = x^3 - (a + 1)x^2 + ax$ and

$g(x) = x^2 - ax$, find the area of the region bounded by the two curves.

$f'(x) = 3x^2 - 2(a + 1)x + a$

$g'(x) = 2x - a$

Since $f'(2) = g'(2)$, $12 - 4(a + 1) + a = 4 - a$

$\therefore \ 4 - 2a = 0 \qquad \therefore \ a = 2$

$\therefore \ f(x) = x^3 - 3x^2 + 2x = x(x^2 - 3x + 2) = x(x - 1)(x - 2)$

$\quad g(x) = x^2 - 2x = x(x - 2)$

The shaded area is

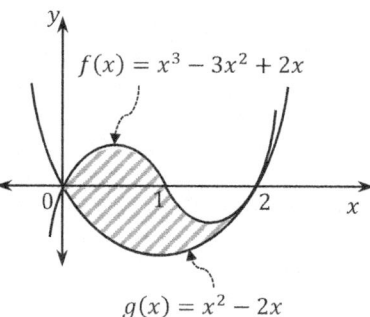

$\int_0^2 \{(x^3 - 3x^2 + 2x) - (x^2 - 2x)\}dx = \int_0^2 (x^3 - 4x^2 + 4x)dx$

$= \left[\frac{1}{4}x^4 - \frac{4}{3}x^3 + 2x^2\right]_0^2 = 4 - \frac{32}{3} + 8 = \frac{4}{3}$

(6) When a curve $y = x\log x + x + a$ and a line $y = 0$ intersect only at one point,

find the area of the region bounded by the curve y, $x = 1$, and $y = 0$.

$y = x\log x + x + a \ \Rightarrow \ y' = \log x + x\frac{1}{x} + 1 = \log x + 2$

$y' = 0 \ \Rightarrow \ \log x = -2 \qquad \therefore \ x = e^{-2}$

$\therefore \ y$ has local extreme value at $x = e^{-2}$.

Since the curve has a point of tangency on the x-axis,

the local minimum value is 0 at $x = e^{-2}$.

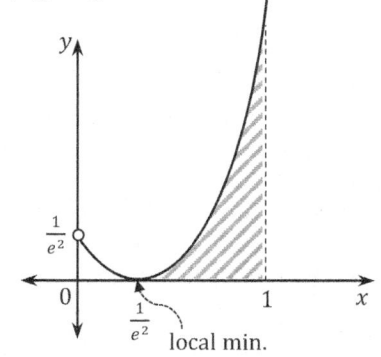

At $x = e^{-2}$, $y = e^{-2}\log e^{-2} + e^{-2} + a = 0$; $-2e^{-2} + e^{-2} + a = 0$

$\therefore \ -e^{-2} + a = 0 ; \quad a = e^{-2} = \frac{1}{e^2}$

Thus, the shaded area is

$\int_{\frac{1}{e^2}}^1 \left(x\log x + x + \frac{1}{e^2}\right) dx = \int_{\frac{1}{e^2}}^1 x\log x \, dx + \int_{\frac{1}{e^2}}^1 \left(x + \frac{1}{e^2}\right) dx$

For $\int x\log x \, dx$,

Let $\quad u = \log x, \quad v' = xdx$

Then, $u' = \frac{1}{x}dx, \quad v = \frac{1}{2}x^2$

$\therefore \ \int x\log x \, dx = \frac{1}{2}x^2\log x - \int \frac{1}{2}x^2 \cdot \frac{1}{x} dx = \frac{1}{2}x^2\log x - \frac{1}{2}\int xdx = \frac{1}{2}x^2\log x - \frac{1}{4}x^2 + C$

$\therefore \ \int_{\frac{1}{e^2}}^1 x\log x \, dx = \left[\frac{1}{2}x^2\log x - \frac{1}{4}x^2\right]_{\frac{1}{e^2}}^1 = -\frac{1}{4} - \left(\frac{1}{2} \cdot \frac{1}{e^4}\log \frac{1}{e^2} - \frac{1}{4} \cdot \frac{1}{e^4}\right)$

$= -\frac{1}{4} - \left(\frac{1}{2} \cdot \frac{1}{e^4}(-2) - \frac{1}{4} \cdot \frac{1}{e^4}\right) = -\frac{1}{4} + \frac{1}{e^4} + \frac{1}{4e^4}$

$\int_{\frac{1}{e^2}}^{1} \left(x + \frac{1}{e^2} \right) dx = \left[\frac{1}{2} x^2 + \frac{1}{e^2} x \right]_{\frac{1}{e^2}}^{1} = \left(\frac{1}{2} + \frac{1}{e^2} \right) - \left(\frac{1}{2} \cdot \frac{1}{e^4} + \frac{1}{e^2} \cdot \frac{1}{e^2} \right) = \frac{1}{2} + \frac{1}{e^2} - \frac{3}{2e^4}$

Therefore, $\int_{\frac{1}{e^2}}^{1} \left(x \log x + x + \frac{1}{e^2} \right) dx = \left(-\frac{1}{4} + \frac{1}{e^4} + \frac{1}{4e^4} \right) + \left(\frac{1}{2} + \frac{1}{e^2} - \frac{3}{2e^4} \right) = \frac{1}{4} + \frac{1}{e^2} - \frac{1}{4e^4}$

(7) When two curves $y = ax^2$ and $y = \log x$ intersect only at one point, find the area of the region bounded by the curves and the x-axis.

Let $y = f(x) = ax^2$ Then, $f'(x) = 2ax$

Let $y = g(x) = \log x$ Then, $g'(x) = \frac{1}{x}$

Let t be the x-coordinate of the point of tangency.

Then, $f(t) = g(t)$ and $f'(t) = g'(t)$

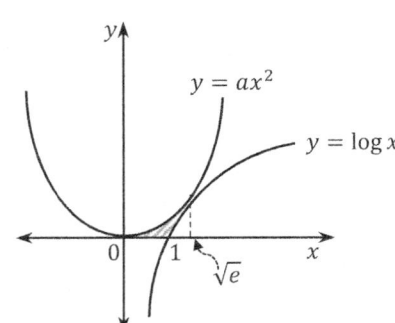

$\therefore$ $at^2 = \log t$ and $2at = \frac{1}{t}$; $2at^2 = 1$; $at^2 = \frac{1}{2}$

$\therefore$ $\frac{1}{2} = \log t$; $t = e^{\frac{1}{2}} = \sqrt{e}$

Since $2at = \frac{1}{t}$, $a = \frac{1}{2t^2} = \frac{1}{2e}$ $\therefore$ $f(x) = \frac{1}{2e} x^2$

Therefore, the shaded area is

$\int_0^{\sqrt{e}} \frac{1}{2e} x^2 dx - \int_1^{\sqrt{e}} \log x \, dx = \frac{1}{2e} \left[\frac{1}{3} x^3 \right]_0^{\sqrt{e}} - [x \log x - x]_1^{\sqrt{e}}$

$= \frac{1}{2e} \left(\frac{1}{3} e\sqrt{e} \right) - \sqrt{e} \log \sqrt{e} + \sqrt{e} - 1 = \frac{1}{6} \sqrt{e} - \frac{1}{2} \sqrt{e} + \sqrt{e} - 1 = \frac{2}{3} \sqrt{e} - 1$

(8) Find the area of the region bounded by the curves $y = e^x$, a tangent line at $(0, 0)$ to the curve, a vertical line $x = -2$, and the x-axis.

$y = e^x \Rightarrow y' = e^x$

The equation of the tangent line at (t, e^t) on the curve $y = e^x$ is

$y - e^t = e^t(x - t)$ ······· ①

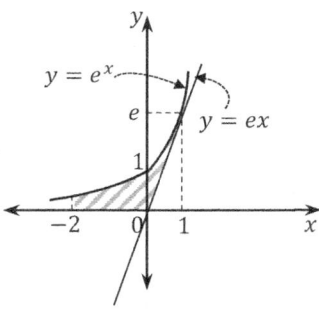

Since ① passes through the origin, $0 - e^t = e^t(0 - t)$; $-e^t = -te^t$

$\therefore$ $t = 1$

Thus, the equation of the tangent line is $y - e = e(x - 1)$; $y = ex$

Therefore, the shaded area is

$\int_{-2}^{1} e^x dx - \int_0^1 ex dx = [e^x]_{-2}^{1} - e \left[\frac{1}{2} x^2 \right]_0^1 = (e - e^{-2}) - \frac{1}{2} e = \frac{1}{2} e - \frac{1}{e^2}$

(9) Find the area of the ellipse $\frac{x^2}{a^2} + \frac{y^2}{b^2} = 1$. $(a > 0, \ b > 0)$

$\frac{x^2}{a^2} + \frac{y^2}{b^2} = 1 \Rightarrow \frac{y^2}{b^2} = 1 - \frac{x^2}{a^2} \Rightarrow y^2 = b^2 \left(1 - \frac{x^2}{a^2} \right) = b^2 \left(\frac{a^2 - x^2}{a^2} \right)$

$\therefore$ $y = \pm \frac{b}{a} \sqrt{a^2 - x^2}$

The shaded area is $4 \times S_1 = 4 \left(\int_0^a \frac{b}{a} \sqrt{a^2 - x^2} \, dx \right)$.

Let $x = a \sin \theta$

Then, $dx = a \cos \theta \, d\theta$

$\theta = 0$ when $x = 0$

$\theta = \frac{\pi}{2}$ when $x = a$

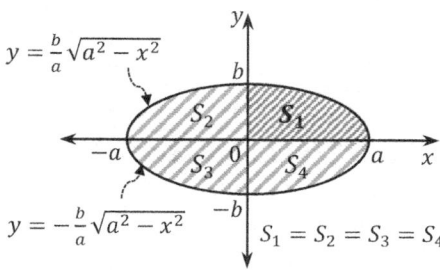

$$\therefore \int_0^a \frac{b}{a} \sqrt{a^2 - x^2} \, dx = \int_0^{\frac{\pi}{2}} \frac{b}{a} \sqrt{a^2 - a^2 \sin^2 \theta} \; (a \cos \theta \, d\theta)$$

$$= \int_0^{\frac{\pi}{2}} ba\sqrt{1 - \sin^2 \theta} \; \cos \theta \, d\theta = ab \int_0^{\frac{\pi}{2}} \cos^2 \theta \, d\theta = ab \int_0^{\frac{\pi}{2}} \frac{1 + \cos 2\theta}{2} \, d\theta$$

$$= \frac{ab}{2} \int_0^{\frac{\pi}{2}} (1 + \cos 2\theta) d\theta = \frac{ab}{2} \left[\theta + \frac{1}{2} \sin 2\theta \right]_0^{\frac{\pi}{2}} = \frac{ab}{2} \left(\frac{\pi}{2} + \frac{1}{2} \sin \pi \right)$$

$$= \frac{ab\pi}{4}$$

Therefore, the shaded area is $4 \times S_1 = 4 \times \frac{ab\pi}{4} = ab\pi$

(10) For a differentiable function $f(x)$ $(x \geq 0)$, $\int_a^x f(t)dt = \frac{4}{3}x\sqrt{x} - \frac{1}{2}x^2 + 1$.

Find the area of the region bounded by $y = f(x)$ and the x-axis.

$\int_a^x f(t)dt = \frac{4}{3}x\sqrt{x} - \frac{1}{2}x^2 + 1$

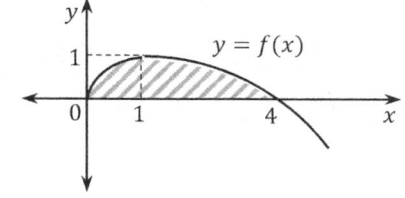

$\Rightarrow \frac{d}{dx}\left(\int_a^x f(t)dt \right) = \frac{d}{dx}\left(\frac{4}{3}x\sqrt{x} - \frac{1}{2}x^2 + 1 \right)$

$\therefore f(x) = \frac{4}{3} \cdot \frac{3}{2}x^{\frac{1}{2}} - \frac{1}{2} \cdot 2x = 2\sqrt{x} - x$

$f(x) = 0 \Rightarrow 2\sqrt{x} = x \; ; \; 4x = x^2 \; ; \; x(x - 4) = 0 \; ; \; x = 0$ or $x = 4$

$\therefore$ The intersection points of the curve and the x-axis are $x = 0$ and $x = 4$.

Therefore, the shaded area is

$$\int_0^4 f(x)dx = \int_0^4 (2\sqrt{x} - x)dx = \left[2 \cdot \frac{2}{3}x^{\frac{3}{2}} - \frac{1}{2}x^2 \right]_0^4 = \frac{4}{3} \cdot 2^3 - 8 = 8\left(\frac{4}{3} - 1 \right) = \frac{8}{3}$$

(11) When a function $f(x) = \lim\limits_{n \to \infty} \dfrac{x^{n+1} + ax^2 + b}{x^n + 1}$ $(x > 0)$ is differentiable at $x = 1$,

find the area of the region bounded by the curve $y = f(x)$, $x = 3$, and the x-axis.

i) When $0 < x < 1$,

$$f(x) = \lim_{n \to \infty} \frac{x^{n+1} + ax^2 + b}{x^n + 1} = ax^2 + b \; ; \quad f'(x) = 2ax$$

ii) When $x = 1$,

$$f(x) = \lim_{n \to \infty} \frac{x^{n+1} + ax^2 + b}{x^n + 1} = \frac{1 + a + b}{2}$$

iii) When $x > 1$,

$$f(x) = \lim_{n \to \infty} \frac{x^{n+1} + ax^2 + b}{x^n + 1} = \frac{x + \dfrac{a}{x^{n-2}} + \dfrac{b}{x^n}}{1 + \dfrac{1}{x^n}}$$

Since $f(x)$ is differentiable at $x = 1$, $f(x)$ is continuous.

$\therefore \quad \lim\limits_{x \to 1^-} f(x) = \lim\limits_{x \to 1^+} f(x) = f(1)$

$\therefore \quad a + b = \dfrac{1+a+b}{2}$; $a + b = 1$

Since $f(x)$ is differentiable at $x = 1$, $\lim\limits_{x \to 1^-} f'(x) = \lim\limits_{x \to 1^+} f'(x)$

$\therefore \quad 2a = 1$; $a = \dfrac{1}{2}$, $b = 1 - a = \dfrac{1}{2}$

$\therefore \quad f(x) = \begin{cases} \dfrac{1}{2}x^2 + \dfrac{1}{2}, & 0 < x < 1 \\ x, & x \geq 1 \end{cases}$

The area is

$$\int_0^3 |f(x)|dx = \int_0^1 f(x)dx + \int_1^3 f(x)dx = \int_0^1 \left(\frac{1}{2}x^2 + \frac{1}{2}\right)dx + \int_1^3 xdx$$

$$= \frac{1}{2}\left[\frac{1}{3}x^3 + x\right]_0^1 + \left[\frac{1}{2}x^2\right]_1^3 = \frac{1}{2}\left(\frac{1}{3} + 1\right) + \frac{1}{2}(3^2 - 1) = \frac{2}{3} + 4 = \frac{14}{3}$$

(12) Find the area of the region bounded by the curve $y = x^2 - 2x + 7$ and the tangent lines at $(1, -3)$ to the curve.

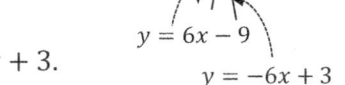

$y = x^2 - 2x + 7 \implies y' = 2x - 2$

Let $P = P(t, \ t^2 - 2t + 7)$ be the point of tangency.

Then, the equation of the tangent line at P is

$y - (t^2 - 2t + 7) = (2t - 2)(x - t)$.

$\therefore \quad y = (2t - 2)x + t^2 - 2t + 7 - 2t^2 + 2t = (2t - 2)x - t^2 + 7$

Since the equation $y = (2t - 2)x - t^2 + 7$ passes through the point $(1, -3)$,

$-3 = (2t - 2) - t^2 + 7$; $t^2 - 2t - 8 = 0$; $(t - 4)(t + 2) = 0$

$\therefore \quad t = 4$ or $t = -2$

$\therefore$ The equations of the tangent lines are $y = 6x - 9$ and $y = -6x + 3$.

$6x - 9 = -6x + 3 \implies 12x = 12$; $x = 1$ (Intersection point of two tangent lines.)

Therefore, the shaded area is

$$\int_{-2}^1 \{(x^2 - 2x + 7) - (-6x + 3)\}dx + \int_1^4 \{(x^2 - 2x + 7) - (6x - 9)\}dx$$

$$= \int_{-2}^1 (x^2 + 4x + 4)dx + \int_1^4 (x^2 - 8x + 16)dx = \left[\frac{1}{3}x^3 + 2x^2 + 4x\right]_{-2}^1 + \left[\frac{1}{3}x^3 - 4x^2 + 16x\right]_1^4$$

$$= \left(\frac{1}{3} + 2 + 4\right) - \left(-\frac{8}{3} + 8 - 8\right) + \left(\frac{1}{3} \cdot 4^3 - 4 \cdot 4^2 + 16 \cdot 4\right) - \left(\frac{1}{3} - 4 + 16\right) = 18$$

(13) For a curve $f(x) = ax^2 + b$ $(x \geq 0)$, $g(x)$ is the inverse of $f(x)$.

When $f(x)$ and $g(x)$ intersect at $x = 1$ and $x = 2$, find the area A of the region bounded by the curves, the x-axis, and the y-axis on $[0, 1]$, and the area B of the region bounded by the curves on $[1, 2]$.

Since $f(x)$ and $g(x)$ are inverse each other, they are symmetric with respect to the line $y = x$.

Thus, the intersection points of the curves $y = f(x)$ and $y = g(x)$ are the intersection points of

the curve $y = f(x)$ and a line $y = x$.

∴ The intersection points are $(1, 1)$ and $(2, 2)$.

∴ $f(1) = 1$ and $f(2) = 2$

Since $f(x) = ax^2 + b$, $a + b = 1$ and $4a + b = 2$

∴ $3a = 1$; $a = \dfrac{1}{3}$, $b = \dfrac{2}{3}$

∴ $f(x) = \dfrac{1}{3}x^2 + \dfrac{2}{3}$

From the graph, the area $A = A_1 + A_2$

Since $A_1 = A_2$ $(\because A_1$ and A_2 are symmetric w.r.t. $y = x)$,

$A = 2A_1 = 2\int_0^1 \left\{ \left(\dfrac{1}{3}x^2 + \dfrac{2}{3} \right) - x \right\} dx = 2\left[\dfrac{1}{9}x^3 + \dfrac{2}{3}x - \dfrac{1}{2}x^2 \right]_0^1 = 2\left(\dfrac{1}{9} + \dfrac{2}{3} - \dfrac{1}{2} \right)$

$= 2 \cdot \dfrac{2+12-9}{18} = \dfrac{5}{9}$

Similarly, the area $B = B_1 + B_2$ and $B_1 = B_2$.

$B = 2B_2 = 2\int_1^2 \left\{ x - \left(\dfrac{1}{3}x^2 + \dfrac{2}{3} \right) \right\} dx = 2\left[\dfrac{1}{2}x^2 - \dfrac{1}{9}x^3 - \dfrac{2}{3}x \right]_1^2$

$= 2\left\{ \left(2 - \dfrac{8}{9} - \dfrac{4}{3} \right) - \left(\dfrac{1}{2} - \dfrac{1}{9} - \dfrac{2}{3} \right) \right\} = 2\left(\dfrac{3}{2} - \dfrac{7}{9} - \dfrac{2}{3} \right) = 2 \cdot \dfrac{27-14-12}{18} = \dfrac{1}{9}$

#5 When the base of a solid is bounded by the circle $x^2 + y^2 = 16$,

(1) Find the volume of the solid with the base that has semicircular cross-sections perpendicular to the y-axis.

$x^2 + y^2 = 16 \Rightarrow x^2 = 16 - y^2 \Rightarrow x = \pm\sqrt{16 - y^2}$

The length of the cross-section is obtained by subtracting

the left boundary of the region from the right boundary;

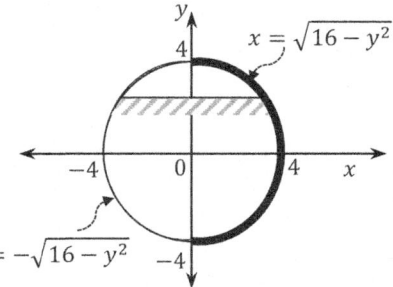

i.e., $\sqrt{16 - y^2} - \left(-\sqrt{16 - y^2} \right) = 2\sqrt{16 - y^2}$.

Since the base is the circle cross-section, the length is the diameter of the cross-section.

∴ The radius of the semicircle is $\sqrt{16 - y^2}$.

∴ The area $A(y)$ of the semicircle is $A(y) = \dfrac{1}{2}\pi r^2 = \dfrac{\pi}{2}(16 - y^2)$.

Therefore, the volume of the solid is

$$\int_{-4}^{4} A(y)dy = \int_{-4}^{4} \frac{\pi}{2}(16 - y^2)dy = \frac{\pi}{2}\int_{-4}^{4}(16 - y^2)dy = \frac{\pi}{2}\left[2\int_{0}^{4}(16 - y^2)dy\right]$$

$$= \pi\int_{0}^{4}(16 - y^2)dy = \pi\left[16y - \frac{1}{3}y^3\right]_{0}^{4} = \pi\left(16 \cdot 4 - \frac{1}{3} \cdot 4^3\right) = \pi\left\{4^3\left(1 - \frac{1}{3}\right)\right\}$$

$$= \frac{2}{3} \cdot 4^3 \cdot \pi = \frac{128}{3}\pi$$

(2) Find the volume of the solid with the base that has semicircular cross-sections perpendicular to the x-axis.

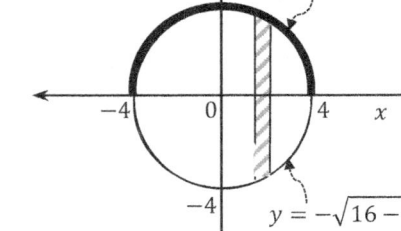

$x^2 + y^2 = 16 \Rightarrow y^2 = 16 - x^2 \Rightarrow y = \pm\sqrt{16 - x^2}$

The upper boundary of the circle is $y = \sqrt{16 - x^2}$ and

the lower boundary is $y = -\sqrt{16 - x^2}$.

∴ The radius of the semicircle is $\sqrt{16 - x^2}$.

∴ The area $A(x)$ of the semicircle is $A(x) = \frac{1}{2}\pi r^2 = \frac{\pi}{2}(16 - x^2)$.

Therefore, the volume of the solid is

$$\int_{-4}^{4} A(x)dx = \int_{-4}^{4} \frac{\pi}{2}(16 - x^2)dx = \frac{\pi}{2}\int_{-4}^{4}(16 - x^2)dx = \frac{\pi}{2}\left[2\int_{0}^{4}(16 - x^2)dx\right]$$

$$= \pi\int_{0}^{4}(16 - x^2)dx = \pi\left[16x - \frac{1}{3}x^3\right]_{0}^{4} = \pi\left(16 \cdot 4 - \frac{1}{3} \cdot 4^3\right) = \pi\left\{4^3\left(1 - \frac{1}{3}\right)\right\}$$

$$= \frac{2}{3} \cdot 4^3 \cdot \pi = \frac{128}{3}\pi$$

#6 Find the volume.

(1) Let the base of a solid be the first quadrant plane region bounded by $4x^2 + 9y^2 = 36$, the x-axis, and the y-axis. When the cross-sections perpendicular to the x-axis are squares, find the volume of the solid.

$4x^2 + 9y^2 = 36 \Rightarrow \frac{x^2}{9} + \frac{y^2}{4} = 1$.

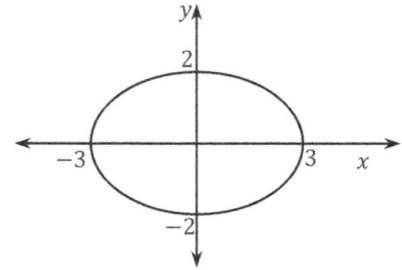

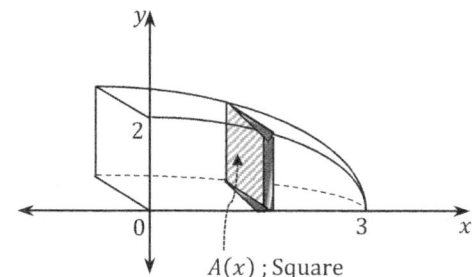

$A(x)$; Square

$A(x) = y^2 = 4\left(1 - \frac{x^2}{9}\right)$

$$\therefore \ V = \int_{0}^{3} A(x)dx = 4\int_{0}^{3}\left(1 - \frac{x^2}{9}\right)dx = 4\left[x - \frac{1}{27}x^3\right]_{0}^{3} = 4\left(3 - \frac{1}{27}3^3\right) = 8$$

(2) Let the base of a solid be the first quadrant plane region bounded by $y = 1 - \dfrac{x^2}{2}$, the x-axis, and the y-axis. Suppose that cross-sections perpendicular to the x-axis are squares. Find the volume of the solid.

$$A(x) = \left(1 - \frac{x^2}{2}\right)^2$$

Therefore, the volume of the solid is

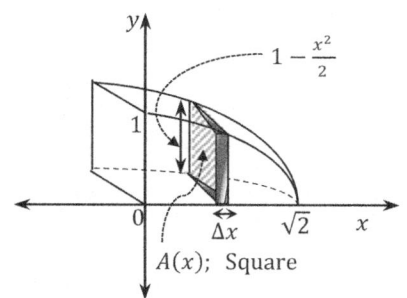

$$V = \int_0^{\sqrt{2}} \left(1 - x^2 + \frac{x^4}{4}\right) dx = \left[x - \frac{1}{3}x^3 + \frac{1}{20}x^5\right]_0^{\sqrt{2}}$$

$$= \sqrt{2} - \frac{1}{3} \cdot 2\sqrt{2} + \frac{1}{20} \cdot 4\sqrt{2}$$

$$= \sqrt{2}\left(1 - \frac{2}{3} + \frac{1}{5}\right) = \frac{15 - 10 + 3}{15}\sqrt{2} = \frac{8}{15}\sqrt{2}$$

(3) A solid has a base in the form of an ellipse with major axis 8 and minor axis 4. Find the volume if every section perpendicular to the major axis is an isosceles triangle with altitude 5.

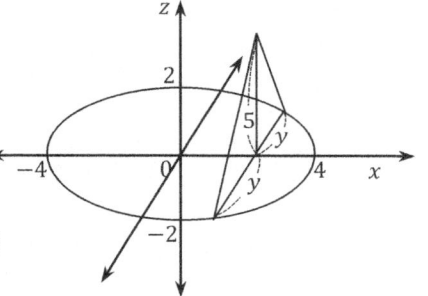

$2a = 8 \; ; \; a = 4$

$2b = 4 \; ; \; b = 2$

The equation of the ellipse is $\dfrac{x^2}{4^2} + \dfrac{y^2}{2^2} = 1 \; ; \; \dfrac{x^2}{16} + \dfrac{y^2}{4} = 1 \; ;$

$\dfrac{y^2}{4} = 1 - \dfrac{x^2}{16} \quad \therefore \; y^2 = 4\left(1 - \dfrac{x^2}{16}\right) = 4\left(\dfrac{16 - x^2}{16}\right) = \dfrac{1}{4}(16 - x^2)$

$\therefore \; y = \pm\dfrac{1}{2}\sqrt{16 - x^2}$

Since the cross-section is an isosceles triangle of base $2y$ and altitude 5,

the area $A(x)$ is $A(x) = \dfrac{1}{2} \cdot 2y \cdot 5 = 5y$

$\therefore \; V = \int_{-4}^{4} A(x)dx = \int_{-4}^{4} 5y\, dx = 5\int_{-4}^{4} \dfrac{1}{2}\sqrt{16 - x^2}dx = \dfrac{5}{2}\int_{-4}^{4}\sqrt{16 - x^2}dx$

Now, consider a circle with radius 4 and center at the origin $(0, 0)$.

The equation of the circle is $x^2 + y^2 = 16$.

$\therefore \; y = \pm\sqrt{16 - x^2}$

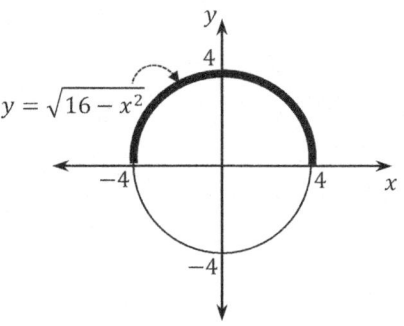

Since the area of the above half circle is $\dfrac{1}{2} \cdot (\pi 4^2) = 8\pi$,

$\int_{-4}^{4}\sqrt{16 - x^2}\, dx = 8\pi.$

Therefore, $V = \dfrac{5}{2} \cdot 8\pi = 20\pi$

(4) The base of a solid is the region between $f(x) = x^2 - 1$ and $g(x) = -x^2 + 1$, and its cross-sections perpendicular to the x-axis are equilateral triangles. When the solid has been truncated to show a triangular cross-section above $x = \frac{2}{3}$, find the volume of the solid.

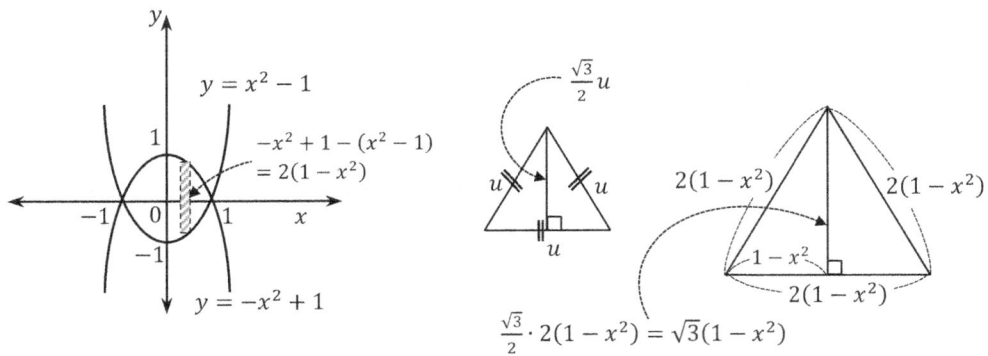

The area of the cross-section is

$$A(x) = \frac{1}{2} \cdot 2(1 - x^2) \cdot \sqrt{3}(1 - x^2) = \sqrt{3}(1 - x^2)^2$$

∴ The volume of the solid is

$$\int_{-1}^{1} \sqrt{3}(1 - x^2)^2 \, dx = \sqrt{3} \int_{-1}^{1} (1 - 2x^2 + x^4) dx = \sqrt{3} \left[x - \frac{2}{3}x^3 + \frac{1}{5}x^5 \right]_{-1}^{1}$$

$$= \sqrt{3} \left\{ \left(1 - \frac{2}{3} + \frac{1}{5} \right) - \left(-1 + \frac{2}{3} - \frac{1}{5} \right) \right\} = \sqrt{3} \left(2 - \frac{4}{3} + \frac{2}{5} \right) = \sqrt{3} \left(\frac{30 - 20 + 6}{15} \right)$$

$$= \frac{16}{15}\sqrt{3}$$

#7 The base of a solid is the region between one arch of $y = \sin x$ ($0 \le x \le \pi$) and the x-axis.

(1) Each cross-section perpendicular to the x-axis is a circle sitting on this base. Find the volume of the solid.

$$A(x) = \pi \left(\frac{1}{2} \sin x \right)^2 = \frac{\pi}{4} \sin^2 x$$

The volume of the solid is

$$\int_0^\pi A(x) dx = \int_0^\pi \frac{\pi}{4} \sin^2 x \, dx = \frac{\pi}{4} \int_0^\pi \sin^2 x \, dx$$

$$= \frac{\pi}{4} \int_0^\pi \frac{1 - \cos 2x}{2} \, dx = \frac{\pi}{8} \int_0^\pi (1 - \cos 2x) dx$$

$$= \frac{\pi}{8} \left[x - \frac{1}{2} \sin 2x \right]_0^\pi = \frac{\pi}{8} \pi = \frac{\pi^2}{8}$$

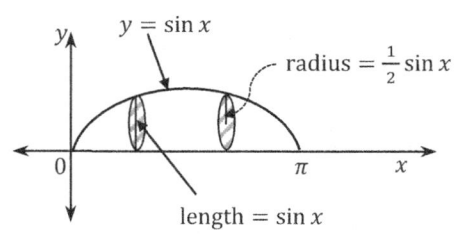

(2) Each cross-section perpendicular to the x-axis is an equilateral triangle sitting on this base. Find the volume of the solid.

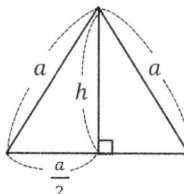

$$a^2 = h^2 + \left(\frac{a}{2}\right)^2$$

$$\therefore\ h^2 = a^2 - \left(\frac{a}{2}\right)^2 = \frac{3}{4}a^2$$

$$\therefore\ h = \frac{\sqrt{3}}{2}a$$

$$A(x) = \frac{1}{2}\cdot a \cdot \frac{\sqrt{3}}{2}a = \frac{\sqrt{3}}{4}a^2$$

The area of an equilateral triangle of side $\sin x$ is $\dfrac{\sqrt{3}}{4}\sin^2 x$.

Therefore, the volume of the solid is

$$\int_0^\pi \frac{\sqrt{3}}{4}\sin^2 x\, dx = \frac{\sqrt{3}}{4}\int_0^\pi \sin^2 x\, dx = \frac{\sqrt{3}}{4}\int_0^\pi \left(\frac{1-\cos 2x}{2}\right) dx$$

$$= \frac{\sqrt{3}}{8}\int_0^\pi (1 - \cos 2x)dx = \frac{\sqrt{3}}{8}\left[x - \frac{1}{2}\sin 2x\right]_0^\pi = \frac{\sqrt{3}}{8}\pi$$

#8 The base of a solid is the region bounded by $y = \sqrt{x}$, the x-axis, and $x = 4$.

(1) Find the volume of the solid with base that has square cross-sections perpendicular to the x-axis.

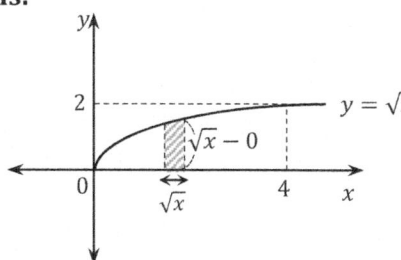
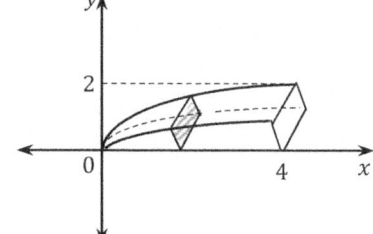

The upper boundary of the region is $y = \sqrt{x}$ and the lower boundary is $y = 0$.

$\therefore$ The length of the cross-section is $\sqrt{x} - 0$.

$\therefore$ The area of the square cross-section is $A(x) = (\sqrt{x})^2 = x$.

$\therefore$ The volume of the solid is $\int_0^4 A(x)dx = \int_0^4 x\,dx = \left[\frac{1}{2}x^2\right]_0^4 = \frac{1}{2}\cdot 16 = 8$

(2) Find the volume of the solid with base that has square cross-sections perpendicular to the y-axis.

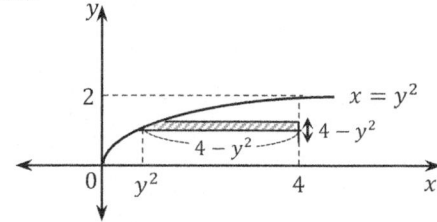

The right boundary of the region is $x = 4$ and the left boundary is $x = y^2$.

$\therefore$ The area of the square cross-section is $A(y) = (4 - y^2)^2 = y^4 - 8y^2 + 16$.

$\therefore$ The volume of the solid is

$$\int_0^2 A(y)dy = \int_0^2 (y^4 - 8y^2 + 16)dy = \left[\frac{1}{5}y^5 - \frac{8}{3}y^3 + 16y\right]_0^2$$

$$= \frac{1}{5} \cdot 2^5 - \frac{8}{3} \cdot 2^3 + 16 \cdot 2$$

$$= \frac{32}{5} - \frac{64}{3} + 32 = \frac{256}{15}$$

(3) Find the volume of the solid with base that has rectangular cross-sections of height 2 that are perpendicular to the x-axis.

The area of a rectangle cross-section is $A(x) = \sqrt{x} \cdot 2 = 2\sqrt{x}$

$\therefore$ The volume of the solid is

$$\int_0^4 2\sqrt{x}\,dx = 2\int_0^4 x^{\frac{1}{2}}\,dx = 2\left[\frac{2}{3}x^{\frac{3}{2}}\right]_0^4 = 2 \cdot \frac{2}{3} \cdot 2^3 = \frac{32}{3}$$

#9 A wedge is cut from a right circular cylinder of radius 3. The upper surface of the wedge is in a plane through a diameter of the circular base and makes an angle 45° with the base. Find the volume of the wedge.

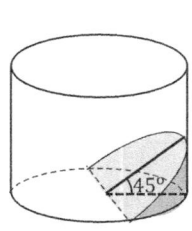 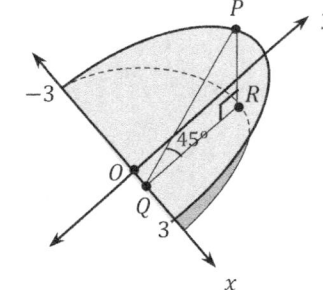 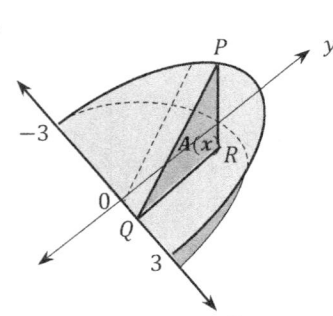

Let $Q = Q(x, 0)$, $-3 \le x \le 3$

From the triangle $\triangle OQR$, $\overline{QR}^2 = \overline{OR}^2 - \overline{OQ}^2 = 3^2 - x^2$

$\therefore \overline{QR} = \sqrt{9 - x^2}$

Since $m(\angle PRQ) = 90°$ and $m(\angle PQR) = 45°$, $\overline{QR} = \overline{PR} = \sqrt{9 - x^2}$

$\therefore A(x) = (\text{The area of } \triangle PQR) = \frac{1}{2} \cdot \overline{QR} \cdot \overline{PR} = \frac{1}{2}(9 - x^2)$

$\therefore$ The volume of the wedge is

$$2\int_0^3 A(x)dx = 2\int_0^3 \frac{1}{2}(9 - x^2)dx = \int_0^3 (9 - x^2)dx = \left[9x - \frac{1}{3}x^3\right]_0^3 = 27 - 9 = 18$$

#10 **For the region bounded by the graphs of the given equations, find the volume V of the solid generated by revolving the region about the x-axis.**

(1) $y = 1 - x^2$, $y = 0$

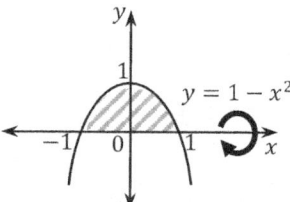

$V = \pi \int_{-1}^{1} y^2 dx = \pi \int_{-1}^{1} (1 - x^2)^2 dx = 2\pi \int_{0}^{1} (1 - 2x^2 + x^4) dx$

$= 2\pi \left[x - \frac{2}{3}x^3 + \frac{1}{5}x^5 \right]_0^1 = 2\pi \left(1 - \frac{2}{3} + \frac{1}{5} \right) = 2\pi \frac{15 - 10 + 3}{15} = \frac{16}{15}\pi$

(2) $y = x^2 + 1$, x-axis, $x = -1$, $x = 1$

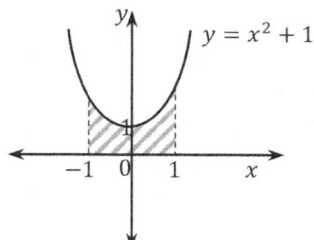

$V = \pi \int_{-1}^{1} (x^2 + 1)^2 dx = \pi \int_{-1}^{1} (x^4 + 2x^2 + 1) dx$

$= \pi \left[\frac{1}{5}x^5 + \frac{2}{3}x^3 + x \right]_{-1}^{1} = \pi \left\{ \frac{1}{5}(1 + 1) + \frac{2}{3}(1 + 1) + (1 + 1) \right\}$

$= \pi \left(\frac{2}{5} + \frac{4}{3} + 2 \right) = \frac{6 + 20 + 30}{15}\pi = \frac{56}{15}\pi$

(3) $y = \sqrt{x + 1}$, $x = 0$, $y = 0$

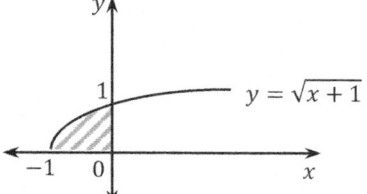

$V = \pi \int_{-1}^{0} \left(\sqrt{x + 1} \right)^2 dx = \pi \int_{-1}^{0} (x + 1) dx$

$= \pi \left[\frac{1}{2}x^2 + x \right]_{-1}^{0} = -\pi \left(\frac{1}{2} - 1 \right) = \frac{1}{2}\pi$

(4) $y = 1 - |x|$, $y = 0$

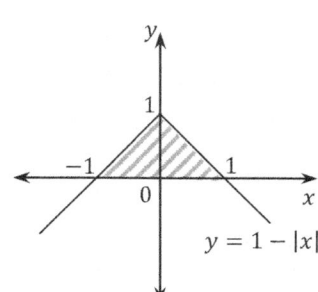

$V = \pi \int_{-1}^{1} (1 - |x|)^2 dx = \pi \left\{ \int_{-1}^{0} (1 + x)^2 dx + \int_{0}^{1} (1 - x)^2 dx \right\}$

$= \pi \left\{ \int_{-1}^{0} (x^2 + 2x + 1) dx + \int_{0}^{1} (x^2 - 2x + 1) dx \right\}$

$= \pi \left\{ \left[\frac{1}{3}x^3 + x^2 + x \right]_{-1}^{0} + \left[\frac{1}{3}x^3 - x^2 + x \right]_{0}^{1} \right\}$

$= \pi \left\{ \left(\frac{1}{3} - 1 + 1 \right) + \left(\frac{1}{3} - 1 + 1 \right) \right\} = \frac{2}{3}\pi$

(5) $x^2 + 4y^2 = 4$, $y = 0$

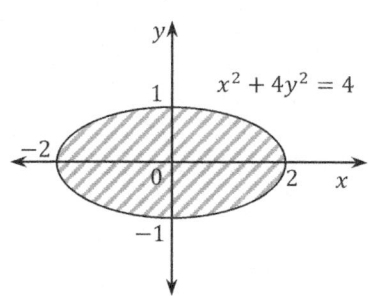

$4y^2 = 4 - x^2 \ ; \ y^2 = \frac{1}{4}(4 - x^2)$

$y^2 = 0 \ \Rightarrow \ x^2 = 4 \ ; \ x = \pm 2$

$V = \pi \int_{-2}^{2} \frac{1}{4}(4 - x^2) dx = 2\pi \int_{0}^{2} \frac{1}{4}(4 - x^2) dx = \frac{\pi}{2} \int_{0}^{2} (4 - x^2) dx$

$= \frac{\pi}{2} \left[4x - \frac{1}{3}x^3 \right]_0^2 = \frac{\pi}{2} \left(8 - \frac{8}{3} \right) = \frac{8\pi}{3}$

(6) $y = x - 2\sqrt{x} \ (0 \le x \le 4)$, $y = 0$

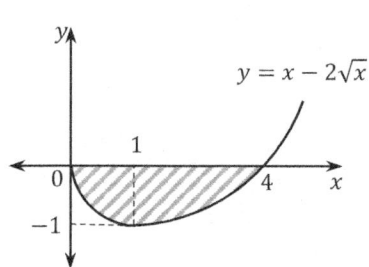

$V = \pi \int_{0}^{4} \left(x - 2\sqrt{x} \right)^2 dx = \pi \int_{0}^{4} \left(x^2 - 4x\sqrt{x} + 4x \right) dx$

$= \pi \left[\frac{1}{3}x^3 - 4 \left(\frac{2}{5}x^{\frac{5}{2}} \right) + 2x^2 \right]_0^4$

$= \pi \left(\frac{1}{3} \cdot 4^3 - 4 \left(\frac{2}{5} \cdot 4^{\frac{5}{2}} \right) + 2 \cdot 4^2 \right) = \frac{32}{15}\pi$

(7) $y = 1 + \sin\frac{x}{2}$, $x = 0$, $y = 0$, $x = 2\pi$

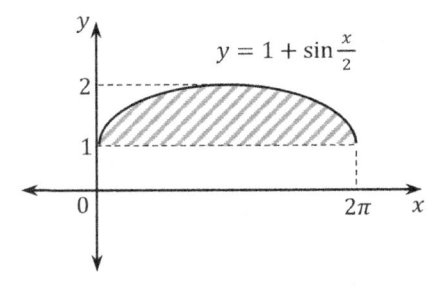

$V = \pi \int_0^{2\pi} \left(1 + \sin\frac{x}{2}\right)^2 dx = \pi \int_0^{2\pi} \left(1 + 2\sin\frac{x}{2} + \sin^2\frac{x}{2}\right) dx$

$= \pi \int_0^{2\pi} \left(1 + 2\sin\frac{x}{2} + \frac{1-\cos\left(2\cdot\frac{x}{2}\right)}{2}\right) dx$

$= \pi \left[x - 2\cdot 2 \cdot \cos\frac{x}{2} + \frac{1}{2}(x - \sin x)\right]_0^{2\pi}$

$= \pi \left[x - 4\cos\frac{x}{2} + \frac{1}{2}(x - \sin x)\right]_0^{2\pi}$

$= \pi \left\{2\pi - 4\cos\pi + \frac{1}{2}(2\pi - \sin 2\pi) - \left(0 - 4\cos 0 + \frac{1}{2}(0 - \sin 0)\right)\right\}$

$= \pi(2\pi + 4 + \pi + 4) = \pi(3\pi + 8)$

(8) $y = e^x - 1$, $x = 1$, $y = 0$

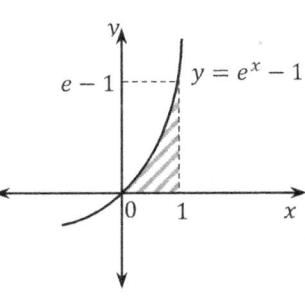

$V = \pi \int_0^1 (e^x - 1)^2 dx = \pi \int_0^1 (e^{2x} - 2e^x + 1) dx$

$= \pi \left[\frac{1}{2}e^{2x} - 2e^x + x\right]_0^1 = \pi \left\{\left(\frac{1}{2}e^2 - 2e + 1\right) - \left(\frac{1}{2}e^0 - 2e^0\right)\right\}$

$= \pi \left(\frac{1}{2}e^2 - 2e + \frac{5}{2}\right) = \frac{\pi}{2}(e^2 - 4e + 5)$

(9) $y = \log x$, $x = e$, $y = 0$

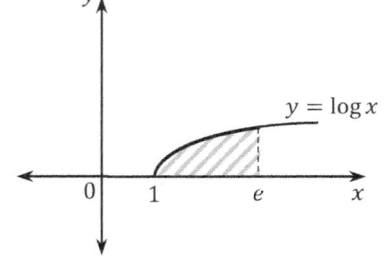

$V = \pi \int_1^e (\log x)^2 dx$

For $\int (\log x)^2 dx$,

Let $\quad u = (\log x)^2$, $\qquad v' = dx$

Then, $\quad u' = 2\log x \cdot \frac{1}{x} dx$, $\quad v = x$

$\therefore \int (\log x)^2 dx = x(\log x)^2 - \int x \cdot 2\log x \cdot \frac{1}{x} dx = x(\log x)^2 - 2(x\log x - x) + C$

Therefore, $V = \pi \int_1^e (\log x)^2 dx = \pi[x(\log x)^2 - 2(x\log x - x)]_1^e = \pi(e - 2)$

(10) $y = 4 - x^2$, $y = 2 - x$

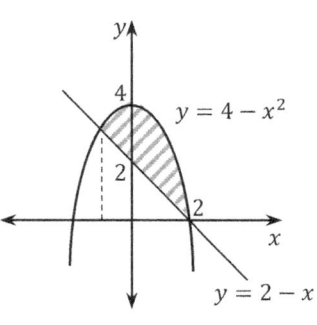

$4 - x^2 = 2 - x \implies x^2 - x - 2 = 0 \implies (x - 2)(x + 1) = 0$

$\implies x = 2$ or $x = -1$

$\therefore$ The intersection points of the curves are $x = 2$ and $x = -1$.

$V = \pi \int_{-1}^2 (4 - x^2)^2 dx - \pi \int_{-1}^2 (2 - x)^2 dx$

$= \pi \int_{-1}^2 \{(x^4 - 8x^2 + 16) - (x^2 - 4x + 4)\} dx$

$= \pi \int_{-1}^2 (x^4 - 9x^2 + 4x + 12) dx = \pi \left[\frac{1}{5}x^5 - 3x^3 + 2x^2 + 12x\right]_{-1}^2$

$= \pi \left(\frac{1}{5}(2^5 + 1) - 3(2^3 + 1) + 2(2^2 - 1) + 12(2 + 1)\right) = \pi \left(\frac{33}{5} - 27 + 6 + 36\right) = \frac{108}{5}\pi$

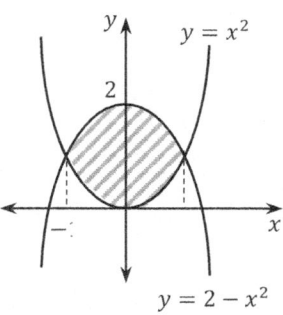

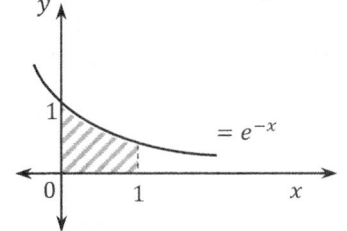

(11) $y = x^2$, $y = 2 - x^2$

$x^2 = 2 - x^2 \Rightarrow 2x^2 = 2 \Rightarrow x^2 = 1 \Rightarrow x = \pm 1$

$\therefore$ The intersection points of the curves are $x = 1$ and $x = -1$.

$V = \pi \int_{-1}^{1}(2 - x^2)^2 dx - \pi \int_{-1}^{1}(x^2)^2 dx = \pi \int_{-1}^{1}(x^4 - 4x^2 + 4 - x^4)dx$

$= 2\pi \int_0^1(-4x^2 + 4)dx = -8\pi \int_0^1(x^2 - 1)dx = -8\pi \left[\frac{1}{3}x^3 - x\right]_0^1$

$= -8\pi\left(\frac{1}{3} - 1\right) = \frac{16}{3}\pi$

(12) $y = e^{-x}$, $x = 0$, $y = 0$, $x = 1$

$V = \pi \int_0^1 y^2 dx = \pi \int_0^1(e^{-x})^2 dx = \pi \int_0^1 e^{-2x} dx$

$= \pi\left[-\frac{1}{2}e^{-2x}\right]_0^1 = -\frac{\pi}{2}(e^{-2} - e^0) = \frac{\pi}{2}\left(1 - \frac{1}{e^2}\right)$

(13) $y = e^x + e^{-x}$, $x = 0$, $y = 0$, $x = 1$

$V = \pi \int_0^1 y^2 dx = \pi \int_0^1(e^x + e^{-x})^2 dx = \pi \int_0^1(e^{2x} + 2 + e^{-2x})dx$

$= \pi\left[\frac{1}{2}e^{2x} + 2x - \frac{1}{2}e^{-2x}\right]_0^1 = \pi\left\{\left(\frac{1}{2}e^2 + 2 - \frac{1}{2}e^{-2}\right) - \left(\frac{1}{2}e^0 - \frac{1}{2}e^0\right)\right\} = \frac{\pi}{2}\left(e^2 + 4 - \frac{1}{e^2}\right)$

(14) $y = \tan x$, $x = \frac{\pi}{4}$, $y = 0$

$V = \pi \int_0^{\frac{\pi}{4}} y^2 dx = \pi \int_0^{\frac{\pi}{4}} \tan^2 x \, dx = \pi \int_0^{\frac{\pi}{4}}(\sec^2 x - 1)dx = \pi[\tan x - x]_0^{\frac{\pi}{4}}$

$= \pi\left(\tan\frac{\pi}{4} - \frac{\pi}{4}\right) = \pi\left(1 - \frac{\pi}{4}\right)$

(15) $y = \sin x$, $y = \sin 2x$ $\left(0 \le x \le \frac{\pi}{2}\right)$

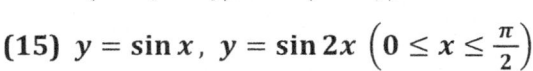

$\sin x = \sin 2x \Rightarrow \sin x = 2\sin x \cos x \Rightarrow \sin x(1 - 2\cos x) = 0$

$\Rightarrow \sin x = 0$ or $\cos x = \frac{1}{2}$

$\therefore x = 0$ and $x = \frac{\pi}{3}$ are the intersection points of the curves.

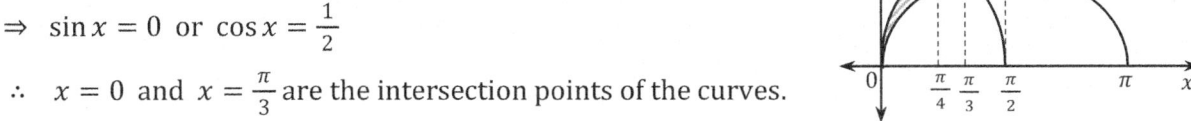

$V = \pi \int_0^{\frac{\pi}{3}} \sin^2 2x \, dx - \pi \int_0^{\frac{\pi}{3}} \sin^2 x \, dx = \pi \int_0^{\frac{\pi}{3}} \left(\frac{1 - \cos 4x}{2} - \frac{1 - \cos 2x}{2}\right) dx$

$= \frac{\pi}{2} \int_0^{\frac{\pi}{3}}(-\cos 4x + \cos 2x)dx = \frac{\pi}{2}\left[-\frac{1}{4}\sin 4x + \frac{1}{2}\sin 2x\right]_0^{\frac{\pi}{3}} = \frac{\pi}{2}\left(-\frac{1}{4}\sin\frac{4\pi}{3} + \frac{1}{2}\sin\frac{2\pi}{3}\right)$

$= \frac{\pi}{2}\left(-\frac{1}{4}\cdot-\frac{\sqrt{3}}{2} + \frac{1}{2}\cdot\frac{\sqrt{3}}{2}\right) = \frac{\pi}{2}\left(\frac{\sqrt{3}}{8} + \frac{\sqrt{3}}{4}\right) = \frac{3\sqrt{3}}{16}\pi$

(16) $y = \cos x + 1$, $x = \pi$, $x = -\pi$, $y = 0$

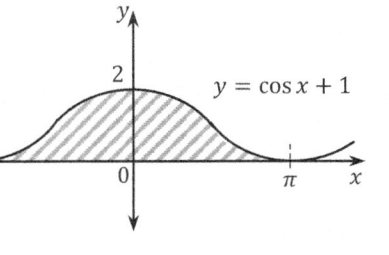

$V = \pi \int_{-\pi}^{\pi}(\cos x + 1)^2 dx = 2\pi \int_0^{\pi}(\cos^2 x + 2\cos x + 1)dx$

$= 2\pi \int_0^{\pi}\left(\frac{1 + \cos 2x}{2} + 2\cos x + 1\right) dx = 2\pi\left[\frac{\sin 2x}{4} + 2\sin x + \frac{3}{2}x\right]_0^{\pi}$

$= 2\pi\left(\frac{\sin 2\pi}{4} + 2\sin \pi + \frac{3}{2}\pi\right) = 3\pi^2$

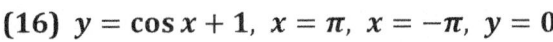

(17) $y = \sin x + \cos x$, $x = 0$, $x = \pi$, $y = 0$

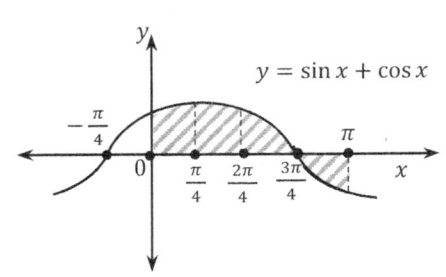

$y = \sin x + \cos x$

$\sin x + \cos x = \sqrt{2} \sin\left(x + \dfrac{\pi}{4}\right)$

$V = \pi \int_0^\pi y^2 dx = \pi \int_0^\pi \{\sin x + \cos x\}^2 dx$

$\quad = \pi \int_0^\pi (1 + 2\sin x \cos x) dx = \pi \int_0^\pi (1 + \sin 2x) dx$

$\quad = \pi \left[x - \dfrac{1}{2}\cos 2x\right]_0^\pi = \pi\left\{\left(\pi - \dfrac{1}{2}\cos 2\pi\right) - \left(-\dfrac{1}{2}\cos 0\right)\right\}$

$\quad = \pi(\pi) = \pi^2$

(18) $y = x^2 - 1$, $y = x + 1$

$x^2 - 1 = x + 1 \;\Rightarrow\; x^2 - x - 2 = 0 \;\Rightarrow\; (x-2)(x+1) = 0 \;\Rightarrow\; x = 2 \text{ or } x = -1$

The intersection points of the curves are $x = 2$ and $x = -1$.

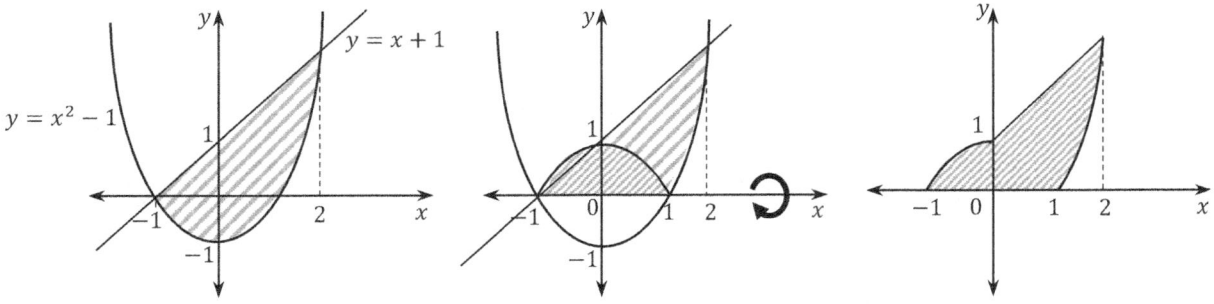

$V = \pi \int_{-1}^0 (x^2 - 1)^2 dx + \pi \int_0^2 (x+1)^2 dx - \pi \int_1^2 (x^2 - 1)^2 dx$

$\quad = \pi \int_{-1}^0 (x^4 - 2x^2 + 1) dx + \pi \int_0^2 (x^2 + 2x + 1) dx - \pi \int_1^2 (x^4 - 2x^2 + 1) dx$

$\quad = \pi \left[\dfrac{1}{5}x^5 - \dfrac{2}{3}x^3 + x\right]_{-1}^0 + \pi \left[\dfrac{1}{3}x^3 + x^2 + x\right]_0^2 - \pi \left[\dfrac{1}{5}x^5 - \dfrac{2}{3}x^3 + x\right]_1^2$

$\quad = \pi \left(\dfrac{1}{5} - \dfrac{2}{3} + 1\right) + \pi \left(\dfrac{8}{3} + 4 + 2\right) - \pi \left(\dfrac{31}{5} - \dfrac{2}{3}\cdot 7 + 1\right)$

$\quad = \dfrac{100}{15}\pi = \dfrac{20}{3}\pi$

(19) $y^2 = x + 2$, $y = x$

$V = \pi \int_{-2}^0 (x+2) dx + \pi \int_0^2 (x+2) dx - \pi \int_0^2 x^2 dx$

$\quad = \pi \left[\dfrac{1}{2}x^2 + 2x\right]_{-2}^0 + \pi \left[\dfrac{1}{2}x^2 + 2x\right]_0^2 - \pi \left[\dfrac{1}{3}x^3\right]_0^2$

$\quad = \pi\left(8 - \dfrac{8}{3}\right) = \dfrac{16}{3}\pi$

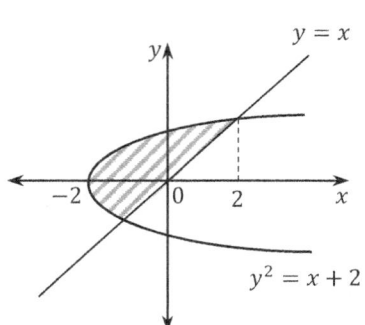

$y = x$

$y^2 = x + 2$

(20) $x^2 + (y-4)^2 = 4$

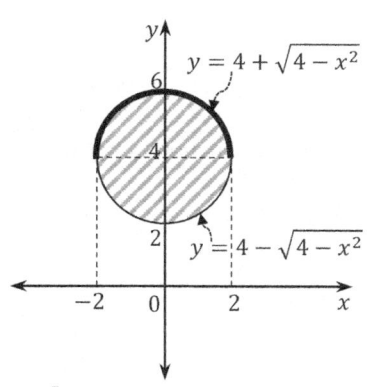

$(y-4)^2 = 4 - x^2$; $y - 4 = \pm\sqrt{4-x^2}$; $y = 4 \pm \sqrt{4-x^2}$

Above half circle : $y = 4 + \sqrt{4-x^2}$

Below half circle : $y = 4 - \sqrt{4-x^2}$

$V = \pi \int_{-2}^{2}\left(4+\sqrt{4-x^2}\right)^2 dx - \pi \int_{-2}^{2}\left(4-\sqrt{4-x^2}\right)^2 dx$

$= 2\pi \left[\int_{0}^{2}\left(4+\sqrt{4-x^2}\right)^2 dx - \int_{0}^{2}\left(4-\sqrt{4-x^2}\right)^2 dx\right]$

$= 2\pi \left[\int_{0}^{2}\{(16+8\sqrt{4-x^2}+4-x^2)-(16-8\sqrt{4-x^2}+4-x^2)\}dx\right]$

$= 2\pi \int_{0}^{2}\left(16\sqrt{4-x^2}\right)dx = 32\pi \int_{0}^{2}\left(\sqrt{4-x^2}\right)dx$

Note that $\int_{0}^{a}\left(\sqrt{a^2-x^2}\right)dx = \frac{1}{4}a^2\pi$

$\therefore \int_{0}^{2}\left(\sqrt{4-x^2}\right)dx = \frac{1}{4}\cdot 2^2\pi = \pi$

Therefore, $V = 32\pi^2$

(21) $y = \sqrt{9-x^2}$, $y = \sqrt{1-\dfrac{x^2}{9}}$, $x = -1$, $y = 2$

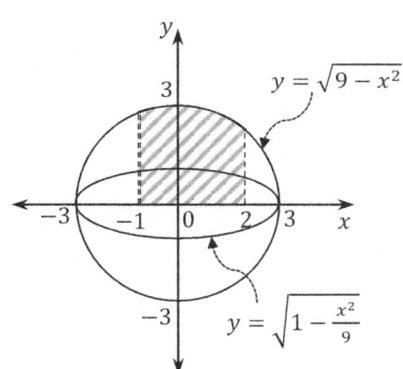

$y = \sqrt{9-x^2} \Rightarrow y^2 = 9-x^2$; $x^2 + y^2 = 9 \ (y \geq 0)$

$y = \sqrt{1-\dfrac{x^2}{9}} \Rightarrow y^2 = 1-\dfrac{x^2}{9}$; $x^2 + 9y^2 = 9 \ (y \geq 0)$

$V = \pi \int_{-1}^{2}\left\{(9-x^2)-\left(1-\dfrac{x^2}{9}\right)\right\}dx$

$= \pi \int_{-1}^{2}\left(8-\dfrac{8}{9}x^2\right)dx = \pi\left[8x - \dfrac{8}{9}\cdot\dfrac{1}{3}x^3\right]_{-1}^{2}$

$= \pi\left\{8(2+1) - \dfrac{8}{9}\cdot\dfrac{1}{3}(8+1)\right\} = \pi\left(24 - \dfrac{8}{3}\right) = \dfrac{64}{3}\pi$

#11 **For the region bounded by the graphs of the given equations, find the volume V of the solid generated by revolving the region about the y-axis.**

(1) $y = 1 - x^2$, $y = 0$

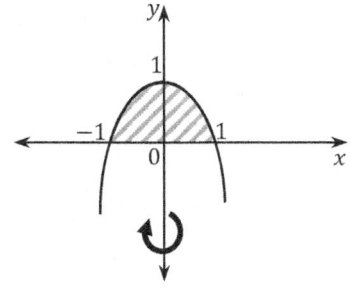

$x^2 = 1 - y$

$V = \pi \int_{0}^{1} x^2 dy = \pi \int_{0}^{1}(1-y)dy = \pi\left[y - \dfrac{1}{2}y^2\right]_{0}^{1}$

$= \pi\left(1 - \dfrac{1}{2}\right) = \dfrac{1}{2}\pi$

(2) $y = \sqrt{x+1}$, $x = 0$, $y = 0$

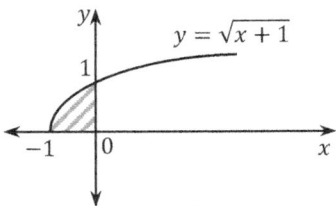

$y^2 = x + 1$; $x = y^2 - 1$

$V = \pi \int_0^1 x^2 dy = \pi \int_0^1 (y^2 - 1)^2 dy = \pi \int_0^1 (y^4 - 2y^2 + 1)dy$

$\quad = \pi \left[\frac{1}{5}y^5 - \frac{2}{3}y^3 + y\right]_0^1 = \pi\left(\frac{1}{5} - \frac{2}{3} + 1\right) = \frac{8}{15}\pi$

(3) $y = \sqrt{x+1}$, $y = x + 1$

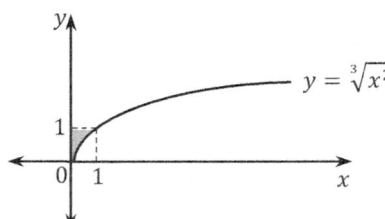

$y^2 = x + 1$; $x = y^2 - 1$

$V = \pi \int_0^1 x^2 dy = \pi \int_0^1 (y^2 - 1)^2 dy - \pi \int_0^1 (y - 1)^2 dy$

$\quad = \pi \int_0^1 \{(y^4 - 2y^2 + 1) - (y^2 - 2y + 1)\}dy = \pi \int_0^1 (y^4 - 3y^2 + 2y)dy$

$\quad = \pi \left[\frac{1}{5}y^5 - y^3 + y^2\right]_0^1 = \pi\left(\frac{1}{5} - 1 + 1\right) = \frac{1}{5}\pi$

(4) $y = \sqrt[3]{x^2}$, $y = 1$

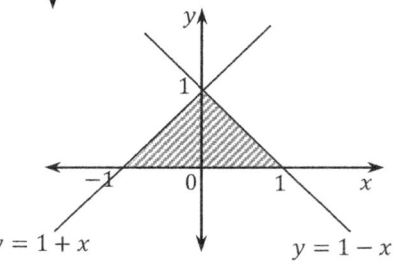

$x^2 = y^3$

$V = \pi \int_0^1 x^2 dy = \pi \int_0^1 y^3 dy = \pi \left[\frac{1}{4}y^4\right]_0^1 = \frac{1}{4}\pi$

(5) $y = 1 - |x|$, $y = 0$

$|x| = 1 - y$; $x^2 = (1-y)^2$

$V = \pi \int_0^1 x^2 dy = \pi \int_0^1 (1-y)^2 dy = \pi \int_0^1 (y^2 - 2y + 1)dy$

$= \pi \left[\frac{1}{3}y^3 - y^2 + y\right]_0^1 = \pi\left(\frac{1}{3} - 1 + 1\right) = \frac{1}{3}\pi$

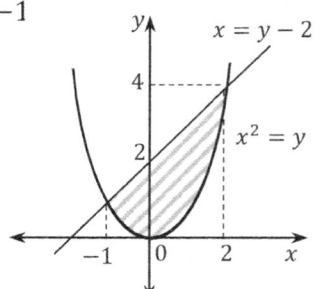

(6) $y = x^2$, $y = x + 2$

$x^2 = x + 2$; $x^2 - x - 2 = 0$; $(x-2)(x+1) = 0$; $x = 2$ or $x = -1$

$V = \pi \int_0^4 y dy - \pi \int_2^4 (y-2)^2 dy = \pi \int_0^4 y dy - \pi \int_2^4 (y^2 - 4y + 4)dy$

$\quad = \pi \left[\frac{1}{2}y^2\right]_0^4 - \pi \left[\frac{1}{3}y^3 - 2y^2 + 4y\right]_2^4$

$\quad = \pi \left(\frac{1}{2} \cdot 4^2\right) - \pi \left\{\frac{1}{3}(4^3 - 2^3) - 2(4^2 - 2^2) + 4(4-2)\right\}$

$\quad = 8\pi - \pi \left\{\frac{1}{3} \cdot 56 - 2 \cdot 12 + 4 \cdot 2\right\}$

$\quad = 8\pi - \frac{8}{3}\pi = \frac{16}{3}\pi$

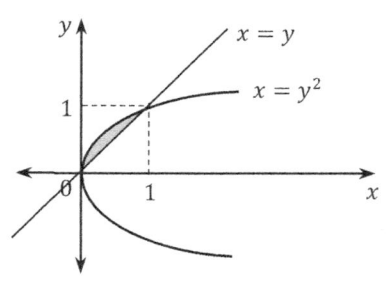

(7) $y^2 = x$, $y = x$

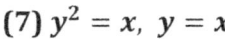

$V = \pi \int_0^1 y^2 dy - \pi \int_0^1 (y^2)^2 dy = \pi \int_0^1 y^2 dy - \pi \int_0^1 y^4 dy$

$\quad = \pi \left[\frac{1}{3}y^3\right]_0^1 - \pi \left[\frac{1}{5}y^5\right]_0^1 = \pi\left(\frac{1}{3}\right) - \pi\left(\frac{1}{5}\right) = \frac{1}{3}\pi - \frac{1}{5}\pi = \frac{2}{15}\pi$

(8) $(x-4)^2 + (y-2)^2 = 1$

$(x-4)^2 = 1 - (y-2)^2$; $x - 4 = \pm\sqrt{1-(y-2)^2}$

$\therefore\ x = 4 \pm \sqrt{1-(y-2)^2}$

Right half circle is $x = 4 + \sqrt{1-(y-2)^2}$

Left half circle is $x = 4 - \sqrt{1-(y-2)^2}$

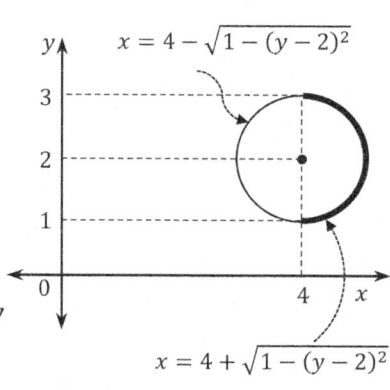

$V = \pi \int_1^3 \left(4 + \sqrt{1-(y-2)^2}\right)^2 dy - \pi \int_1^3 \left(4 - \sqrt{1-(y-2)^2}\right)^2 dy$

$= \pi \int_1^3 \left\{\left(4 + \sqrt{1-(y-2)^2}\right)^2 - \left(4 - \sqrt{1-(y-2)^2}\right)^2\right\} dy$

$= 2\pi \int_1^2 \left\{\left(4 + \sqrt{1-(y-2)^2}\right)^2 - \left(4 - \sqrt{1-(y-2)^2}\right)^2\right\} dy$

$= 2\pi \int_1^2 \left\{16\sqrt{1-(y-2)^2}\right\} dy$

$= 32\pi \int_1^2 \left\{\sqrt{1-(y-2)^2}\right\} dy$

Note that $\int_1^2 \left\{\sqrt{1-(y-2)^2}\right\} dy = \dfrac{\pi}{4}$ ($\because$ Since $\int_1^2 \left\{\sqrt{1-(y-2)^2}\right\} dy$ is $\dfrac{1}{4}$ of the area of

a circle with radius 1,

$\int_1^2 \left\{\sqrt{1-(y-2)^2}\right\} dy = \dfrac{1}{4}(\pi \cdot 1^2) = \dfrac{\pi}{4}$)

Therefore, $V = 32\pi \cdot \dfrac{\pi}{4} = 8\pi^2$

(9) $y = \log(x+1)$, $x = 0$, $y = 1$

$y = \log(x+1) \ \Rightarrow\ x + 1 = e^y$; $x = e^y - 1$

$\log(x+1) = 1 \ \Rightarrow\ x + 1 = e^1$; $x = e - 1$

$\therefore\ (e-1, 1)$ is the intersection point

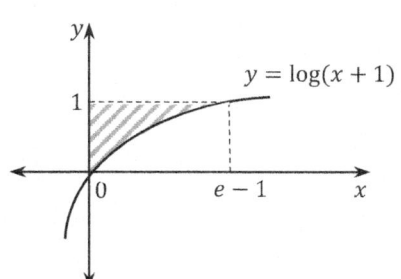

$V = \pi \int_0^1 x^2 dy = \pi \int_0^1 (e^y - 1)^2 dy = \pi \int_0^1 (e^{2y} - 2e^y + 1) dy$

$= \pi \left[\dfrac{1}{2}e^{2y} - 2e^y + y\right]_0^1 = \pi \left\{\left(\dfrac{1}{2}e^2 - 2e + 1\right) - \left(\dfrac{1}{2}e^0 - 2e^0\right)\right\}$

$= \pi \left(\dfrac{1}{2}e^2 - 2e + \dfrac{5}{2}\right) = \dfrac{\pi}{2}(e^2 - 4e + 5)$

(10) $y = \log x$, $y = \log 2$, $x = 0$, $y = 0$

$y = \log x \ \Rightarrow\ x = e^y$

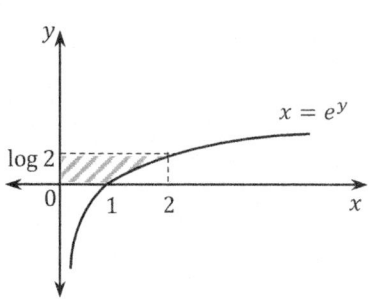

$V = \pi \int_0^{\log 2} x^2 dy = \pi \int_0^{\log 2} (e^y)^2 dy = \pi \int_0^{\log 2} e^{2y} dy$

$= \pi \left[\dfrac{1}{2}e^{2y}\right]_0^{\log 2} = \pi \left(\dfrac{1}{2}e^{2\log 2} - \dfrac{1}{2}e^0\right) = \pi \left(\dfrac{1}{2} \cdot 2^2 - \dfrac{1}{2}\right)$

$= \dfrac{3}{2}\pi$

(11) $y = \log(2 - x)$, $x = 0$, $y = 0$

$y = \log(2 - x) \Rightarrow 2 - x = e^y \; ; \; x = 2 - e^y$

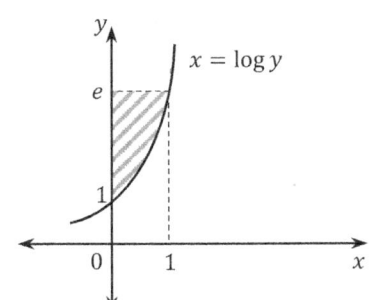

$V = \pi \int_0^{\log 2} x^2 dy = \pi \int_0^{\log 2} (2 - e^y)^2 dy$

$= \pi \int_0^{\log 2} (e^{2y} - 4e^y + 4) dy$

$y = \log(2 - x)$

$= \pi \left[\frac{1}{2} e^{2y} - 4e^y + 4y \right]_0^{\log 2} = \pi \left\{ \left(\frac{1}{2} e^{2\log 2} - 4e^{\log 2} + 4 \log 2 \right) - \left(\frac{1}{2} e^0 - 4e^0 \right) \right\}$

$= \pi \left\{ \left(\frac{1}{2} \cdot 4 - 4 \cdot 2 + 4 \log 2 \right) - \left(\frac{1}{2} - 4 \right) \right\} = \pi \left(4 \log 2 - \frac{5}{2} \right)$

(12) $y = e^x$, $x = 0$, $y = e$

$y = e^x \Rightarrow x = \log y$

$V = \pi \int_1^e x^2 dy = \pi \int_1^e (\log y)^2 dy$

For $\int (\log x)^2 dx$,

Let $\quad u = (\log x)^2$, $\qquad v' = dx$

Then, $\quad u' = 2 \log x \cdot \frac{1}{x} dx$, $\quad v = x$

$\therefore \; \int (\log x)^2 dx = x (\log x)^2 - \int x \cdot 2 \log x \cdot \frac{1}{x} dx = x (\log x)^2 - 2(x \log x - x) + C$

Therefore, $V = \pi \int_1^e (\log x)^2 dx = \pi [x(\log x)^2 - 2(x \log x - x)]_1^e = \pi(e - 2)$

(13) $y = e^x - x - 1$, $x = 0$, $x = 1$

$x = 0 \Rightarrow y = e^0 - 1 = 0 \; ; \quad x = 1 \Rightarrow y = e - 2$

Since $y' = e^x - 1 \geq 0$, y is increasing function in $[0, 1]$.

$V = \pi \int_0^{e-2} x^2 dy$

Since $y = e^x - x - 1$, $\quad dy = (e^x - 1) dx$

$\therefore \; V = \pi \int_0^{e-2} x^2 dy = \pi \int_0^1 x^2 (e^x - 1) dx = \pi \int_0^1 x^2 e^x dx - \pi \int_0^1 x^2 dx$

For $\int x^2 e^x dx$,

Let $\quad u = x^2$, $\qquad v' = e^x dx$

Then, $\quad u' = 2x dx$, $\quad v = e^x$ $\qquad \therefore \; \int x^2 e^x dx = x^2 e^x - \int e^x \cdot 2x dx = x^2 e^x - 2 \int x e^x dx$

For $\int x e^x dx$,

Let $\quad u = x$, $\qquad v' = e^x dx$

Then, $\quad u' = dx$, $\quad v = e^x$ $\qquad \therefore \; \int x e^x dx = x e^x - \int e^x dx = x e^x - e^x + C_1$

Thus, $\int x^2 e^x dx = x^2 e^x - 2(x e^x - e^x + C_1) = x^2 e^x - 2(x e^x - e^x) + C$

Therefore, $V = \pi \int_0^1 x^2 e^x dx - \pi \int_0^1 x^2 dx$

$$= \pi [x^2 e^x - 2(x e^x - e^x)]_0^1 - \pi \left[\frac{1}{3} x^3 \right]_0^1 = \pi(e - 2) - \frac{1}{3} \pi = \pi \left(e - \frac{7}{3} \right)$$

(14) $y = e^{-x^2}$ $(0 \le x \le 1)$, $x = 0$, $y = e^{-1}$

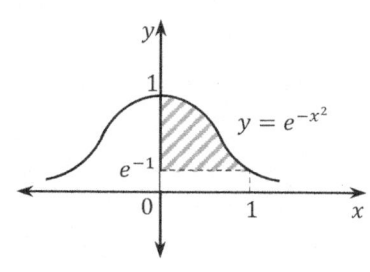

$y = e^{-x^2} \Rightarrow -x^2 = \log y$; $x^2 = -\log y$

$V = \pi \int_{e^{-1}}^1 x^2 dy = \pi \int_{e^{-1}}^1 (-\log y) dy = -\pi \int_{e^{-1}}^1 \log y\, dy$

$\quad = -\pi [y \log y - y]_{e^{-1}}^1 = -\pi \{(0-1) - (e^{-1} \log e^{-1} - e^{-1})\}$

$\quad = -\pi(-1 + 2e^{-1}) = \pi \left(1 - \dfrac{2}{e}\right)$

#12 Find the value.

(1) For the region bounded by $y = ax$ $(a > 0)$, $y = 0$, $x = 1$, and $x = 2$, the volume of the solid generated by revolving the region about x-axis is equal to 14π. Find the value of a.

The volume of the solid is

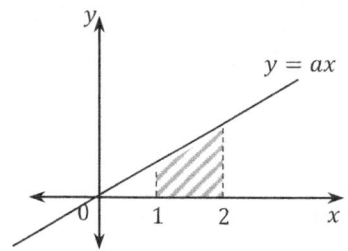

$V = \pi \int_1^2 y^2 dx = \pi \int_1^2 a^2 x^2 dx = a^2 \pi \int_1^2 x^2 dx = a^2 \pi \left[\dfrac{1}{3} x^3\right]_1^2$

$\quad = \dfrac{1}{3} a^2 \pi (2^3 - 1^3) = \dfrac{7}{3} a^2 \pi$

Since $V = 14\pi$, $\dfrac{7}{3} a^2 \pi = 14\pi$

$\therefore \ a^2 = 6$; $a = \pm\sqrt{6}$

Since $a > 0$, $a = \sqrt{6}$

(2) When the volume of the solid generated by revolving the region bounded by the ellipse $x^2 + ay^2 = 1$ $(a > 0)$ about the x-axis is equal to $\dfrac{\pi}{3}$, find the value of a.

$x^2 + ay^2 = 1 \ \Rightarrow \ y^2 = \dfrac{1}{a}(1 - x^2)$

The volume of the solid is

$V = \pi \int_{-1}^1 y^2 dx = 2\pi \int_0^1 y^2 dx = 2\pi \int_0^1 \dfrac{1}{a}(1-x^2)dx = \dfrac{2}{a}\pi \int_0^1 (1-x^2)dx = \dfrac{2}{a}\pi \left[x - \dfrac{1}{3}x^3\right]_0^1$

$\quad = \dfrac{2}{a}\pi \left(1 - \dfrac{1}{3}\right) = \dfrac{4}{3a}\pi$

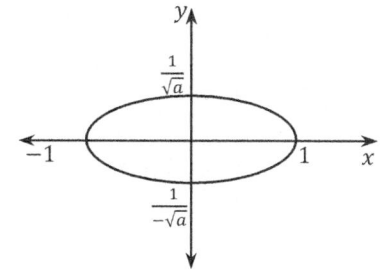

Since $V = \dfrac{\pi}{3}$, $\dfrac{4}{3a}\pi = \dfrac{\pi}{3}$

$\therefore \ \dfrac{4}{3a} = \dfrac{1}{3}$; $3a = 12$

Therefore, $a = 4$

(3) When a point $(1, 0)$ is the point of tangency of the curve $y = ax^4 + bx^2 + 1$, find the volume of the solid generated by revolving the region bounded by the curve y and $y = 0$ about the x-axis.

$y = ax^4 + bx^2 + 1 \ \Rightarrow \ y' = 4ax^3 + 2bx$

Since $(1, 0)$ lies on the curve y, $\ 0 = a + b + 1 \ \cdots\cdots \ ①$

Since $(1, 0)$ is the point of tangency of the curve y, $\ y' = 0$ when $x = 1$

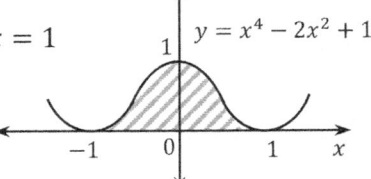

$\therefore \ 4a + 2b = 0 \ ; \quad 2a + b = 0 \ \cdots\cdots ②$

By ① and ②, $a = 1$, $b = -2$

$\therefore \ y = x^4 - 2x^2 + 1 = (x^2 - 1)^2 = (x + 1)^2(x - 1)^2$

$V = \pi \int_{-1}^{1} y^2 dx = 2\pi \int_0^1 y^2 dx = 2\pi \int_0^1 (x^4 - 2x^2 + 1)^2 dx$

$= 2\pi \int_0^1 (x^8 + 4x^4 + 1 - 4x^6 + 2x^4 - 4x^2) dx = 2\pi \int_0^1 (x^8 - 4x^6 + 6x^4 - 4x^2 + 1) dx$

$= 2\pi \left[\frac{1}{9}x^9 - \frac{4}{7}x^7 + \frac{6}{5}x^5 - \frac{4}{3}x^3 + x \right]_0^1 = 2\pi \left(\frac{1}{9} - \frac{4}{7} + \frac{6}{5} - \frac{4}{3} + 1 \right)$

$= 2\pi \left(\frac{7-36}{63} + \frac{18-20+15}{15} \right) = 2\pi \left(\frac{-29}{63} + \frac{13}{15} \right) = 2\pi \left(\frac{-435+819}{945} \right) = 2\pi \left(\frac{384}{945} \right)$

$= 2\pi \left(\frac{128}{315} \right) = \frac{256}{315} \pi$

(4) Find the value of a at which the volume of the solid generated by revolving the region bounded by the curve $x^2 + a(a - 1)y^2 = a^2 \ (a > 1)$ about the x-axis has minimum value.

$x^2 + a(a-1)y^2 = a^2 \ \Rightarrow \ y^2 = \frac{1}{a(a-1)}(a^2 - x^2)$

$V = 2\pi \int_0^a y^2 dx = 2\pi \int_0^a \frac{1}{a(a-1)}(a^2 - x^2) dx = \frac{2\pi}{a(a-1)} \int_0^a (a^2 - x^2) dx = \frac{2\pi}{a(a-1)} \left[a^2 x - \frac{1}{3}x^3 \right]_0^a$

$= \frac{2\pi}{a(a-1)} \left(a^2 a - \frac{1}{3}a^3 \right) = \frac{2\pi}{(a-1)} \cdot \frac{2}{3}a^2 = \frac{4\pi a^2}{3(a-1)} = \frac{4\pi\{(a-1)^2 + 2a - 1\}}{3(a-1)} = \frac{4\pi}{3} \left(a - 1 + \frac{2a-1}{a-1} \right)$

$= \frac{4\pi}{3} \left(a - 1 + \frac{2(a-1)+1}{a-1} \right) = \frac{4\pi}{3} \left(a - 1 + \frac{1}{a-1} + 2 \right)$

By the relationship between arithmetic and geometric means,

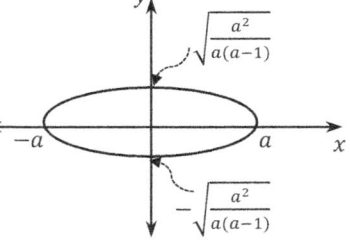

$a - 1 + \frac{1}{a-1} \geq 2\sqrt{(a-1) \cdot \left(\frac{1}{a-1} \right)}$

$= 2 \quad$ (when $a - 1 = \frac{1}{a-1}$, the equality holds.)

Since $a - 1 + \frac{1}{a-1} \geq 2$, the volume has minimum value when $a - 1 = 1$.

Therefore, $a = 2$

(5) When the volume of the solid generated by revolving the region bounded by the curve $y^2 = x + a^2$ ($a > 0$) and the y-axis is equal to $\frac{16}{15}\pi$, find the value of a.

When $x = 0$, $y^2 = a^2$ $\quad \therefore \ y = \pm a$

The intersection points of the curve and the y-axis are $y = a$ and $y = -a$.

$V = \pi \int_{-a}^{a} x^2 dy = 2\pi \int_{0}^{a} x^2 dy = 2\pi \int_{0}^{a} (y^2 - a^2)^2 dy = 2\pi \int_{0}^{a} (y^4 - 2a^2 y^2 + a^4) dy$

$= 2\pi \left[\frac{1}{5} y^5 - \frac{2}{3} a^2 y^3 + a^4 y \right]_{0}^{a} = 2\pi \left(\frac{1}{5} a^5 - \frac{2}{3} a^2 a^3 + a^4 a \right) = 2\pi \left(\frac{1}{5} a^5 - \frac{2}{3} a^5 + a^5 \right)$

$= \frac{16}{15} a^5 \pi$

Since $\frac{16}{15} a^5 \pi = \frac{16}{15} \pi$, $a^5 = 1$

Since $a > 0$, $a = 1$

#13 Prove it.

(1) Prove the volume of a sphere with radius r is $\frac{4}{3}\pi r^3$.

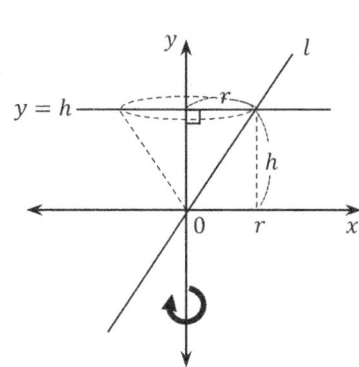

Consider a circle with radius r and center at the origin.

Then , the equation of the circle is $x^2 + y^2 = r^2$.

$\therefore \ y^2 = r^2 - x^2 \ ; \ \ y = \pm\sqrt{r^2 - x^2}$

$V = \pi \int_{-r}^{r} y^2 dx = \pi \int_{-r}^{r} (r^2 - x^2) dx = 2\pi \int_{0}^{r} (r^2 - x^2) dx = 2\pi \left[r^2 x - \frac{1}{3} x^3 \right]_{0}^{r}$

$= 2\pi \left(r^2 r - \frac{1}{3} r^3 \right) = 2\pi \cdot \frac{2}{3} r^3 = \frac{4}{3} \pi r^3$

(2) Prove the volume of a right circular cone with radius r and height h is $\frac{1}{3}\pi r^2 h$.

(r, h; non-zero real numbers.)

Consider the first quadrant region bounded by the x-axis, the y-axis, and a line $y = h$.

Let l be the line passing through $(0, 0)$ and (r, h).

Then, the equation of the line l is $y = \frac{h}{r} x$.

Considering right boundary equation,

express the equation in terms of y; i.e., $x = \frac{r}{h} y$

$V = \pi \int_{0}^{h} x^2 dy = \pi \int_{0}^{h} \left(\frac{r}{h} y \right)^2 dy = \frac{r^2 \pi}{h^2} \int_{0}^{h} y^2 dy$

$= \frac{r^2 \pi}{h^2} \left[\frac{1}{3} y^3 \right]_{0}^{h} = \frac{r^2 \pi}{h^2} \left(\frac{1}{3} h^3 \right)$

$= \frac{1}{3} \pi r^2 h$

#14 Find the volume of the solid generated by rotating the following region about the given line.

    (1) Rotating the region bounded by a curve $y = 2 - x^2$ and the x-axis about the line $y = -1$.

$$2 - x^2 = 0 \implies x = \pm\sqrt{2}$$

$$V = \pi \int_{-\sqrt{2}}^{\sqrt{2}} \{(3 - x^2)^2 - (1)^2\}dx = \pi \int_{-\sqrt{2}}^{\sqrt{2}} (x^4 - 6x^2 + 8)dx$$

$$= \pi \left[\frac{1}{5}x^5 - 2x^3 + 8x\right]_{-\sqrt{2}}^{\sqrt{2}}$$

$$= \pi \left\{\frac{1}{5}\left(\sqrt{2}^5 + \sqrt{2}^5\right) - 2\left(\sqrt{2}^3 + \sqrt{2}^3\right) + 8(\sqrt{2} + \sqrt{2})\right\}$$

$$= \pi \left\{\frac{1}{5}(8\sqrt{2}) - 2(4\sqrt{2}) + 8(2\sqrt{2})\right\}$$

$$= 2\sqrt{2}\pi \left(\frac{4}{5} - 4 + 8\right) = 2\sqrt{2}\pi \cdot \frac{24}{5} = \frac{48\sqrt{2}}{5}\pi$$

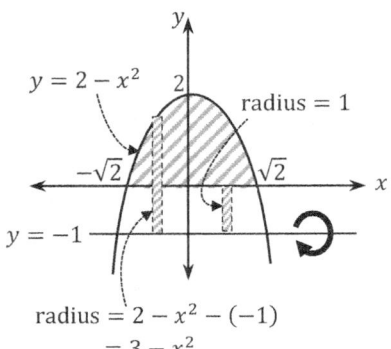

radius $= 2 - x^2 - (-1)$
$= 3 - x^2$

    (2) Rotating the region bounded by a curve $y = 2 - x^2$ and the x-axis about the line $y = 2$.

$$V = \pi \int_{-\sqrt{2}}^{\sqrt{2}} \{(2)^2 - (x^2)^2\}dx = \pi \int_{-\sqrt{2}}^{\sqrt{2}} (4 - x^4)dx$$

$$= \pi \left[4x - \frac{1}{5}x^5\right]_{-\sqrt{2}}^{\sqrt{2}} = \pi \left\{\left(4\sqrt{2} - \frac{1}{5}\cdot 4\sqrt{2}\right) - \left(-4\sqrt{2} + \frac{1}{5}\cdot 4\sqrt{2}\right)\right\}$$

$$= \pi \left(8\sqrt{2} - \frac{8\sqrt{2}}{5}\right) = \frac{32\sqrt{2}}{5}\pi$$

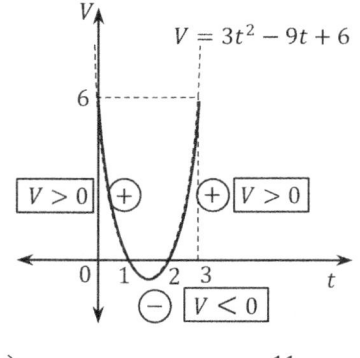

radius $= 2 - (2 - x^2)$
$= x^2$

#15 An object P is moving along a coordinate line subject to the indicated velocity V (in feet per second). Find the directed distance S (in feet) during the given time interval.

    (1) $V = 3t^2 - 9t + 6$ (from $t = 0$ to $t = 3$)

$V = 3t^2 - 9t + 6 = 3(t^2 - 3t + 2) = 3(t - 1)(t - 2)$

$\therefore \quad t = 1$ and $t = 2$ are the intersection points

    between the curve and the x-axis.

$$S = \int_0^3 |V(t)|dt = \int_0^1 V(t)dt + \int_1^2 -V(t)dt + \int_2^3 V(t)dt$$

$$= \left[t^3 - \frac{9}{2}t^2 + 6t\right]_0^1 - \left[t^3 - \frac{9}{2}t^2 + 6t\right]_1^2 + \left[t^3 - \frac{9}{2}t^2 + 6t\right]_2^3$$

$$= \left(1 - \frac{9}{2} + 6\right) - (8 - 18 + 12) + \left(1 - \frac{9}{2} + 6\right) + \left(27 - \frac{81}{2} + 18\right) - (8 - 18 + 12) = \frac{11}{2}$$

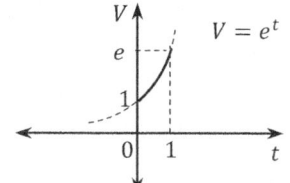

(2) $V = e^t$ **(from** $t = 0$ **to** $t = 1$**)**

$S = \int_0^1 |V(t)| dt = \int_0^1 e^t dt = [e^t]_0^1 = e^1 - e^0 = e - 1$

(3) $V = (t - 1)e^{-t}$ **(from** $t = 0$ **to** $t = 2$**)**

$S = \int_0^2 |V(t)| dt = \int_0^2 |(t-1)e^{-t}| dt = \int_0^2 |(t-1)| e^{-t} dt$

$= \int_0^1 (1-t)e^{-t} dt + \int_1^2 (t-1)e^{-t} dt = \int_0^1 (1-t)e^{-t} dt - \int_1^2 (1-t)e^{-t} dt$

For $\int (1-t)e^{-t} dt$,

Let $\qquad u = 1 - t, \qquad v' = e^{-t} dt$

Then, $\qquad u' = -dt, \qquad v = -e^{-t}$

$\therefore \quad \int (1-t)e^{-t} dt = -(1-t)e^{-t} - \int (-e^{-t})(-dt) = (t-1)e^{-t} - \int e^{-t} dt$

$\qquad\qquad = (t-1)e^{-t} + e^{-t} + C = te^{-t} + C$

Thus, $\int_0^1 (1-t)e^{-t} dt = [te^{-t}]_0^1 = e^{-1}$ and $\int_1^2 (1-t)e^{-t} dt = [te^{-t}]_1^2 = 2e^{-2} - e^{-1}$

Therefore, $S = e^{-1} - (2e^{-2} - e^{-1}) = 2e^{-1} - 2e^{-2} = 2(e^{-1} - e^{-2}) = 2\left(\dfrac{1}{e} - \dfrac{1}{e^2}\right)$

#16 **When a ball is thrown upward from an initial height of 10 feet with a velocity of 20 feet per second, its velocity t seconds later is given by $V = 20 - 10t$.**

(1) Find its height from ground level 3 seconds later.

(2) Find the maximum height that the ball reaches from the ground.

(3) Find the total distance during the first 3 seconds.

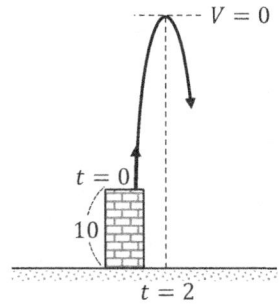

$V(t) = 20 - 10t = 10(2 - t)$

$V(t) = 0 \quad \Rightarrow \quad t = 2$

$0 \le t < 2 \quad \Rightarrow \quad$ The ball is going upward.

$t = 2 \quad \Rightarrow \quad$ The ball reaches the maximum height.

$t > 2 \quad \Rightarrow \quad$ The ball is going downward.

(1) The height from the ground after 3 seconds is

$S = 10 + \int_0^3 |V(t)| dt = 10 + \int_0^3 (20 - 10t) dt = 10 + [20t - 5t^2]_0^3$

$= 10 + (20 \cdot 3 - 5 \cdot 3^2) = 25$ feet

(2) When $V = 0$, the ball reaches its maximum height.

$\therefore$ The maximum height h is $h = 10 + \int_0^2 V(t) dt = 10 + \int_0^2 (20 - 10t) dt$

$= 10 + [20t - 5t^2]_0^2 = 10 + (20 \cdot 2 - 5 \cdot 2^2) = 30$ feet

(3) Directed distance is

$$S = \int_0^3 |V(t)|dt = \int_0^2 V(t)dt + \int_2^3 -V(t)dt$$

$$= \int_0^2 (20 - 10t)dt + \int_2^3 -(20 - 10t)dt$$

$$= [20t - 5t^2]_0^2 - [20t - 5t^2]_2^3 = (20 \cdot 2 - 5 \cdot 2^2) - (20 \cdot 3 - 5 \cdot 3^2) + (20 \cdot 2 - 5 \cdot 2^2)$$

$$= 2(20 \cdot 2 - 5 \cdot 2^2) - (20 \cdot 3 - 5 \cdot 3^2) = 40 - 15 = 25 \text{ feet}$$

#17 A point $P(x, y)$ moves in the coordinate plane, starting at the origin.

When the velocity vector of the point at time t is given by $v(t) = \langle e^t + 1, \ e^{-t} - 1\rangle$,

find the coordinates of P at $t = 1$.

Since $v(t) = \langle e^t + 1, \ e^{-t} - 1\rangle$, $\frac{dx}{dt} = e^t + 1$ and $\frac{dy}{dt} = e^{-t} - 1$

At $t = 1$, $x = \int_0^1 (e^t + 1)dt = [e^t + t]_0^1 = (e^1 + 1) - e^0 = e$

$$y = \int_0^1 (e^{-t} - 1)dt = [-e^{-t} - t]_0^1 = (-e^{-1} - 1) + e^0 = -e^{-1}$$

Therefore, $P(x, y) = \langle e, \ -\frac{1}{e}\rangle$

#18 An object is moving along a coordinate line with the initial velocity 30 feet/sec.

If the object decreases speed at $\frac{1}{2}$ feet/sec, find the distance from the starting to

stopping point.

Let V be the velocity after t seconds.

Then, $V = 30 - \frac{1}{2}t$

When $V = 0$, the object will stop.

$30 - \frac{1}{2}t = 0 \ \Rightarrow \ t = 60$

Therefore, the distance is

$$\int_0^{60} |V(t)|dt = \int_0^{60} \left(30 - \frac{1}{2}t\right)dt = \left[30t - \frac{1}{4}t^2\right]_0^{60} = 30 \cdot 60 - \frac{1}{4} \cdot 60^2$$

$$= 60(30 - 15) = 900 \text{ feet}$$

#19 A tank in the shape of a right circular cone is being filled with water.

When the height of the tank is h, the area of its top is $S(h) = 4\pi h^2$.

If the velocity of filling the tank is given by $V(t) = 2t(3 - t)$,

find the maximum amount of water over the edge of the tank.

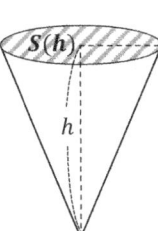

When $V(t) = 0$, the tank is full of water.

$V(t) = 0 \Rightarrow 2t(3 - t) = 0 \qquad \therefore \ t = 0$ or $t = 3$

When $t = 3$, $\ h = \int_0^3 |V(t)| dt = \int_0^3 2t(3 - t) dt = \int_0^3 (6t - 2t^2) dt$

$$= \left[3t^2 - \frac{2}{3}t^3\right]_0^3 = 3 \cdot 3^2 - \frac{2}{3} \cdot 3^3 = 3^3\left(1 - \frac{2}{3}\right) = \frac{1}{3} \cdot 3^3 = 3^2 = 9$$

Therefore, the volume of the tank is

$$\int_0^9 S(h) dh = \int_0^9 (4\pi h^2) dh = 4\pi \int_0^9 h^2 dh = 4\pi \left[\frac{1}{3}h^3\right]_0^9 = \frac{4\pi}{3} \cdot 9^3 = 4\pi \cdot 243 = 972\pi$$

#20 **A water tank in the shape of a rectangle with the surface area of 3 feet^2 is draining at a speed of $V(t) = 2t$ at t seconds. Find the amount of water that has drained during the first 10 seconds.**

The distance of the path of drained water for 10 seconds is

$\int_0^{10} |V(t)| dt = \int_0^{10} 2t \, dt = [t^2]_0^{10} = 100$ feet

Since the surface area is 3 ft^2, the amount of water that has drained is

$3 \times 100 = 300$ feet^3

#21 **When a certain object shoots straight up from the ground level with an initial velocity of a feet per second ($a > 0$), the velocity after t seconds is given by**

$V(t) = a - 10t$ feet/sec.

What velocity will cause the object to reach its maximum height more than 20 feet?

When $V(t) = 0$, the object reaches its maximum height.

$a - 10t = 0 \Rightarrow t = \dfrac{a}{10}$

The distance of the movement until the object reaches the maximum height is

$$S = \int_0^{\frac{a}{10}} V(t) dt = \int_0^{\frac{a}{10}} (a - 10t) dt = [at - 5t^2]_0^{\frac{a}{10}} = a \cdot \frac{a}{10} - 5\left(\frac{a}{10}\right)^2 = \frac{a}{10}\left(a - 5 \cdot \frac{a}{10}\right)$$

$$= \frac{a}{10} \cdot \frac{a}{2} = \frac{a^2}{20}$$

Since $\dfrac{a^2}{20} \geq 20$, $\ a^2 \geq (20)^2$

$\therefore \ a \geq 20 \quad (\because \ a > 0)$

Therefore, the velocity more than 20 feet/sec will cause it to happen.

#22 **Starting from rest, an object is moving along a coordinate line with an initial velocity**

$V(t) = \frac{3}{4}t^2 - \frac{1}{2}t + \frac{1}{2}$ **(feet/sec) for the first 6 feet, and then travels at a constant speed.**

Find the directed distance during the first 11 seconds.

Suppose the object travels 6 feet for x seconds.

Then, $\int_0^x V(t)dt = \int_0^x \left(\frac{3}{4}t^2 - \frac{1}{2}t + \frac{1}{2}\right) dt = \left[\frac{1}{4}t^3 - \frac{1}{4}t^2 + \frac{1}{2}t\right]_0^x$

$\qquad\qquad = \frac{1}{4}x^3 - \frac{1}{4}x^2 + \frac{1}{2}x = 6$

$\therefore\ x^3 - x^2 + 2x - 24 = 0$

$\therefore\ (x-3)(x^2 + 2x + 8) = 0$

$$
\begin{array}{r|rrrr}
 & 1 & -1 & 2 & -24 \\
3 & & 3 & 6 & 24 \\
\hline
 & 1 & 2 & 8 & 0
\end{array}
$$

Since $x^2 + 2x + 8 = (x+1)^2 + 7 > 0, \ x = 3$

That is, the object travels 6 feet for 3 seconds.

At $t = 3$, $V(t) = \frac{3}{4} \cdot 3^2 - \frac{1}{2} \cdot 3 + \frac{1}{2} = \frac{27}{4} - \frac{6}{4} + \frac{2}{4} = \frac{23}{4}$

$\therefore$ After 3 seconds, the object travels at a constant speed of $\frac{23}{4}$ feet/sec.

Therefore, the distance for the first 11 seconds is

$\int_0^{11} V(t)dt = \int_0^3 \left(\frac{3}{4}t^2 - \frac{1}{2}t + \frac{1}{2}\right) dt + \int_3^{11} \left(\frac{23}{4}\right) dt = 6 + \frac{23}{4}[t]_3^{11}$

$\qquad\qquad = 6 + \frac{23}{4}(11 - 3) = 52 \text{ feet}$

#23 **Find the length l of the indicated curve.**

(1) $y = 2x + 1 \quad (0 \le x \le 1)$

$l = \int_0^1 \sqrt{1 + \left(\frac{dy}{dx}\right)^2} \, dx = \int_0^1 \sqrt{1 + 2^2} dx = \int_0^1 \sqrt{5} dx = \sqrt{5}[x]_0^1 = \sqrt{5}$

(2) $x = 6t^2, \ y = t^3 - 12t \quad (0 \le t \le 1)$

$l = \int_0^1 \sqrt{\left(\frac{dx}{dt}\right)^2 + \left(\frac{dy}{dt}\right)^2} \, dt = \int_0^1 \sqrt{(12t)^2 + (3t^2 - 12)^2} \, dt$

$= \int_0^1 \sqrt{(12t)^2 + 9(t^2 - 4)^2} \, dt = \int_0^1 \sqrt{(12t)^2 + 9(t^4 - 8t^2 + 16)} \, dt$

$= \int_0^1 \sqrt{9t^4 + 72t^2 + 9 \cdot 16} \, dt = \int_0^1 \sqrt{9(t^4 + 8t^2 + 16)} \, dt = 3 \int_0^1 \sqrt{(t^2 + 4)^2} \, dt$

$= 3 \int_0^1 (t^2 + 4) \, dt = 3 \left[\frac{1}{3}t^3 + 4t\right]_0^1 = 3 \left(\frac{1}{3} + 4\right) = 13$

(3) $x = t - \sin t$, $y = 1 - \cos t$ $(0 \leq t \leq 2\pi)$

$l = \int_0^{2\pi} \sqrt{\left(\frac{dx}{dt}\right)^2 + \left(\frac{dy}{dt}\right)^2}\, dt = \int_0^{2\pi} \sqrt{(1 - \cos t)^2 + (\sin t)^2}\, dt$

$= \int_0^{2\pi} \sqrt{(1 - 2\cos t + \cos^2 t) + \sin^2 t}\, dt = \int_0^{2\pi} \sqrt{(2 - 2\cos t)}\, dt$

$= \int_0^{2\pi} \sqrt{2(1 - \cos t)}\, dt = \int_0^{2\pi} \sqrt{2 \cdot 2\sin^2\left(\frac{t}{2}\right)}\, dt = 2\int_0^{2\pi} \sin\left(\frac{t}{2}\right) dt$

$= 2\left[-2\cos\left(\frac{t}{2}\right)\right]_0^{2\pi} = -4(\cos\pi - \cos 0) = 8$

> $\cos 2x = 1 - 2\sin^2 x$
> $\therefore\ \cos x = 1 - 2\sin^2\left(\frac{x}{2}\right)$
> $\therefore\ 1 - \cos x = 2\sin^2\left(\frac{x}{2}\right)$

(4) $y = x\sqrt{x}$ $(0 \leq x \leq 2)$

$y = x\sqrt{x} = x^{\frac{3}{2}} \Rightarrow y' = \frac{3}{2}x^{\frac{1}{2}}$

$l = \int_0^2 \sqrt{1 + (y')^2}\, dx = \int_0^2 \sqrt{1 + \left(\frac{3}{2}x^{\frac{1}{2}}\right)^2}\, dx = \int_0^2 \sqrt{1 + \frac{9}{4}x}\, dx = \int_0^2 \sqrt{\frac{4 + 9x}{4}}\, dx =$

$\frac{1}{2}\int_0^2 \sqrt{4 + 9x}\, dx = \frac{1}{2}\left[\frac{2}{3}(4 + 9x)^{\frac{3}{2}} \cdot \frac{1}{9}\right]_0^2$

$= \frac{1}{27}\left\{\left((4 + 9\cdot 2)^{\frac{3}{2}}\right) - 4^{\frac{3}{2}}\right\} = \frac{1}{27}\left(22^{\frac{3}{2}} - 2^3\right) = \frac{1}{27}(22\sqrt{22} - 8) = \frac{2}{27}(11\sqrt{22} - 4)$

(5) $y = \frac{e^x + e^{-x}}{2}$ $(-1 \leq x \leq 1)$

$y' = \frac{e^x - e^{-x}}{2}$

$l = \int_{-1}^1 \sqrt{1 + (y')^2}\, dx = \int_{-1}^1 \sqrt{1 + \left(\frac{e^x - e^{-x}}{2}\right)^2}\, dx = \int_{-1}^1 \sqrt{1 + \left(\frac{e^{2x} - 2 + e^{-2x}}{4}\right)}\, dx$

$= \int_{-1}^1 \sqrt{\frac{e^{2x} + 2 + e^{-2x}}{4}}\, dx = \frac{1}{2}\int_{-1}^1 \sqrt{e^{2x} + 2 + e^{-2x}}\, dx = \frac{1}{2}\int_{-1}^1 \sqrt{(e^x + e^{-x})^2}\, dx$

$= \frac{1}{2}\cdot 2\int_0^1 \sqrt{(e^x + e^{-x})^2}\, dx = \int_0^1 (e^x + e^{-x})\, dx = [e^x - e^{-x}]_0^1 = (e^1 - e^{-1}) - (e^0 - e^0)$

$= e - \frac{1}{e}$

(6) $x = \log t$, $y = \frac{1}{2}\left(t + \frac{1}{t}\right)$ $\left(\frac{1}{e} \leq t \leq e\right)$

$\frac{dx}{dt} = \frac{1}{t}$ and $\frac{dy}{dt} = \frac{1}{2}(1 - t^{-2})$

$l = \int_{\frac{1}{e}}^e \sqrt{\left(\frac{dx}{dt}\right)^2 + \left(\frac{dy}{dt}\right)^2}\, dt = \int_{\frac{1}{e}}^e \sqrt{\left(\frac{1}{t}\right)^2 + \left(\frac{1}{2}(1 - t^{-2})\right)^2}\, dt = \int_{\frac{1}{e}}^e \sqrt{\frac{1}{t^2} + \frac{1}{4}\left(1 - \frac{2}{t^2} + t^{-4}\right)}\, dt$

$= \frac{1}{2}\int_{\frac{1}{e}}^e \sqrt{1 + \frac{2}{t^2} + \frac{1}{t^4}}\, dt = \frac{1}{2}\int_{\frac{1}{e}}^e \sqrt{\left(1 + \frac{1}{t^2}\right)^2}\, dt = \frac{1}{2}\int_{\frac{1}{e}}^e \left(1 + \frac{1}{t^2}\right) dt = \frac{1}{2}[t - t^{-1}]_{\frac{1}{e}}^e$

$= \frac{1}{2}\left\{(e - e^{-1}) - \left(\frac{1}{e} - e\right)\right\} = \frac{1}{2}\{2(e - e^{-1})\} = e - e^{-1} = e - \frac{1}{e}$

(7) $x = e^{-t} \cos t, \quad y = e^{-t} \sin t \quad (0 \le t \le \pi)$

$\dfrac{dx}{dt} = -e^{-t} \cos t - e^{-t} \sin t = -e^{-t}(\cos t + \sin t)$

$\dfrac{dy}{dt} = -e^{-t} \sin t + e^{-t} \cos t = e^{-t}(\cos t - \sin t)$

$l = \int_0^\pi \sqrt{\left(\dfrac{dx}{dt}\right)^2 + \left(\dfrac{dy}{dt}\right)^2}\, dt = \int_0^\pi \sqrt{\left(-e^{-t}(\cos t + \sin t)\right)^2 + (e^{-t}(\cos t - \sin t))^2}\, dt$

$\quad = \int_0^\pi \sqrt{e^{-2t}(1 + 2\sin t \cos t) + e^{-2t}(1 - 2\sin t \cos t)}\, dt = \int_0^\pi \sqrt{2e^{-2t}}\, dt$

$\quad = \int_0^\pi \sqrt{2(e^{-t})^2}\, dt = \sqrt{2}\int_0^\pi e^{-t}\, dt = \sqrt{2}[-e^{-t}]_0^\pi = -\sqrt{2}(e^{-\pi} - e^0) = \sqrt{2}\left(1 - \dfrac{1}{e^\pi}\right)$

(8) $y = \displaystyle\int_0^x \dfrac{x - t}{\cos^2 t}\, dt \quad \left(0 \le x \le \dfrac{\pi}{6}\right)$

$y = \int_0^x \dfrac{x-t}{\cos^2 t}\, dt = \int_0^x \dfrac{x}{\cos^2 t}\, dt - \int_0^x \dfrac{t}{\cos^2 t}\, dt = x\int_0^x \dfrac{1}{\cos^2 t}\, dt - \int_0^x \dfrac{t}{\cos^2 t}\, dt$

$\therefore\ y' = 1 \cdot \int_0^x \dfrac{1}{\cos^2 t}\, dt + x \cdot \dfrac{d}{dx}\left\{\int_0^x \dfrac{1}{\cos^2 t}\, dt\right\} - \dfrac{d}{dx}\left\{\int_0^x \dfrac{t}{\cos^2 t}\, dt\right\}$

$\quad = \int_0^x \dfrac{1}{\cos^2 t}\, dt + x \cdot \dfrac{1}{\cos^2 x} - \dfrac{x}{\cos^2 x} = \int_0^x \dfrac{1}{\cos^2 t}\, dt = \int_0^x \sec^2 t\, dt = [\tan t]_0^x = \tan x$

$l = \int_0^{\frac{\pi}{6}} \sqrt{1 + (y')^2}\, dx = \int_0^{\frac{\pi}{6}} \sqrt{1 + \tan^2 x}\, dx = \int_0^{\frac{\pi}{6}} \sqrt{\sec^2 x}\, dx = \int_0^{\frac{\pi}{6}} \sec x\, dx = \int_0^{\frac{\pi}{6}} \dfrac{1}{\cos x}\, dx$

$\quad = \int_0^{\frac{\pi}{6}} \dfrac{\cos x}{\cos^2 x}\, dx = \int_0^{\frac{\pi}{6}} \dfrac{\cos x}{1 - \sin^2 x}\, dx$

Let $\sin x = u$

Then $\cos x\, dx = du$

$u = 0$ when $x = 0$

$u = \dfrac{1}{2}$ when $x = \dfrac{\pi}{6}$

$\therefore\ l = \int_0^{\frac{\pi}{6}} \dfrac{\cos x}{1 - \sin^2 x}\, dx = \int_0^{\frac{1}{2}} \dfrac{1}{1 - u^2}\, du = \int_0^{\frac{1}{2}} \dfrac{1}{2}\left(\dfrac{1}{1+u} + \dfrac{1}{1-u}\right) du = \dfrac{1}{2}[\log|1 + u| - \log|1 - u|]_0^{\frac{1}{2}}$

$\quad = \dfrac{1}{2}\left\{\left(\log\left|1 + \dfrac{1}{2}\right| - \log\left|1 - \dfrac{1}{2}\right|\right) - (\log|1 + 0| - \log|1 - 0|)\right\}$

$\quad = \dfrac{1}{2}\left\{\log\left(\dfrac{3}{2}\right) - \log\left(\dfrac{1}{2}\right)\right\} = \dfrac{1}{2}\log\left(\dfrac{\frac{3}{2}}{\frac{1}{2}}\right) = \dfrac{1}{2}\log 3$

#24 When the length of the curve $y = f(x)$ from $(0, 1)$ to (x, y) on the curve is given by

$l = e^{2x} + y - 1$, find the slope of the tangent line at $x = 2$ on the curve.

($y = f(x)$ is a differentiable function defined on $x \geq 0$.)

Since $\int_0^x \sqrt{1 + \{f'(t)\}^2}\, dt = e^{2x} + y - 1$, $\quad \dfrac{d}{dx}\left[\int_0^x \sqrt{1 + \{f'(t)\}^2}\, dt\right] = \dfrac{d}{dx}[e^{2x} + y - 1]$

$\therefore \ \sqrt{1 + \{f'(x)\}^2} = 2e^{2x} + y' = 2e^{2x} + f'(x)$

Squaring both sides, $\ 1 + \{f'(x)\}^2 = 4e^{4x} + 4e^{2x}f'(x) + \{f'(x)\}^2$

$\therefore \ 1 = 4e^{4x} + 4e^{2x}f'(x)$

$\therefore \ f'(x) = \dfrac{1}{4e^{2x}}(1 - 4e^{4x})$

Therefore, $f'(2) = \dfrac{1}{4e^4}(1 - 4e^8) = \dfrac{1}{4e^4} - e^4$

Printed in Great Britain
by Amazon

43530597R00150